Studies in Big Data

Volume 67

Series Editor

Janusz Kacprzyk, Polish Academy of Sciences, Warsaw, Poland

The series "Studies in Big Data" (SBD) publishes new developments and advances in the various areas of Big Data- quickly and with a high quality. The intent is to cover the theory, research, development, and applications of Big Data, as embedded in the fields of engineering, computer science, physics, economics and life sciences. The books of the series refer to the analysis and understanding of large, complex, and/or distributed data sets generated from recent digital sources coming from sensors or other physical instruments as well as simulations, crowd sourcing, social networks or other internet transactions, such as emails or video click streams and other. The series contains monographs, lecture notes and edited volumes in Big Data spanning the areas of computational intelligence including neural networks, evolutionary computation, soft computing, fuzzy systems, as well as artificial intelligence, data mining, modern statistics and Operations research, as well as self-organizing systems. Of particular value to both the contributors and the readership are the short publication timeframe and the world-wide distribution, which enable both wide and rapid dissemination of research output.

** Indexing: The books of this series are submitted to ISI Web of Science, DBLP, Ulrichs, MathSciNet, Current Mathematical Publications, Mathematical Reviews, Zentralblatt Math: MetaPress and Springerlink.

More information about this series at http://www.springer.com/series/11970

Prasant Kumar Pattnaik ·
Raghvendra Kumar · Souvik Pal
Editors

Internet of Things and Analytics for Agriculture, Volume 2

Springer

Editors
Prasant Kumar Pattnaik
School of Computer Engineering
KIIT Deemed to be University
Bhubaneswar, Odisha, India

Raghvendra Kumar
Department of Computer Science
and Engineering
LNCT College
Jabalpur, Madhya Pradesh, India

Souvik Pal
Department of Computer Science
and Engineering
Brainware University
Kolkata, West Bengal, India

ISSN 2197-6503 ISSN 2197-6511 (electronic)
Studies in Big Data
ISBN 978-981-15-0665-9 ISBN 978-981-15-0663-5 (eBook)
https://doi.org/10.1007/978-981-15-0663-5

This Springer imprint is published by the registered company Springer Nature Singapore Pte Ltd.
The registered company address is: 152 Beach Road, #21-01/04 Gateway East, Singapore 189721, Singapore

Preface

The main aim of this book is to bring together leading academic scientists, researchers, and research scholars to exchange and share their experiences and research results on all aspects of wireless IoT and analytics for agriculture. It also provides a premier interdisciplinary platform for researchers, practitioners, and educators to present and discuss the most recent innovations, trends, and concerns as well as practical challenges encountered and solutions adopted in the fields of IoT and analytics for agriculture. This book is organized into 13 chapters.

Chapter "IoT-Based Soil Health Monitoring and Recommendation System" discusses a soil health monitoring system in which a farmer will be able to monitor soil moisture, soil temperature, and soil pH in his android smartphone. The farmer will also get the recommendations of lime and sulfur on the basis of pH of the soil. The area of study is Jaipur, Rajasthan (26.9124° N, 75.7873° E). The proposed system is tested with the results taken from authorized laboratories. The proposed system is validated with T-test having no significant difference between calibrated values and laboratory recorded values. The proposed system is implemented on android smartphone, so it is useful to farmers, agriculture scientists, agriculture professionals, and IoT experts.

Chapter "Implementing IoT and Wireless Sensor Networks for Precision Agriculture" affords the deployment of an Internet of things (IoT)-based precision agriculture technology using wireless sensor networks (WSNs) in crop management. An online crop management system (OCMS) was developed to complement the IoT platform layer. The application aimed at providing the farm environmental conditions including the temperature, water level, pH value, and dissolved oxygen measurements from the paddy field with food crops and horticulture crops being grown. Further studies could be done by analyzing the inter- and intra-field variabilities such as soil herbicide weed control, side dressing, late blight control, and haulm killing.

Chapter "Bayesian Network for Development of Expert System in Pest Management" discusses the application of Bayesian network which may greatly assist in development of innovative methods of selecting appropriate management option based on uncertain or tentative information of identified agroecological factors besides crop condition that farmers provide. Application of the methods thus developed in information and communication technology (ICT)-based expert systems (ESs) of agriculture would automate and speed up the process of providing real-time decision support on insect pest management to the farmers.

Chapter "An IoT and Wireless Sensor Network-Based Technology for a Low-Cost Precision Apiculture" presents the foundational space of apiculture, challenges, and emerging techniques of the apiarists in beekeeping including precision apiculture and a low-cost IoT and remote sensing technology deployed for data gathering and monitoring the vigor and productivity of beehive colonies. The system has redundancy components and utilized the three-level hierarchical model including wireless node, local information server, and the cloud data server. The system also consists of sensors deployed across distinct locations to capture the beehives' temperature, relative humidity, and loads of the hives and honeycombs, and the information is stored to the local and cloud server. This system will complement the urgent need to countervail the honeybee colonies' collapse by helping the apiarists, apiculture sub-sector, and agriculture sector to embark emerging precision apiculture technologies.

Chapter "Adaptive Encryption Technique for Collaborative Cloud Environments" presents a realistic solution to the problem where numeric data are transformed into alphanumeric types, and hence, there is minimal probability of storing encrypted data in the existing numeric field. The proposed algorithm allows translucent verification of intensity encryption that does not modify the data field category or the fixed time span.

Chapter "Role of Internet of Things and IT in Health Care" explores in depth the importance of the Internet of things devices in health care and also the role of information technology in managing the massive volume of high-security medical information of patients. This chapter proposes few IoT applications in rural health care and ways to boost primary health desires of the developing nations.

Chapter "Adopting Big Data Analysis in the Agricultural Sector: Financial and Societal Impacts" summarizes the state-of-the-art analytical methods that have been recently used in the agriculture industry, along with their financial and societal impacts.

Chapter "AI-Based Yield Prediction and Smart Irrigation" acknowledges the past breakthroughs and emerging artificial intelligence-based techniques in precision farming specifically for yield prediction and smart irrigation. Artificial intelligence-based system provides sufficient information about crop yields at an early stage, and its associated smart irrigation management system is effective in judicious use of essential resources such as water and energy for agriculture.

Chapter "IoT-Based Smart Tree Management Solution for Green Cities" discusses key challenges faced by transformation of smart cities and integration of green technologies, initiatives, and existing systems that are in place to solve these challenges. With overpopulation, fast urbanization, and environmental changes, powerful administration of trees and plants is required like never before. We present an Internet of things (IoT)-based smart tree management architecture for cities which is able to track various characteristics of a given tree such as the air quality, sunlight level, sound pollution level, and other important characteristics for efficient city planning, afforestation plans, and higher-quality produce.

Chapter "Data Analysis and Prediction Using Big Data Analytics in Agriculture" is composed of four sections. The first section provides an overview of analytics on big data in agriculture. This is followed by the second section which highlights the different big data techniques in agriculture. The third section describes two case studies that apply techniques for big data analysis in the agricultural domain. The fourth section analyzes the various research areas related to agriculture that can be explored by the academicians and researchers. The summary of the chapter is provided at the end.

Chapter "Generic IoT Platform for Analytics in Agriculture" aims in creating a universal platform used especially in the field of agriculture for data transmission and reception from various sensors and plotting the statistical graphs on a dashboard. Apart from this as a purpose of providing easy usage and extensibility, smart devices can also be installed, which provide the plug-and-play capability of our system.

Chapter "Mission SDGs Through Food Waste Management: Nature and Approaches" attempts to assess and account the loss of economy for wastage of food items from global canvas to national perspective. This study has also focused on how to use IoT platform so that the food wastage can be reduced up to a considerable amount both in the supply chain and even in the household practices. This research work is based on secondary information like research papers, reports, and results of other relevant studies. The chapter has attempted to develop and devise a conceptual and strategic model where the IoT ecosystem can be incorporated to ensure real-time solutions and to curb on massive food wastage practices. If the model is implemented and practiced with appropriate case-specific modifications and customizations, it would make the economy more efficient and address different perspectives and dimensions of UN Sustainable Development Goals (UNSDGs) to a larger extent primarily in the Indian context.

Chapter "Frost Prediction in Highland Crops Management Using IoT-Enabled System and Multiple Regression" discusses the concepts of precision agriculture, applications of the Internet of things (IoT) in agriculture, the design and challenges of an IoT-enabled system for highland crop management, and utilized multiple regression as a frost prediction technique. The remote sensing devices deployed in the farm collect the frost climatic events including air velocity, temperature, and humidity, and transmit the information to the Web server. The system provides early warnings and frost forecast for the farmers using SMS. The system helped farmers as a useful resource to conduct frost protection activities on the farm, thereby reducing the frost harm to the crops.

We are sincerely thankful to the Almighty for supporting and standing at all times with us, whether it is good or tough times and given ways to concede us. Starting from the call for chapters till the finalization of chapters, all the editors have given their contributions amicably, which is a positive sign of significant teamwork.

The editors are sincerely thankful to all the members of Springer, especially Prof. Aninda Bose for the providing constructive inputs and allowing an opportunity to edit this important book. We are equally thankful to reviewers who hailed from different places in and around the globe, shared their support, and stood firm toward quality chapter submission.

Bhubaneswar, India — Prasant Kumar Pattnaik
Jabalpur, India — Raghvendra Kumar
Kolkata, India — Souvik Pal

Contents

About the Editors

Prasant Kumar Pattnaik Ph.D (Computer Science), Fellow IETE, Senior Member IEEE is a Professor at the School of Computer Engineering, KIIT Deemed University, Bhubaneswar. He has more than a decade of teaching and research experience. Dr. Pattnaik has published numbers of Research Papers in peer-reviewed International Journals and Conferences. He also published many edited book volumes in Springer and IGI Global Publication. His areas of interest include Mobile Computing, Cloud Computing, Cyber Security, Intelligent Systems and Brain Computer Interface. He is one of the Associate Editor of Journal of Intelligent & Fuzzy Systems, IOS Press and Intelligent Systems Book Series Editor of CRC Press, Taylor Francis Group.

Raghvendra Kumar is working as Associate Professor in Computer Science and Engineering Department at L.N.C.T Group of College Jabalpur, M.P. India. He received B. Tech, M.Tech and Ph.D. in Computer Science and Engineering, India, and Postdoc Fellow from Institute of Information Technology, Virtual Reality and Multimedia, Vietnam National University (VNU), Vietnam. He serves as Series Editor Internet of Everything (IOE): Security and Privacy Paradigm, Green Engineering and Technology: Concepts and Applications, publishes by CRC press, Taylor & Francis Group, USA and Bio-Medical Engineering: Techniques and Applications, Publishes by Apple Academic Press, CRC Press, Taylor & Francis Group, USA. He also serves as acquisition editor for Computer Science by Apple Academic Press, CRC Press, Taylor & Francis Group, USA.

Souvik Pal Ph.D, MCSI; MCSTA/ACM, USA; MIAENG, Hong Kong; MIRED, USA; MACEEE, New Delhi; MIACSIT, Singapore; MAASCIT, USA is Associate Professor at the Department of Computer Science and Engineering, Brainware University, Kolkata, India. Dr. Pal received his B. Tech degree in Computer Science and Engineering from West Bengal University of Technology, Kolkata. He has received his M. Tech and PhD. degree in Computer Engineering from KIIT University, Bhubaneswar, India. He has worked as Assistant Professor in Nalanda Institute of Technology, Bhubaneswar, and JIS College of Engineering,

Kolkata (NAAC "A" Accredited College). He has also worked as head of the Computer Science Department in Elitte College of Engineering, Kolkata. Dr. Pal has published several research papers in Scopus-indexed International journals and conferences.

IoT-Based Soil Health Monitoring and Recommendation System

Vaibhav Bhatnagar and Ramesh Chandra

Abstract Soil is the base of agriculture. Soil provides nutrients that increase the growth of a crop. Some chemical and physical properties of soil, such as its moisture, temperature and its pH, heavily affect the yield of a crop. These properties can be sensed by the open-source hardware, and they can be used in the field. In this chapter, a soil health monitoring system is proposed in which farmer will be able to monitor soil moisture, soil temperature and soil pH in his android smart-phone. The farmer will also get the recommendations of lime and sulphur on the basis of pH of the soil. The area of study is Jaipur, Rajasthan (26.9124° N, 75.7873° E). The proposed system is tested with the results taken from authorized laboratories. The proposed system is validated with *t*-test having no significance difference between calibrated values and laboratory-recorded values. The proposed system is implemented on android smart-phone so it is useful for farmers, agriculture scientists, agriculture professionals and IoT experts.

Keywords Internet of things · Fertilization · Soil health

1 Introduction

Soil is the upper most layer of the earth. Agriculture, which is the prime component of livelihood, cannot be expected without soil. There are so many chemical and physical properties of soil that affect the growth of crops such as soil moisture, soil pH, soil temperature, soil electrical conductivity and its humidity. Water present in the soil is known as soil moisture. Soil moisture acts as carrier of nutrients for crop and plant growth. Moreover, it itself acts as nutrient for crops and plants; it also helps

V. Bhatnagar (✉)
Department of Computer Science, S.S. Jain Subodh PG College, Jaipur, India
e-mail: vaibhav.bhatnagar15@gmail.com

R. Chandra
Cyber-Physical Systems Laboratory, Department of ICT and Natural Sciences, Faculty of Information Technology and Electrical Engineering NTNU - Norwegian University of Science and Technology, Ålesund, Norway

P. K. Pattnaik et al. (eds.), *Internet of Things and Analytics for Agriculture, Volume 2*, Studies in Big Data 67,
https://doi.org/10.1007/978-981-15-0663-5_1

to maintain the soil temperature [1]. Soil temperature is referred as warmth in the soil. It helps to maintain the life in the soil of germs. The ideal for soil organisms lies between 25 and 350 °C [1]. Soil humidity is the water present in the soil in vapour form. Soil humidity plays an important role in evapotranspiration that directly affects crop's water need. Soil pH is most prominent factor responsible for successful growth of crops and plants. As it is commonly known that range of pH lies between 0 and 14, and distilled water has exactly 7 pH (neutral). The ideal pH for crop growth lies between 6.0 and 7.0. If a sample of soil has pH less than 6.0, then it is acidic in nature and needs to be raised with the help of different liming material, on the other hand if sample of soil has pH greater, then it is alkaline in nature and needs to be lowered with the help of sulphur and gypsum. Table 1a and b shows fertilizer requirement for maintaining the pH level [2].

The objectives of the chapter are to show the current status of soil moisture, soil temperature, humidity and soil pH in the smart-phone of farmer. To show the pH of the soil, the pH sensor is needed to be calibrated with the soil which is the second objective, and the last objective is to recommend the fertilizer on the basis of pH of soil. The area of study is Jaipur, Rajasthan (26.9124° N, 75.7873° E).

This chapter is divided into five sections—it begins with introduction, the second part consists of literature review and research motivation, third part consists of design of proposed system, fourth part consists of results and discussion, and fifth part consists of conclusion and future work.

Table 1 **a** Limestone needed to raise pH. **b** Sulphur needed to lowering pH

(**a**) Approximate amount of finely ground limestone needed to raise the pH of a 7-inch layer of soil, tones per acre

Soil texture	From pH 4.5 to 5.5	From pH 5.5 to 6.5
Sand and loam sand	0.5	0.6
Sandy loam	0.8	1.3
Loam	1.2	1.7
Silt loam	1.5	2.0
Clay loam	1.9	2.3
Muck	3.8	4.3

(**b**) Tons of sulphur needed per acre to lower pH to 6.5 (1 ton of sulphur)

Original pH	Sandy soil	Clay soil
8.5	0.7–1.0	1.0–1.3
8.0	0.5–0.7	0.7–1.1
7.5	0.2–0.3	0.4–0.5

2 Related Work

In the year 2017, Encinas et al. [3] presented a prototype model of distributed monitoring system of different properties of aquaculture water quality, such as pH, temperature and availability of oxygen. The proposed prototype model is useful for improving environmental control, reduction of production cost, etc. Authors used Arduino sensors, Zigbee technology, Arduino module, web services, mobile application and database to design the system. The prototype is shown in Fig. 1 [3].

Using the Arduino module, sensors sense pH, temperature and availability of oxygen and sensed value which are transmitted to external cloud with the help of database and different web services. Smart-phones are connected through the cloud by which these sensed data can be visualized. As the future work, alarm conditions are to be applied in the system with help of artificial intelligence.

In the year 2009, Coyle et al. [4] designed and calibrated textile-based wearable sensor for sports performance. These sensors are comfortable in wearing in that manner a person can do exercise and can play a sport. To analyse the rehydration, the fabric-based pH sensor is used. This pH sensor used sweat of the body to sense the rehydration. This special sensor is depicted in Fig. 2 [4].

This above-mentioned pH sensor is calibrated with the help of artificial sweat of range 4–8 pH. Authors calibrated three times and illustrated that this pH sensor is sensitive to changes in pH of less 0.2 units and that the response is repeatable. LilyPad Arduino is used to control the sensors attached to the body.

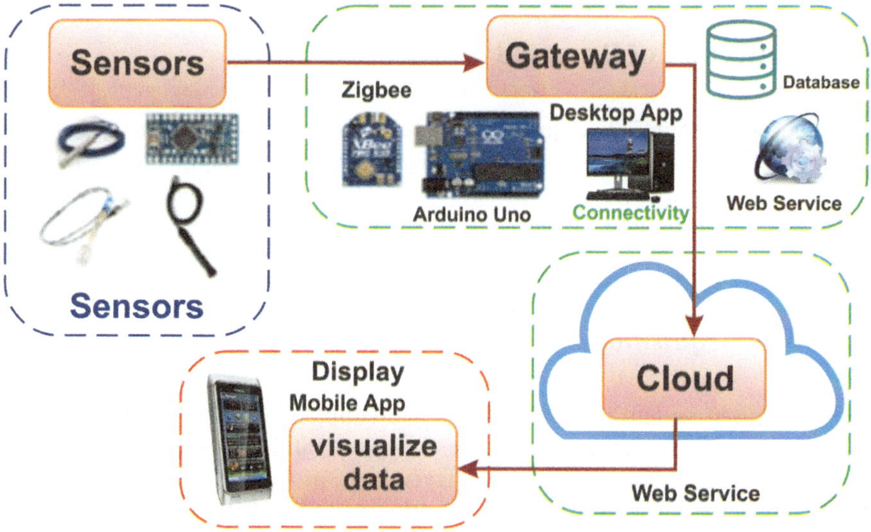

Fig. 1 Prototype model of distributed monitoring system of different properties of aquaculture water quality

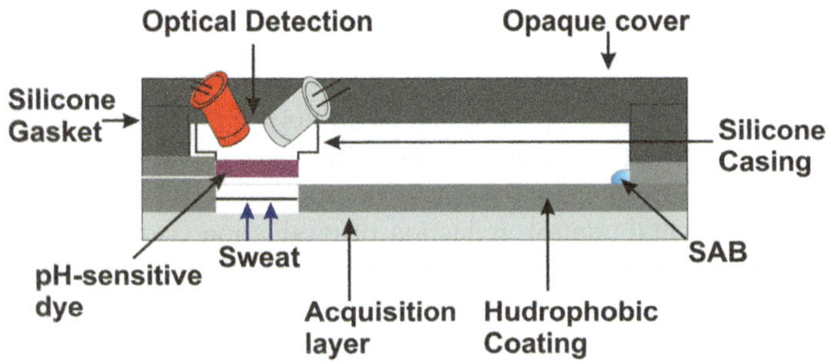

Fig. 2 Special pH sensor for sweat measurement

Yamaguchi et al. [5] presented a sensor network-based agriculture project called as e-kakashi project using different sensors such as camera, Zigbee module, temperature and humidity sensor. It was basically designed for improvement of productivity in farming system and reduction of vermin disease. This application was designed for iOS operating system. Lee et al. [6] presented IoT-designed agricultural production system for stabilizing the demand and supply of agricultural related products while developing the required sensors and prediction system for the growth and production amount of crops. The authors used HSDPA/Dual CDMA, TCP/UDP/IP/ICMP Protocol Support, 2.4 GHz IEEE 802.15.4 compatible, temperature sensor, humidity sensor, soil EC senor and pH sensor.

3 Architecture of Proposed System

The proposed system is two-layer architecture. First layer is data acquisition and calibration layer, and second layer is fertilization recommendation layer. The connectivity between these two layers is done by Bluetooth communication system (IEEE 802.15) [7]. Figure 3 depicts the overview of proposed system.

3.1 Data Acquisition and Calibration Layer

In the proposed system, data acquisition and calibration layer is responsible for acquiring raw data from the field using sensors, calibration of the input and sending the data to fertilization recommendation layer. In the proposed system, Arduino Uno [8] microprocessor is used. DHT11, SEN: 0161 and SEN: 0193 sensors are used in the system.

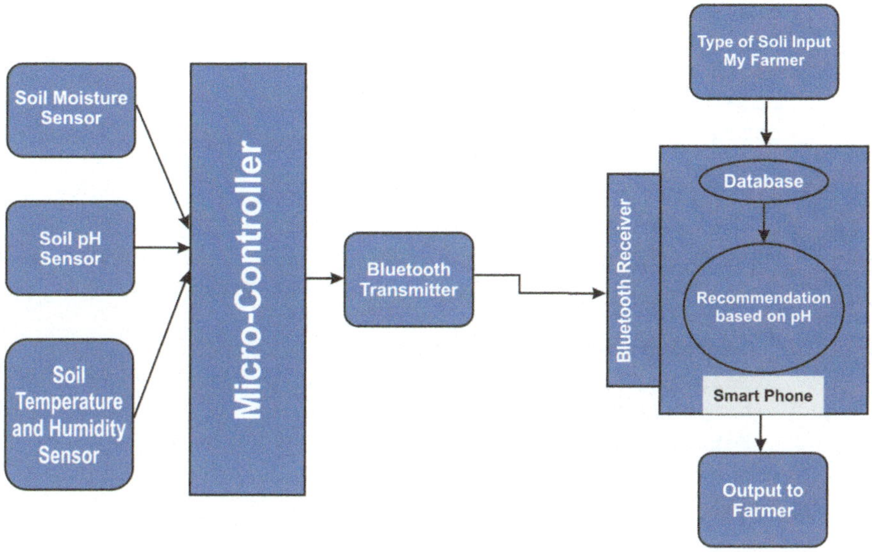

Fig. 3 Overview of proposed system

3.2 Soil Temperature and Humidity Sensor

Here DHT11 [9] sensor is used. DHT11 sensor is a analogue sensor which is used for sensing the soil temperature and soil humidity. Since library of DHT11 is directly available, this sensor generates the calibrated values, so manual calibration is not required. This sensor gives the values of both soil humidity and temperature. The sketch code of this sensor shown in Figs. 4 and 5 illustrates the DHT11 sensor [9].

3.3 pH Sensor

SEN: 0161 pH sensor [10] is used. SEN: 0161 sensor is typical analogue sensor which is used to sense the pH of soil. This sensor has been calibrated for soil of Rajasthan. The sensor is calibrated from 0 to 12 values. SEN: 0161 sensor is shown in Fig. 6 [10].

```
#include <dht11.h>
int dht11Pin= A0;
float temp;
float humidity;
dht11 myDHTSensor;
void setup() {
Serial.begin(9600);
}
void loop() {
  // put your main code here, to run repeatedly:
  humidity = getDhtHumidity(dht11Pin);
  Serial.println(humidity);
  delay(1000);
  temp = getDhtTemp(dht11Pin);
  Serial.println(humidity);
}
float getDhtHumidity(int dht11Pin)
{
          myDHTSensor.read(dht11Pin);
        return (float)myDHTSensor.humidity;
}
float getDhtTemp(int dht11Pin)
{
          myDHTSensor.read(dht11Pin);
        return (float)myDHTSensor.temperature;
}
```

Fig. 4 Sketch code of DHT11 sensors

Fig. 5 DHT11 sensor

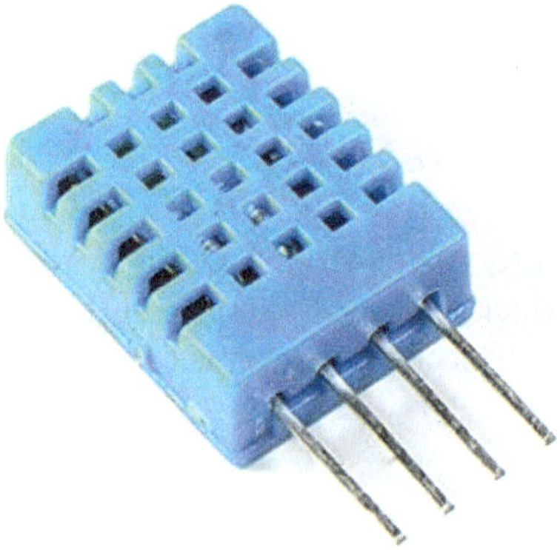

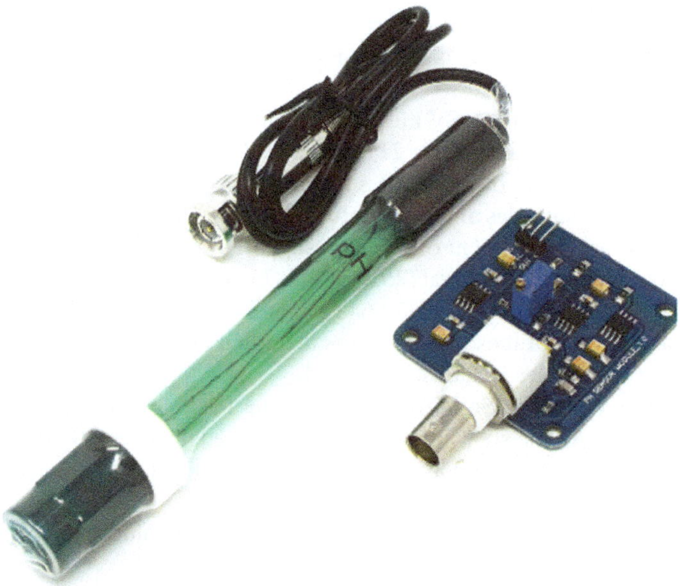

Fig. 6 SEN: 0161 sensor

The following steps are used for calibration:

- **Selection of Materials and Instruments**: To perform the calibration the pH rapid tester (Yeslike Digital PH Meter), oxalic acid [11] is used as acid agent, and potassium bicarbonate [12] is used as alkaline agent. pH buffer tablets of 7.00 are used along with 100 ml water with the help of measuring cylinder. A total of 21 samples ranging 0–12 pH are drawn for calibration.
- **Capturing the Reading**: The next step is to read the sensor values from the samples that are arranged in last step. For acquiring the inputs, Table 2 depicts the requirements.
  SD-card data logger [13] is used to save the values sensed by both the sensors. The reason behind using data logger is that the readings of sensors fluctuate with

Table 2 Requirement for acquiring input

S. No.	Name of device	Use
1	Arduino Uno	Microcontroller
2	D-cable	Connect Arduino to computer
3	SD-card data logger	To save data sensed by the sensors
4	Jumper wire	Male–male, female–male, female–female
5	Sketch	IDE for development
6	Breadboard	For connecting two or more input sensors

certain range of values. In this experiment, 500 readings of every experiment are being recorded. Same Sketch is used to develop the code for sensing and recording the values of both the sensors.

All the values are saved in.CSV file that are convenient to open in MS Excel. This sketch code has been executed for 21 times, and 500 values have been saved. Mean values of all 500 values of 21 samples are depicted in Table 3.

- **Selection of Soil Samples and Laboratory Testing**: The next step is to collect the soil samples for development of decision support system. The soil samples have been collected from Jaipur Rural (26.9124° N, 75.7873° E), Sambhar (26.9096° N, 75.1859° E), Kota (25.2138° N, 75.8648° E) and Jhalawar (24.5973° N, 76.1610° E) regions. Soil pH of these samples was identified in authorized soil testing lab. The pH of soil samples is shown in Table 4.
- **Reading the Values of Soil from Sensor**: The next step is to read the sensor value from soil. All the six soil samples are dissolved in water, 100 readings were taken from each sample, and average has been calculated. The values are shown in Table 5.

  Now the next step is to match these values with actual values received in above tables. These values are shown in Table 6 and Fig. 7.

Table 3 Mean values of pH sensor

S. No.	pH value	Mean value received
1	7	520
2	7.5	480
3	8	440
4	8.5	400
5	9	360
6	9.5	320
7	10	300
8	10.5	310
9	11	305
10	11.5	305
11	12	300
12	6.5	565
13	6	610
14	5.5	655
15	5	700
16	4.5	745
17	4	785
18	3.5	827
19	3	872
20	2.5	871
21	2	873

Table 4 pH of soil samples

S. No.	Soil type	Soil location	Soil pH
1	Clay	Jhalawar	7.9
2	Clay (yellow)	Kota	8.2
3	Clay (black)	Jhalawar	8.0
4	Loam	Sambhar	8.0
5	Silt sand	Kalwar, Jaipur	8.2
6	Pure sand	Jagatpura	7.8

Table 5 Mean values of pH of soil taken from pH sensor

S. No.	Soil type	Mean value
1	Clay	445
2	Clay (yellow)	435
3	Clay (black)	439
4	Loam	437
5	Silt sand	432
6	Pure sand	452

Table 6 Laboratory values and sensor values of pH of soil

S. No.	Soil type	Actual pH	Sensor value	Nearest pH value from
1	Clay	7.9	445	8
2	Clay (yellow)	8.2	435	8
3	Clay (black)	8.0	439	8
4	Loam	8.0	437	8
5	Silt sand	8.2	432	8
6	Pure sand	7.8	452	8

- **Setting the Values**: these values are set with the pH values by taking interval of 10. These values will be sent to the smart-phone and this table will be saved in the database of smart-phone (SQLite). The set values are shown in Table 7.

3.4 Soil Moisture

Here capacitance-based soil moisture sensor is used, i.e. SEN: 0193 [14]. This sensor is also a analogue sensor which is used to sense soil moisture. It gives value from 0 to 1023; therefore, calibration is required. The calibration of this sensor is also done with the same soil samples and with same data logger. In order to prepare the

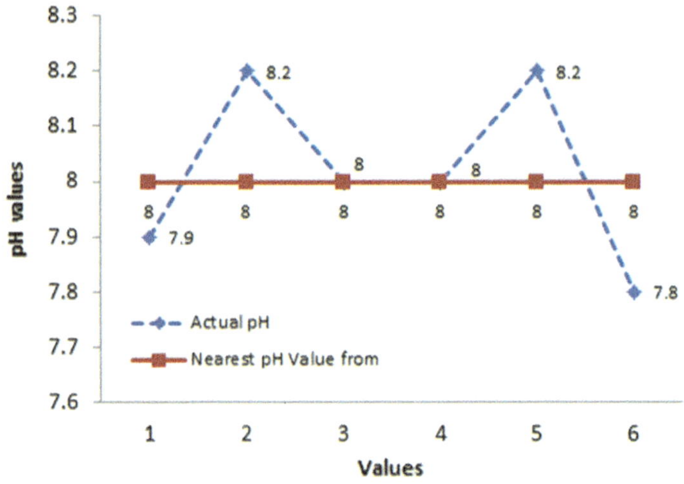

Fig. 7 Laboratory values and sensor values of pH of soil

Table 7 Set values of pH of soil

S. No.	pH value	Range
1	9.5 and greater	320 below
2	9.1–9.4	321–349
3	9	350–370
4	8.6–8.9	371–389
5	8.5	390–410
6	8.1–8.4	411–429
7	8	430–450
8	7.6–7.9	451–469
9	7.5	470–490
10	7.1–7.4	491–509
11	7	510–530
12	6.6–6.9	531–554
13	6.5	555–575
14	6.6–6.1	576–599
15	6	600–620
16	5.6–5.9	621–644
17	5.5	645–665

first sample, a random soil sample of large amount is collected from the field of more than 20 kg in weight. This 20 kg of soil is put in the oven so that the moisture already present inside the soil is evaporated by using gravimetric method [15]. Now from this sample of 20 kg soil, 500 g soil has been taken. Now this sample has 0% moisture, and the soil and sensors are inserted for reading the values. Further again, 500 g of soil has been taken from the collected sample, and 25 ml water is added to this sample. Now this sample has 5% moisture content, and sensors are put in the soil for reading the values. This same process has been carried out for 21 times. The last sample has 500 g of soil with 500 ml water content and has 100% moisture, and sensors are put for reading the values as shown in Tables 8, 9 and 10.

After the calibration, the physical properties sensed by different sensors would be sent to the smart-phone using Bluetooth communication [16] with the help HC-05. HC-05 is a Bluetooth chip that will be used to bridge the gap between data acquisition and calibration layer and fertilization recommendation layer. This Bluetooth chip is

Table 8 Soil moisture calibration of loam soil

S. No.	Size of sample (g)	Water content added	Moisture content, % water content	Sensor value	Range value
1	500	Nil	0	560	1000–560
2	500	25 ml	5	500	559–500
3	500	50 ml	10	400	499–400
4	500	75 ml	15	349	399–349
5	500	100 ml	20	342	348–342
6	500	125 ml	25	280	341–280
7	500	150 ml	30	279	279–279
8	500	175 ml	35	274	278–277
9	500	200 ml	40	269	276–269
10	500	225 ml	45	265	268–265
11	500	250 ml	50	254	264–264
12	500	275 ml	55	253	263–253
13	500	300 ml	60	251	252–251
14	500	325 ml	65	250	250–250
15	500	350 ml	70	248	249–248
16	500	375 ml	75	248	247–248
17	500	400 ml	80	247	247–247
18	500	425 ml	85	247	247–247
19	500	450 ml	90	240	246–240
20	500	475 ml	95	240	246–240
21	500	500 ml	100	240	246–240

Table 9 Soil moisture calibration of clay soil

S. No.	Size of sample (g)	Water content added	Moisture content, % water content	Sensor value	Range value
1	500	Nil	0	555	1000–555
2	500	25 ml	5	506	554–506
3	500	50 ml	10	449	505–449
4	500	75 ml	15	436	448–436
5	500	100 ml	20	430	435–429
6	500	125 ml	25	425	428–425
7	500	150 ml	30	420	424–420
8	500	175 ml	35	401	419–401
9	500	200 ml	40	363	400–363
10	500	225 ml	45	360	362–360
11	500	250 ml	50	352	359–352
12	500	275 ml	55	301	351–301
13	500	300 ml	60	285	300–285
14	500	325 ml	65	284	284–284
15	500	350 ml	70	282	283–282
16	500	375 ml	75	282	282–0
17	500	400 ml	80	282	282–0
18	500	425 ml	85	282	282–0
19	500	450 ml	90	282	282–0
20	500	475 ml	95	282	282–0
21	500	500 ml	100	282	282–0

connected with Arduino which will send the data to the smart-phone. In this case, android-based smart-phone is used. Figure 8 depicts the sketch code.

The experimental setup is shown in Fig. 9.

3.5 Fertilization Recommendation Layer

The next layer is fertilization recommendation layer which is responsible for depicting physical properties of soil and lime requirement to the farmer. This layer is developed in the form of android-based smart-phone application. The physical properties of soil sensed by sensors are depicted here using Bluetooth available in the smart-phone. As discussed in the above section, soil temperature and humidity sensor are pre-calibrated so direct sensed values are used. Soil moisture sensor and pH

Table 10 Soil moisture calibration of sand soil

S. No.	Size of sample (g)	Water content added	Moisture content, % water content	Sensor value	Range value
1	500	Nil	0	578	1000–578
2	500	25 ml	5	405	576–405
3	500	50 ml	10	348	404–348
4	500	75 ml	15	343	347–343
5	500	100 ml	20	340	342–340
6	500	125 ml	25	309	339–309
7	500	150 ml	30	301	308–301
8	500	175 ml	35	293	300–293
9	500	200 ml	40	284	292–284
10	500	225 ml	45	275	283–275
11	500	250 ml	50	273	274–273
12	500	275 ml	55	271	272–271
13	500	300 ml	60	270	270–270
14	500	325 ml	65	270	270–270
15	500	350 ml	70	270	270–270
16	500	375 ml	75	269	269–269
17	500	400 ml	80	268	268–268
18	500	425 ml	85	268	268–268
19	500	450 ml	90	261	267–261
20	500	475 ml	95	260	260–260
21	500	500 ml	100	261	260–260

sensor are already calibrated, and these calibrated values are already saved in SQLite database. Two fertilizers are recommended on the basis of pH, i.e. sulphur and limestone. The standard recommendations for fertilizer use have already been discussed in Sect. 1 (Introduction). The flow diagram of fertilization recommendation layer is shown in Fig. 10. The pseudo-code below gives the sulphur and lime requirements on the basis of pH and land area given by the farmer. The average value is taken for sulphur and lime requirements as not a specific value is recommended but a range is recommended.

```
#include <SoftwareSerial.h>
#include <dht11.h>
SoftwareSerial BluetoothSerial(8,9);
int f_sensor_pin=A0; int s_sensor_pin=A1; int t_sensor_pin=A2;
String final_data_send;
dht11 myDHTSensor;
String v1; String v2; String v3;
void setup() {
  pinMode(f_sensor_pin, INPUT);
  pinMode(s_sensor_pin, INPUT);
  pinMode(t_sensor_pin, INPUT);
  Serial.begin(9600);
  BluetoothSerial.begin(9600);
}
void loop() { band_fun();}
void band_fun()
{
  float value1 = analogRead(f_sensor_pin);
  Serial.print("value1--"); Serial.println(value1);
   delay(1000);
  float value2 = analogRead(s_sensor_pin);
  myDHTSensor.read(s_sensor_pin);
  value2 =myDHTSensor.temperature;
  Serial.print("value2--"); Serial.println(value2);
  delay(1000);
  float value3 = analogRead(t_sensor_pin);
  Serial.print("value3--"); Serial.println(value3);
   delay(1000);
 int iv1=int (value1); int iv2=int (value2) int iv3=int (value3);
  v1=String(iv1); v2=String(iv2); v3=String(iv3);
  final_data_send= String(v1+","+v2 +","+ v3);
  Serial.println(final_data_send);
  BluetoothSerial.println(final_data_send);
  delay(1000);
}
```

Fig. 8 Sketch code of data calibration and acquisition layer

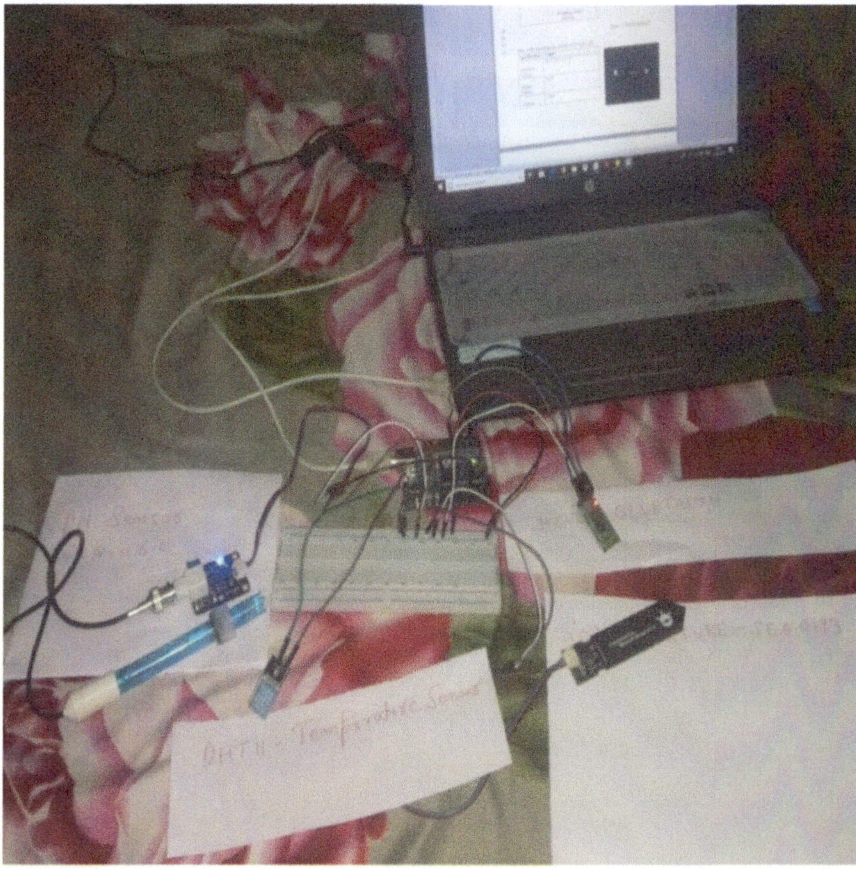

Fig. 9 Experimental setup of proposed system with pH sensor, soil moisture sensor, soil temperature sensor and Bluetooth chip

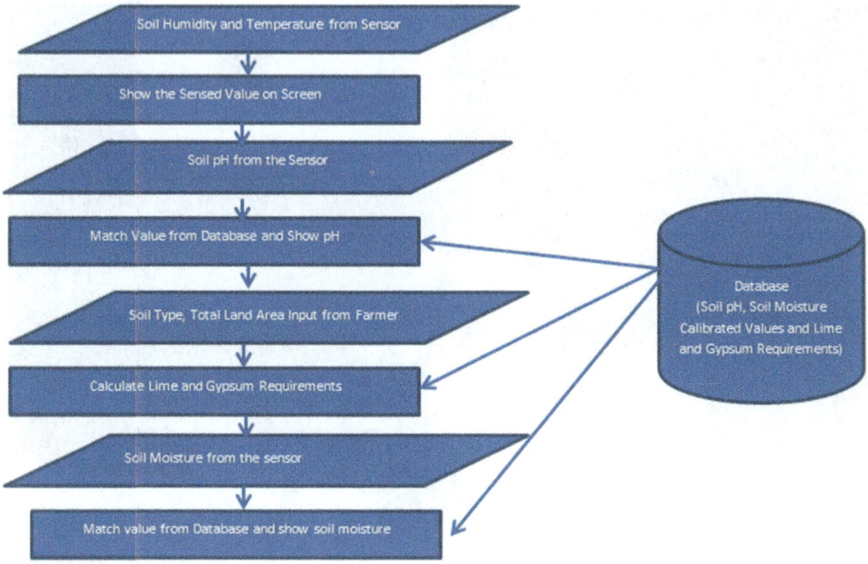

Fig. 10 Flow chart of fertilization recommendation system

```
Begin
    1.  Read type_soil        \\ To be Given by Farmer
    2.  Read land_sizeB       \\ To be Given by Farmer in Bigha
    3.  Read pH_r             \\ Raw value received from pH Sensor
    4.  Set land_sizeA :- land_sizeB * 0.25   \\ Convert Bigha into Acre
    5.  Set pH :- pH_r        \\ Match raw sensor value with Database
    6.  If (type_soil = "Sand") then
        6.1. If(pH>=4.5 && pH<5.5) then
             6.1.1.   Lime_ Requirement :- 0.5 * land_sizeA
             6.1.2.   Else if (pH>=5.5 && pH<=6.5) then
             6.1.3.   Lime_ Requirement :- 0.6 * land_sizeA
    7.  Else if (type_soil = "Loam") then
        7.1. If(pH>=4.5 && pH<5.5) then
             7.1.1.   Lime_ Requirement :- 1.2 * land_sizeA
             7.1.2.   Else if (pH>=5.5 && pH<=6.5) then
             7.1.3.   Lime_ Requirement :- 1.7 * land_sizeA
    8.  Else if (type_soil = "Clay") then
        8.1. If(pH>=4.5 && pH<5.5) then
             8.1.1.   Lime_ Requirement :- 1.9 * land_sizeA
             8.1.2.   Else if (pH>=5.5 && pH<=6.5) then
             8.1.3.   Lime_ Requirement :- 2.3* land_sizeA
End IF
    9.  IF (type_soil = "Sand" && pH=8.5) then
        9.1. Sulphur_Requirement :- 0.8* land_sizeA
        9.2. Else IF (type_soil = "Clay" && pH=8.5)
        9.3. Sulphur_Requirement :- 1.2* land_sizeA
        End IF
    10. IF (type_soil = "Sand" && pH=8.0) then
        10.1.Sulphur_Requirement :- 0.6* land_sizeA
        10.2.Else IF (type_soil = "Clay" && pH=8.0)
        10.3.Sulphur_Requirement :- 0.9* land_sizeA
             End IF
    11. IF (type_soil = "Sand" && pH=7.5) then
        11.1.Sulphur_Requirement :- 0.25* land_sizeA
        11.2.Else IF (type_soil = "Clay" && pH=7.5)
        11.3.Sulphur_Requirement :- 0.45* land_sizeA
             End IF
End
```

4 Results and Discussion

The proposed system is validated with authorized soil laboratory (shown in Appendix E). New soil samples are taken from the same place and tested again with different parameters. The values are shown in Table 11 and Fig. 11.

Table 11 Comparison between calibrated values and laboratory values

S. No.	Sample soil	Time	Sensor value	Calibrated value (%)	Laboratory value
1	Sandy soil	9.00 A.M.	341	20	23
2	Sandy soil	1.00 P.M.	291	35	31
3	Sandy soil	6.00 P.M.	390	10	9
4	Loam soil	9.00 A.M.	293	25	19
5	Loam soil	1.00 P.M.	279	30	28
6	Loam soil	6.00 P.M.	458	10	13
7	Clay soil	9.00 A.M.	510	5	7
8	Clay soil	1.00 P.M.	526	5	9
9	Clay soil	6.00 P.M.	546	5	11

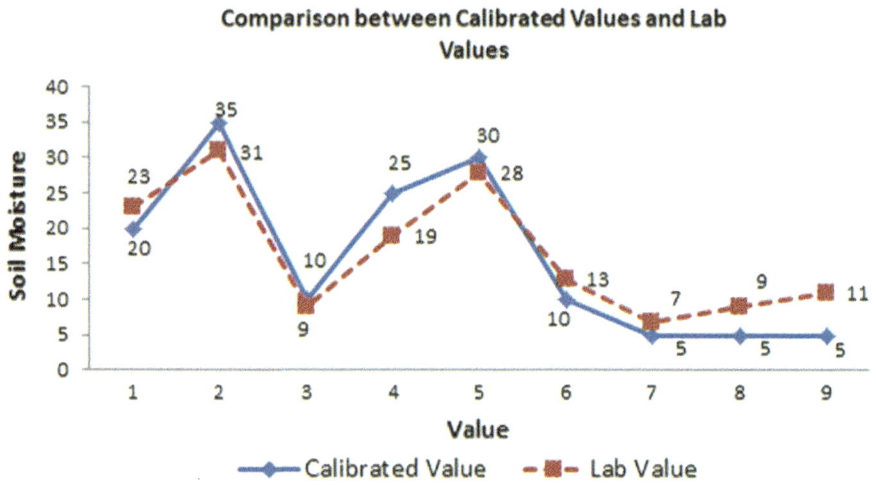

Fig. 11 Comparison between calibrated values and laboratory values

Table 12 Comparison of calibrated values and laboratory values of pH sensor

S. No.	Soil type	Actual pH	Sensor value	Nearest pH value from
1	Clay	7.9	451	8
2	Loam	8.0	430	8
3	Pure sand	7.8	449	8

To validate the calibrated value of soil moisture sensor from laboratory-tested values, Two-tailed paired t-test is applied [17] between calibrated values and laboratory-tested values. The t-test can determine whether the means of two sets of data are significantly different from each other. The value of T is obtained with the help of MS Excel, and the value is 0.688, that is less than tabular value of T at eight degrees of freedom at 5% level of significance 1.860; there is no significant difference between calibrated value and laboratory value. Therefore, it can be stated that the proposed model can be used in the field.

The calibration values of pH sensor are also validated with laboratory values. New soil samples are taken from the same place. The values are shown in Table 12. To validate the calibrated value of pH sensor from laboratory-tested values, two-tailed paired t-test is applied between calibrated values and laboratory-tested values. The value of T is obtained with the help of MS Excel, and the value is 0.225, that is

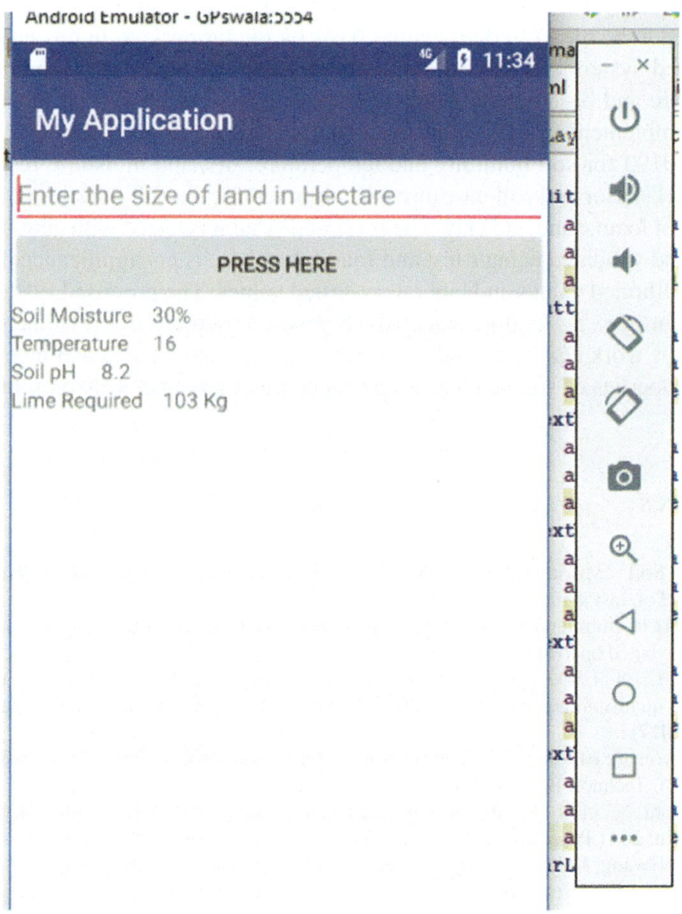

Fig. 12 Output of proposed system

less than tabular value of T at eight degrees of freedom at 5% level of significance 1.860; there is no significant difference between calibrated value and laboratory value. Therefore, it can be stated that the proposed model can be used in the field. The proposed output is shown Fig. 12.

5 Conclusion and Future Work

Soil is essential for life. It habitat for many insects, act as a filtration system for surface water and maintenance of atmospheric gases. It is an important component for agriculture. It is responsible for crop yield. Crops and plants get irrigation and fertilization through the soil only. There are some physical and chemical properties of soil that affect the crop yield, such as soil moisture, soil temperature, soil humidity, electrical conductivity, organic compounds and soil pH. These properties are required to be monitored on regular basis by the farmers, so in this chapter, IoT sensor-based system is proposed that monitors the soil temperature, humidity, pH and moisture and based on pH, lime and sulphur are recommended. The proposed system is implemented on android smart-phone. Three sensors DHT11, SEN: 0161 and SEN: 0193 for soil humidity and temperature, pH, and moisture, respectively, are used. pH sensor and soil moisture are calibrated for the soil of Rajasthan of three categories of loam, sand and clay. The proposed system is tested with authorized laboratories and validated using t-test and found that there is no significance difference between calibrated values and laboratory-tested values. The proposed system is beneficial for farmers, agriculture scientist, chemists, agriculture professionals and IoT experts. This work can be extended with more geographical area and incorporation of other properties of soil, such as electrical conductivity and NPK present in soil.

References

1. Topic: Soil Moisture, available at http://www.agriinfo.in/?page=topic&superid=4&topicid=274, last visited on 1 Mar 2019
2. Topic: pH Recommendations, available athttp://vric.ucdavis.edu/pdf/Soil/ChangingpHinSoil.pdf, last visited on 1 Mar 2019
3. Encinas, C., et al.: Design and implementation of a distributed IoT system for the monitoring of water quality in aquaculture. In: 2017 Wireless Telecommunications Symposium (WTS). IEEE (2017)
4. Coyle, S., et al.: BIOTEX—Biosensing textiles for personalised healthcare management. IEEE Trans. Inf. Technol. Biomed. **14**(2), 364–370 (2010)
5. Yamaguchi, N., et al.: E-kakashi project, an agri sensor network using ad hoc network technology. In: 2011 Proceedings of SICE Annual Conference (SICE). IEEE (2011)
6. Lee, M., Hwang, J., Yoe, H.: Agricultural production system based on IoT. In: 2013 IEEE 16th International Conference on Computational Science and Engineering (CSE). IEEE (2013)
7. Molisch, A.F., et al.: IEEE 802.15. 4a channel model-final report. IEEE P802 **15**(04), 0662 (2004)

8. Badamasi, Y.A.: The working principle of an Arduino. In: 2014 11th International Conference on Electronics, Computer and Computation (ICECCO). IEEE (2014)
9. Tianlong, N.: Application of single bus sensor DHT11 in temperature humidity measure and control system. Microcontrollers Embed. Syst. **6**, 026 (2010)
10. Topic: PH SEN: 0161, available at https://wiki.dfrobot.com/PH_meter_SKU__SEN0161, last visited on 10 Mar 2019
11. Hodgkinson, A.: Oxalic Acid in Biology and Medicine. New York, p 416 (1977)
12. Nitta, I., Tomiie, Y., Koo, C.H.: The crystal structure of potassium bicarbonate, KHCO3. Acta Crystallogr. **5**(2), 292–292(1952)
13. Mahzan, N.N., et al.: Design of data logger with multiple SD cards. In: 2013 IEEE Conference on Clean Energy and Technology (CEAT). IEEE (2013)
14. Topic: PH SEN: 0161, available at https://wiki.dfrobot.com/Capacitive_Soil_Moisture_Sensor_SKU_SEN0193, last visited on 10 Mar 2019
15. Vance, Eric D., Brookes, Peter C., Jenkinson, David S.: An extraction method for measuring soil microbial biomass C. Soil Biol. Biochem. **19**(6), 703–707 (1987)
16. Pahuja, Ritika, Kumar, Narender: Android mobile phone controlled bluetooth robot using 8051 microcontroller. Int. J. Sci. Eng. Res. **2**(7), 14–17 (2014)
17. Ruxton, Graeme D.: The unequal variance t-test is an underused alternative to Student's t-test and the Mann-Whitney U test. Behav. Ecol. **17**(4), 688–690 (2006)

Implementing IoT and Wireless Sensor Networks for Precision Agriculture

D. D. Dasig Jr.

Abstract The burgeoning effects of climate change to the agricultural systems have resulted in reduced agricultural productivity. These extreme weather events, changes, and increased in temperatures and precipitation patterns can be countervailed by deploying precision farming technologies in this milieu of a cyber-physical system. Among the precision farming technologies is the application of Internet of things (IoT) devices and sensors in farming management by observation, measurement, response, and monitoring of the food and horticulture crops inter- and intra-field variability. This chapter presents the rudimentary and recent advances in Agriculture 4.0. This chapter also afforded the deployment of an Internet of things (IoT)-based precision agriculture technology using wireless sensor networks (WSN) in crop management. An online crop management system (OCMS) was developed to complement the IoT platform layer. The application aimed at providing the farm environmental conditions including the temperature, water level, pH value, and dissolved oxygen measurements from the paddy field with food crops and horticulture crops being grown. Further studies could be done by analyzing the inter- and intra-field variabilities such as soil herbicide weed control, side dressing, late blight control, and haulm killing.

Keywords Agricultural system · Internet of things · Precision farming · Tomato crop management · Wireless sensor network

1 Overview

Fundamental to creating a functional economy and sound societies is the agricultural and food production, hence, sustainable agriculture. Today, the Internet of things (IoT) progresses toward becoming limestone on the planet, where every object or everything is interconnected and interacting with a specific task or functions.

D. D. Dasig Jr. (✉)
Graduate Studies, College of Science and Computer Studies, De La Salle University—Dasmarinas, DBB-B, 4115 West Ave, Dasmariñas, Cavite, Philippines
e-mail: dhuniel2006@yahoo.com

P. K. Pattnaik et al. (eds.), *Internet of Things and Analytics for Agriculture, Volume 2*, Studies in Big Data 67,
https://doi.org/10.1007/978-981-15-0663-5_2

In this chapter, the issues and advances in the agricultural sector are discussed. The effects of Industry 4.0 on ramifications and industrial modifications in the industrial and agricultural production systems have open new paradigm and opportunities for the researchers and scientists to conduct empirical investigations and technological innovations to suppress and address the increasing world population to feed, and the changing demands of the society.

There exists a growing demand for ubiquitous connectivity as the world forward with the Fourth Industrial Revolution (4IR). The 4IR has changed the many faces of society, hence, Agriculture 4.0 era. The recent advances in the primary sector of the economy such as in agriculture strike the notion of more productive agriculture, modern, and innovative.

These characteristics associated with Agriculture 4.0 have coined smart agriculture, diversified farming, digital farm, disruptive models in agriculture, and many others. Agriculture 4.0 will have a pivotal role in agribusiness and agriculture economics. These agricultural technologies will help improve food production, reduced costs, and promote process efficiency.

2 Introduction

The primary sector of the economy including the agriculture sector plays a vital role in the national and economic development of a country [1]. The activities being associated to the primary sector of the economy includes the extractions and harvesting of basic goods and raw materials from the earth such as agriculture and forestry, fishing, quarrying, mining, gathering, and hunting [2]. Agriculture is the science with regard to cultivating or farming, including cultivation of the soil for the growing and management of crops and the rearing of animals to produce and provide food, fleece, and other variety of agricultural produce. Agricultural development and an increase in agricultural production are crucial to broader national and economic development [3].

Agricultural crops production is primarily important for human use and domestic consumption. Crop production is highly dependent and affected by the type of soil, pests, irrigation, weeds, bacterial endophytes, residues, and yields. In recent years, agricultural crop production has drastically reduced due to short episodes of high temperatures during the reproductive period [4]. Also, a decreased in production was attributed to the extent of application of fertilizers since and intensive high-yield agriculture is heavily dependent on fertilizers [5].

In some studies, it was posited the water salinity in an irrigated agriculture, and the problem of solidity [6] and the endophytic bacteria latently residing in the plant tissue have impacted crops production [7], although agriculturists recommended the use of cover crops to improve the soil fertility and crop performance [8]. The need to remodel crop management should be done to improve the growth and development of the crop, the expected yields through modifications in the management practices of irrigation, biological interactions, countervail diseases, and weeds and pests using environment-friendly solutions.

Most studies have concurred deploying well-advanced technologies for future development in the world food, nutrition, and agriculture [9] to sustain the supply and demand. In most agricultural countries, lower agricultural productivity is associated with traditional farming methods and climate [10]. The burgeoning and long-term effects of climate change to the agricultural systems have resulted in reduced to adverse agricultural productivity, a problem burnt by agricultural-dependent communities and countries [11]. These extreme weather events, changes, and increased in temperatures and precipitation patterns can be countervailed by utilizing quantitative tools and techniques [12], climate model forecasting, and deploying precision farming technologies in this milieu of a cyber-physical system.

Among the precision farming technologies is the application of Internet of things (IoT) devices and sensors in farming management by observation, measurement, response and monitoring of the crops inter- and intra-field variability. This study presents the deployment of wireless sensor networks (WSN) in crop management to observe, measure, and monitor the farm temperature, humidity, and pressure using P89V51RD2 microcontroller, ZigBee module XBee module IEEE 802.15.4 standard, and RF transceiver.

3 Agroecology and Sustainable Agriculture

Fundamental to creating a functional economy and sound societies is the agricultural and food production, hence, sustainable agriculture. Sustainable agriculture is a prominent and a very vague discourse [13] defined as cultivating or the conduct of farming in a modest and environment-friendly approach to meet the present needs of society without compromising the needs of tomorrow's generation, given their comprehension of the relationships and interactions of the biological system.

Today, the agricultural sector should transform from a traditional farming system to e-agroecology to achieve sustainable agriculture. This e-agroecology integrates information and communications technology (ICT) to the ecological processes in agricultural production systems. The e-agroecology recognizes the importance of ICT in agriculture. Agroecology is the investigation of natural processes related to farming techniques as the ecology of food systems [14]. This farming approach recognizes environmental, economic, and social dimensions centered on food production which allows the future generations to be able to feed themselves [15, 16].

According to the study which utilized participatory learning, aside from having well-advanced and top of the line technologies in the agricultural production systems, there is a need to have experts in the agricultural fields who are passionate to share their knowledge, expertise, and empower communities, and agricultural stakeholders [17]. Participatory learning in agriculture developed the community of farmers and agriculturists equipped with ecological processes and technological skills to manage engineering and ICT-based solutions toward sustainable agriculture.

Agriculturists, scientists, and engineers should work together to improve the design, development, production strategies, and better decision-making.

3.1 Challenges and Developments in Sustainable Agriculture

Fundamental to global demands, agricultural work has always had been considered the most prevalent employment opportunities [18]. Agriculture, however, threatens the water supply since it is the major contributors to wastewater. Wastewater from agriculture is a global issue which has directly and indirectly adversely impacted the ecosystem and biological diversity. According to the United Nations Environment, water is principal and fundamental to life on our planet, yet this valuable asset has progressively been sought after and under risk [19] and water shortages in many regions [20]. Water covers multiple thirds of the planet, and our agricultural sector has always had produced wastewater including those chemicals from industrialized agricultural spaces which flows directly to waterways, lakes, wetlands, and springs.

Another persistent issue in agriculture is agricultural biotechnology. Although its well-advanced and scientifically investigated application has presented several promising benefits, it also brought a new dimension of challenges to agricultural sector [21]. Land clearing, land conversions, and industrial agricultural production have been seen as a factor to radiative forcing of climate because of secretions of greenhouse gasses, and ozone on crops and agro-ecosystems [22, 23].

3.1.1 Industrial Agriculture

Industrial agriculture is now the prevailing food production system; a large-scale production system heavily utilizes and promotes substantial utilization of compound composts, chemical fertilizers, and pesticides. Industrial agriculture posed soil depletion a made it vulnerable to soil erosion. This kind of farming technique develops "dead zones" or the oxygen-deprived spaces.

Industrial agriculture posed health risks as a source of air contamination, one of the reasons for agricultural production laborers respiratory and cardiovascular conditions. Although industrialized farming has been effective in creating huge amounts of sustenance, the eventual fate of food production is in peril because of loss of agricultural lands and reduced variations in the production of crops and livestock even more.

3.1.2 Conservation Agriculture

Today, several agricultural production systems have been integrated into the framework of large-scale food production including conservation agriculture. In this context, it is characterized as a sustainable agriculture production framework including land cultivating practices with an embedded adaptation technique based on the yields and Ambiental and environmental conditions of the region or locality. CA is utilized as a soil management practice to shield the soil from disintegration, degradation, and erosion [24], as well as to improve its quality.

CA advocated the minimal soil disturbance, permanent soil cover combined with rotations as a sustainable agricultural production system [25]. CA will help the agriculture sector to efficiently and effectively provide the need of the human population while making the agricultural land productive [26] to help economy advance toward agricultural sustainability and biodiversity conservation [27, 28].

3.1.3 Climate-Smart Agriculture

Another transformative approach in the agricultural production system is sustainable intensification and climate-smart agriculture (CSA). Climate-smart agriculture (CSA) is a methodology that aides' activities expected to change and recalibrate agricultural systems to successfully bolster improvement and guarantee sustenance security in a changing climate. CSA approach is characterized as one of the rural practices that economically viable and helped increment the profitability in agriculture with space for flexibility while diminishing ozone-depleting substance emissions.

Climate-smart agriculture can be characterized as a transformative approach toward agricultural efficiency and livelihoods by transforming and reorienting agricultural systems [29]. Climate-smart agriculture (CSA) compliments sustainable intensification which essentially important in adapting to climate change [30], increased in food security and adaptive climate mitigation capacity in a sustainable way [31, 32] while providing the increasing demands of food, fuel, and fiber [33].

3.1.4 The e-Agroecology

Due to the increasing threat to the natural resources brought by modern agriculture, there have been approaches and diversified farming methods applied by the rural social movements to attain sustainable agriculture [34, 35]. One of the frameworks used to address the issues and challenges in agricultural production is the agroecology. Primarily, agroecology was used in crop production [15, 36]. Agroecological approach dealt with farming systems to preserve and feed future generations considering the economic, social, and environmental dimensions [15, 37].

In this study, the e-agroecology was proposed to integrate technological ramifications into the agricultural production systems centered on food production considering the environmental, social, economic, and technological dimensions (ESET) as illustrated in Fig. 1.

This farming approach will utilize the cyber-physical age instrumentation and communications techniques including the Internet of things (IoT) devices and wireless sensor networks.

Fig. 1 e-Agroecology (ESET) model

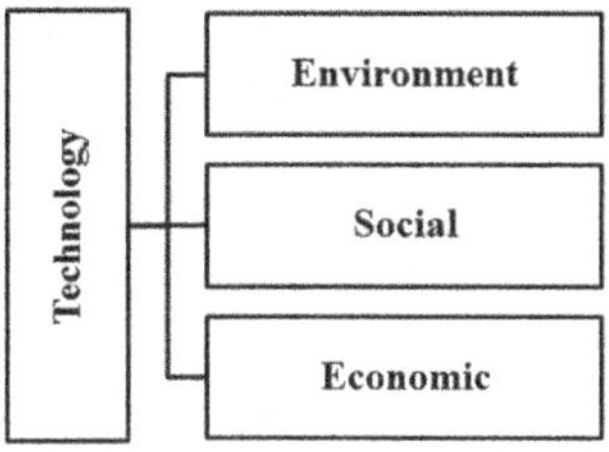

3.1.5 Precision Agriculture

The industrial ramifications of the Fourth Industrial Revolution (4IR) have reshaped the society with the technological advances, disruptive technologies, and emerging business models. It has disrupted the conventional industries with new technical models and technologies. In the agricultural sector, the use of drones, the Internet of things (IoT), and blockchain have been notably considered as the agritech sector welcomes Agriculture 4.0. The use of these new technologies to farming management is referred to as precision agriculture. Aside from the deployment of technological innovations in the farming and crop management, precision agriculture is seen as a technified model to farming management to increase crops productions and making sure of environmental, social, and economic sustainability.

The agricultural sector has been primed to develop and produce food to feed the world. Over the years, there have been innovations in the agritech using emerging technologies including aeroponics and hydroponics. These methods utilized engineering and information technology the same as precision agriculture. Precision agriculture is a farm or crop management approach utilizing information technology and engineering to ensure the optimal health of the soil and increase productivity. With the use of information technology services, the farmers could be able to conduct the processes of observing and measuring as well as able to immediately respond to the inter- and intra-field crops variability. Precision agriculture has been seen to boost the agricultural sector as an effective method in farm and crop management.

Notably, in the mid of 1980s, precision agriculture was initiated to blend with agriculturists and farmers way of doing the farming processes with the available and emerging technologies [38]. Precision agriculture is an inventive approach, coordinated and universally standardized method intends to further the effectiveness of technologies and soil use and to decrease the vulnerability of decisions required to control minor departure from farms [33]. In deploying technology for precision farming, it would involve a well-designed technology utilizing sensors, information systems, improved apparatus, and information management system [39, 40]. These spatially planned and deployed sensing devices will be the gathering tools for sensing the environmental and Ambiental conditions of the farms, hence improving the farmer management system and production, as well as ecological quality [41].

Precision agriculture also advocated farm-site management from a specific approach to large-scale site management to improve soil quality, pesticides distribution to support farmers to increase the profitability of yield creation, improve crops and farm produce quality, and protect the environment [42]. Other precision agriculture technologies used optical lists from hyperspectral remote detecting in the evaluation of vegetation biophysical factors both in forestry and agriculture management [43]. Remote sensing techniques and detection applications in precision agriculture began with sensors for soil natural issue and have rapidly diversified toward satellite, flying, and handheld or tractor-mounted sensors [44]. Spatial estimations of yield utilizing mechanical advances like in a hurry yield monitoring systems [45], while the use of the airborne remote-detecting system has been recognized worldwide as a promising strategy for identifying and mapping weeds in yields [46].

4 Agriculture 4.0

With the onslaught of the Fourth Industrial Revolution (4IR), there have been ramifications and disruption to the sectors of the economy. The 4IR can be characterized as the era of cyber-physical systems where automation and advancements to industries were shaped and heavily influenced by information technology. With the automation and machine-driven production and technology-driven processes, there have been an increased in technology proficiency requirements, higher demand of engineers and information technology professionals, and enabled service-centric and customer-centric organizations.

The 4IR has brought emerging and innovations to current technologies including Internet of things, artificial intelligence and machine learning, robotics and automation, virtual, and commercial reality. These modifications and ramifications are attributed to advancements in the process and technology in the production to better the lives of humanity. The Fourth Industrial Revolution or commonly known as Industry 4.0 has changed the society with technological concepts and solutions, top of the line technologies to answer the recurring problems [47, 48]. It was opined to realize mesh economy of scale and the economy of scope.

With the Industry 4.0, the agricultural sector has been seen to be innovative to be able to adapt to the changing demands of society today with the concept of Agriculture 4.0. Agriculture 4.0 has invigorated agricultural systems to address challenges and issues in soil management, fertilizer, pests, crop management, yields, plant stresses, etc. [48, 49]. The synergies with combination and cooperative energies of the most recent ICT advancements such as embedded system, real-time system, telecommunication, web automation, mobile communications and ICT infrastructure, remote sensing, and wireless communications have pinned into the agricultural sector geared toward smart agriculture and precision agriculture [49–51].

Agriculture 4.0 was coined as the function and work of agricultural engineering in the design, development, and deployment of systems including drones, Internet of things (IoT), smart and vertical farming, artificial intelligence, robotics, and solar energy [52] to advocate low-cost farming and increasing yields and productivity [53]. In the study for agro-technology, it was emphasized that Agriculture 4.0 utilizes a decision support system to address climate and weather forecasting, estimation of complex, and ill effects of climate to crops and soil [54].

The Agriculture 4.0 has opened opportunities toward the application of drones in farm and crop management, using TCP/IP protocols, image processing for plant disease identification and management, machine learning and classification techniques, and a complete data-driven agriculture value chain. The uses of the Internet of things and machine learning have produced an optimal management approach in agricultural production system [50]. Researchers concur that Industry 4.0 has boosted the economies and industrial upheaval. These advances in the distributed computing were pivotal to foster the design, development, and deployment of smart agricultural systems, precision farming [55]. Industry 4.0 has immensely revolutionized the agribusiness models and transformed the agricultural sector [55, 56].

In the context of Agriculture 4.0, the agri-food value chain considers and interconnected stakeholders [57] who advocate an enhanced productivity, and with the support of policy-makers in the technology revolution [58], this era will achieve higher productivity in agriculture, better marketing strategies using technology, reduced food wastes, and greater eco-efficiency [57–59].

5 Internet of Things (IoT)

The Internet of things (IoT) is one of the ramifications in this Industry 4.0 era. IoT is a system of interrelated and connected computing devices, a cyber-physical system, and digital machines with unique attributes and identifiers with the ability to transfer data over a network. Internet of things is a system which combines the Internet and the emerging and future technologies [60]. Because of the Internet of things (IoT), the ability to gather, transfer, and process data without human intervention has been made possible; hence, human–human and human–computer interactions (HCI) have been eliminated from the architecture [60, 61]. IoT is characterized as a global Internet-based architecture which utilizes IoT devices, IoT gateway, and IoT platform layers [61, 62] with the networked interconnection of communicating things equipped with ubiquitous intelligence [62, 63].

The Internet of things (IoT) as a technological paradigm is also considered as the Industrial Internet because of its applications to the industries and realm of the physical world [64] to provide secured and powerful industrial systems [62, 64, 65]. Internet of things (IoT) also known as the Internet of everything is a system that utilizes spatially planned and distributed remote sensing devices and actuators, wireless, and mobile devices with an embedded system to seamlessly collect and transfer data over an open access system in the IoT environment [66–68]. With the technological revolution of 4IR, Internet of things has been seen to provide numerous opportunities to industrial and agricultural sectors given the support of academic and technical communities toward sophisticated and a more transparent and secured ubiquitous world [69].

Fig. 2 IoT architecture

The Internet of things as an Internet-based technical architecture utilizes the IoT architecture as illustrated in Fig. 2. The IoT architecture affords the three layers including the IoT device layer, IoT gateway layer, and the IoT platform layer. Each layer of this architecture is characterized according to their functions to collect and transfer data from the source to the physical world of the IoT solution.

The Internet of the future or the Internet of things (**IoT**) has been seen to have grown rapidly in recent years. It has become the domain of discourse and research focus to many which embed IoT devices such as RFID, the sensors and actuators into the design and development of an IoT architecture IoT solution [64]. The expeditious developments of the Internet of things and the cyber-physical systems have even increased the demands of the IoT and smart devices [65] as these devices are main system components to serve as the bridge between the physical and virtual environments [70].

Looking at the IoT architecture's IoT device layer, the IoT devices are spatially distributed purposely embedded in the design for sensing and actuating the physical environment such as Ambiental conditions [71] and collecting information [72] from the locale for transfer to the IoT gateway layer using a protocol [65, 70]. These IoT devices are also connected to multiple devices to connect to the Internet to facilitate communication to the cloud server and utilize the cloud computing services in the IoT platform layer [73]. These IoT devices are geographically deployed in heterogeneous platforms, topographically dispersed over heterogeneous stages, traversing various administration and functional areas [71].

In the Internet of things (IoT), the IoT gateway layer is considered as the software program or a physical device that bridges the connection between the IoT device layer and IoT platform layer. This layer is responsible for connecting the sensing and actuating devices and the heterogeneous platforms or applications in the IoT platform layer. It is exemplified by providing additional and improved data security for the intelligent system or network as well as the data that it transports [74]. Primarily, IoT gateway is the enabler of local communication or a short-range communication between the sensors, actuators, and other IoT devices commonly applied in industrial and production systems, public security system, smart communities and buildings, logistics [74], and even in agriculture.

In a design of an IoT gateway-centric architecture, the IoT architecture was used to develop novel M2M services by providing real-time interactions between the smart things and the mobile clients [75]. Another implementation of smart IoT gateway has emphasized the applicability of join point of a heterogeneous network [76], Zigbee

Adhoc network, cable network, and wireless LAN to countervail design issues and demand for ubiquitous connectivity [77].

The third layer of the IoT architecture is the IoT platform layer. It is a multi-layer technology or commonly referred to as middleware which facilitates data communication, flow, device management, and application support and management. IoT platform is the enabler for provisioning of the IoT solution services, management of IoT services, and the automation of connected things and devices. The Internet of things universe has increased the demands of IoT platforms to reduce IoT solutions development cost, risks, and scope. These commercially available and open sources IoT platforms have even help to foster IoT solutions development's time-to-market. These IoT platforms maybe an end-to-end solution, connectivity, cloud platform, or data platform.

These IoT platforms play a pivotal role in the IoT solution development for industrial applications since it enables value for a sector or business and its customer. In the context of the agricultural sector, IoT platforms help developers and farmers to select appropriate platform and services which heavily rely on the type of attributes and defined system functions. IoT platform allows solutions users to be able to collect the needed data, store and manage the same data, and process into actionable insights using analytical models and visualization techniques for a data-driven platform [78]. In the study enabling IoT ecosystem in the deployment of the Internet of the future applications, and ramifications, the researchers explored the functional design of the system using a platform-specific API and IoT solutions information models [79].

The proliferations of ICT and engineering solutions have our cities transformed into smart cities and smart communities utilizing IoT hubs [80]. These IoT hubs subscribed to the cloud-based platform or cloud computing services with the advent of Platform-as-a-Service (PaaS) architecture [78, 80] to support IoT hubs for the smart cities and smart communities' services. Among other revolutionary IoT platforms to manage IoT devices and applications in Industry 4.0 is the blockchain technology's bitcoin deployed to cashless communities, financial, and banking sector [81]. In digital agriculture, a "Phenonet" was proposed by enhancing an IoT platform semantically enhanced IoT platform to address issues of interoperability, IoT silos, among others [81, 82].

6 The Wireless Sensor Network

In the wireless sensor network, the main building block is the sensor node. A sensor node or a "mote" is a node in a sensor network. A sensor node will observe the climatic and environmental conditions, process, and gather the sensory information, and it communicates with other nodes in the network of sensors. A sensor node in a sensor network consists of the sensor unit, processing, communication, and power units. A wireless sensor network is a group of geographically deployed and dispersed sensors for gathering, monitoring, and recording the environmental and Ambiental conditions [83], then organized these collected data at a central location.

A wireless sensor network has immensely reinforced numerous industrial and application-specific use since it has been re-engineered into its smaller, cheaper, and intelligent features [83] as an integrated micro-devices and micro-technology, digital electronics, wireless, and telecommunication devices [83, 84]. This compact design and feature-rich embedded in the sensor and actuators have been given importance in the areas of research, embedded system, real-time system, measurements, and network implementations [85]. The sensor nodes' sensing capabilities, power management, routing, and transfer protocols are specifically designed based on the purpose of the application [85, 86].

Technical communities and researchers have deployed wireless sensor networks using different design, protocols, methods, algorithm, estimations, and enhancements to align with specific system characteristics. A T-MAC WSN was developed to deploy a network with low message rate and lower sensitivity to latency [87]. Another study advocated Energy-LEACH protocol to improve the choice method of cluster head [88]. Another study on estimation of convex position was proposed to estimate the sensor network unknown node position [89]. A medium access control (MAC) protocol was proposed for a WSN that allows the network of devices to collaborate for common applications [90]. A low-power transceiver was effectively used to advance an energy-efficient WSN using hierarchical cluster algorithm [91].

Another WSN deployed to a vineyard setting utilized a 65-node multi-hop network to identify ways and identify the farm conditions [92]. Another system deployed to monitor different parameters of distributed crop aimed at improving the agricultural production by utilizing a multiplicity of emerging and innovative technologies [93]. Another innovation in WSN was a pervasive wireless sensor network consists of connected cheap and simple devices capable of self-configuration network to transform agriculture utilizing precision technologies for agriculture [94]. A Zigbee-based technology was used in precision agriculture with a sensor network based on the IEEE 802.15.4 standard aimed at collecting the data of climatologically and other environmental properties [15, 94].

7 The System

This study presents an IoT-based precision agriculture technology using wireless sensor networks (WSN). The technology aimed at collecting, processing the Ambiental and environmental condition on the farm and provides notifications and recommended courses of actions when needed.

The technology removes the human-to-human and human-to-computer interaction utilizing the interconnected sensor nodes deployed spatially over the distributed mixed crop farm with food and horticulture crops (rice, corn, fruits, and vegetables). The next sections discuss the design and detail of the system.

7.1 IoT-Architectured Precision Agriculture Technology

In this study, the IoT device layer includes the sensors and IoT gateway layer, and the IoT platform layer. The architecture of the developed technology is illustrated in Fig. 3. All system components are connected to a P89V51RD2 microcontroller including the sensors (for four sensor nodes), Zigbee module, and RF transceiver. The IoT architecture was employed in the design and development of the technology with the IoT device layer, IoT gateway layer, and IoT platform layer. In the IoT device layer, the sensor node is composed of temperature sensor, pH sensor, dissolved oxygen sensor, and the water level sensor. The sensor node will observe the climatic and environmental conditions in the farm where food and horticulture crops are grown.

The sensing devices will process the gathered sensory information, and it will be communicated to the other nodes or the platform layer using the TCP/IP and Zigbee protocol. These sensors were deployed in the paddy field or an arable land primarily used to grow aquatic and semi-aquatic rice. A wireless sensor network and Zigbee, and TCP/IP protocol were used in the IoT gateway layer.

The cloud computing services were used with an online crop management system—a web and mobile compatible application. The data observed and measured is collected and displayed in a dashboard which signals the system to respond based on the needed parameterized measurements.

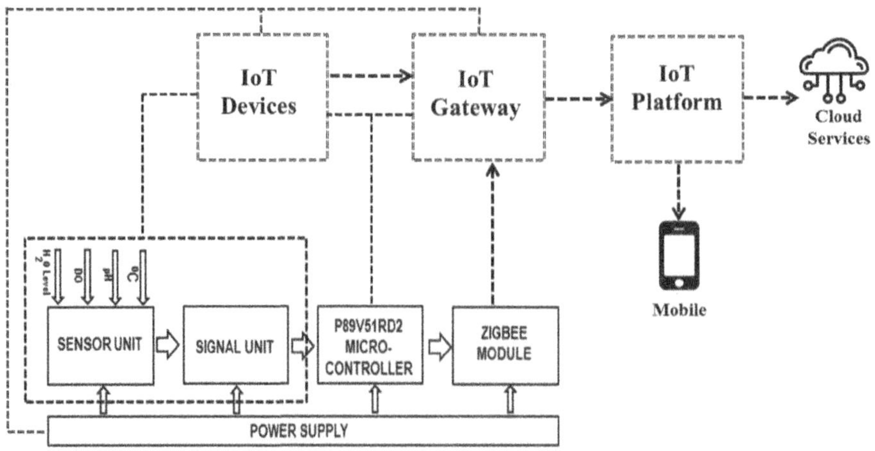

Fig. 3 Block diagram of precision agriculture technology

7.2 *The Precision Agriculture Wireless Sensor Network*

In this design, the sensor unit is composed of temperature, pH, dissolved oxygen, and the water level sensors. In measuring the water temperature, a thermometer sensing technology was used, and a glass electrode method was utilized for pH value. On the other hand, a membrane electrode technique was utilized for a dissolved oxygen sensor and a localized sensor designed out of an operational amplifier as a functional water level sensor.

In this system, there were four (4) sensor nodes spatially deployed in the farm of food and horticulture crops as illustrated in Fig. 4. Each sensor unit has a temperature sensor, pH sensor, dissolved oxygen sensor, and water level sensors. These sensors will also gather sensory information from the paddy field to measure and monitor the value of the parameter specified for each sensor. In measuring the water temperature in the field of aquatic and semi-aquatic rice, a thermometer sensing technique is used. For the pH value of water in the field an electrode, a method was used to identify

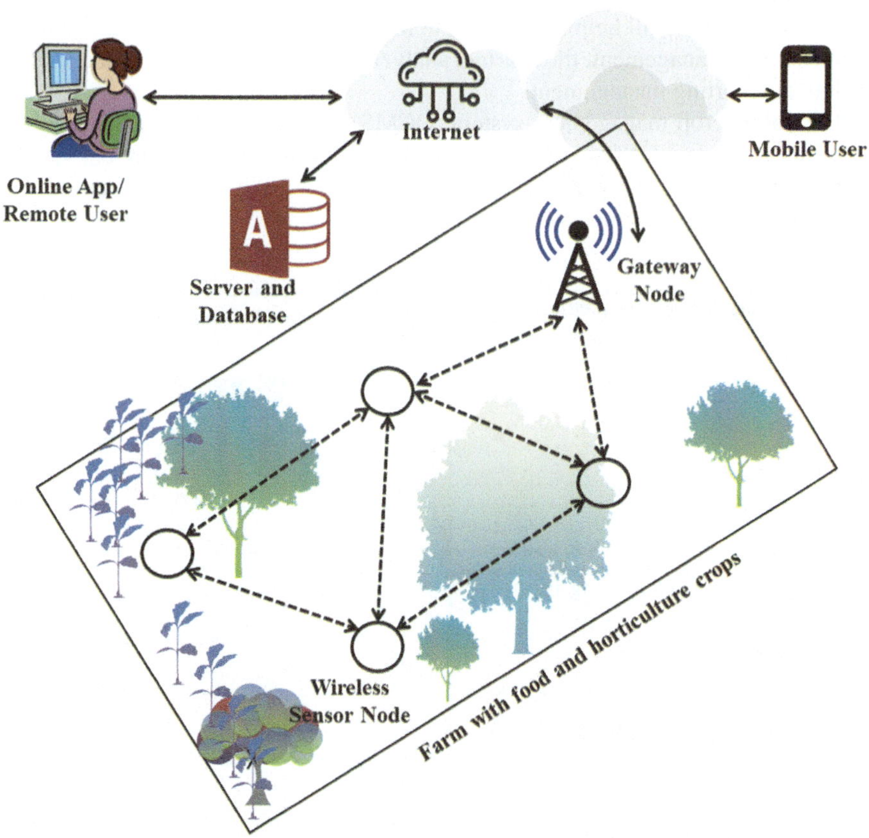

Fig. 4 Wireless sensor network (WSN)

and measure the current pH level of the water or acidity. This is to make sure that the plants or crops will have optimal and healthy conditions, although some of the crops thrive to survive even the pH value is below the range, 7; however, these plants and crops are heavily affected.

For aquatic and semi-aquatic crops, the dissolved oxygen in their habitat is primed for their survival. The membrane electrode technique was utilized for dissolved oxygen sensing and measurement. For the water level sensor, there has been a localized sensor designed out of an operational amplifier as a functional water level sensor.

7.3 Online Crop Management System

In this study, an online crop management system (OCMS) was developed as part of the IoT platform layer. The system aimed at providing the farm environmental conditions including the temperature, water level, pH value, and dissolved oxygen measurements from the arable farm with food crops and horticulture crops being grown. The OCMS will help the farmers to the optimized farm operation and management, data management, file sharing, analyzing the farm Ambiental data, and for analytical reporting management.

The online crop management system (OCMS) graphical user interface as illustrated in Fig. 5 provides descriptive visualization in terms of temperature in (°C), pH or water acidity in moles per liter, dissolved oxygen in milligrams per liter (mg/L), and water level measurements (ft).

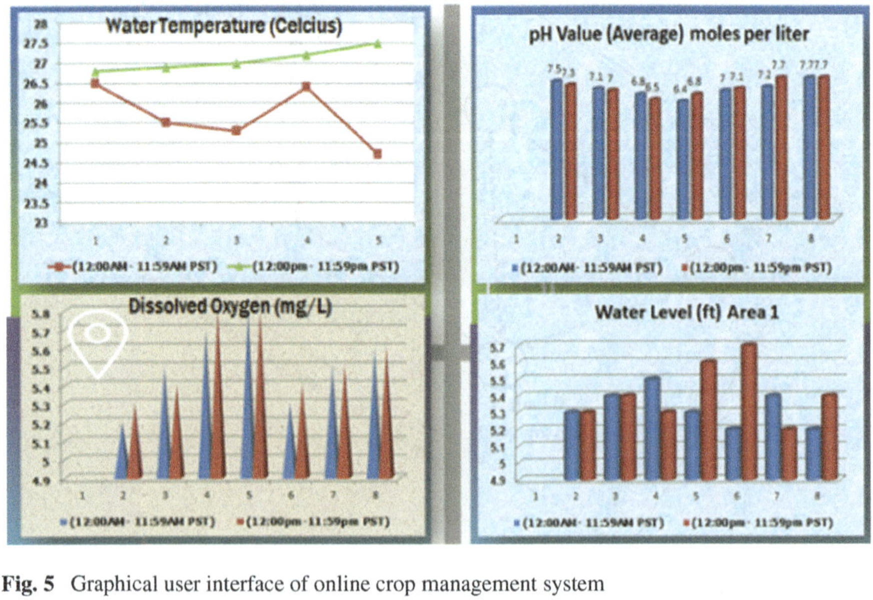

Fig. 5 Graphical user interface of online crop management system

8 Results and Discussion

The succeeding subsections will discuss the data gathered from the 6-day sensory observations during the field tests conducted in the farm with food crops and horticulture crops. The gathered data includes temperature in (°C), pH or water acidity in moles per liter, dissolved oxygen in milligrams per liter (mg/L), and water level measurements (ft).

8.1 Farm/Water Temperature

As illustrated in Table 1, the average water temperature in the farm was within the normal temperature of 25 °C except during the fifth day of testing. It also shows that field or the water temperature is higher than the air temperature. In farm management, when the water temperature is above 25 °C, there is a need for moisture to about five mm per day.

It was warm and stems elongation growth stage during the testing days; hence, each of the food and horticulture crops will utilize water as well in the paddy field. In this soil system of the paddy fields where the food crops and horticulture crops are grown, the soil temperature change also happens due to the competition in the ecosystem on water consumption [95].

8.2 pH Level

The pH level is the negative logarithm of hydrogen ions concentration in water. The pH is a scale being used to specify if the solution or water is basic or acidic. Accordingly, the pH in water is heavily influenced by the concentration of carbon dioxide. The farmers sometimes utilized dolomitic limestone and phytoremediation so that the pH level can be adjusted.

Table 1 Average temperature

Test day	Temperature (°C)	
	Soil/water	Air
1	24	22
2	24	23
3	24	22
4	25	23
5	26	24
6	25	23

°C Celsius

Table 2 Average pH level

Test day	pH level (moles/L)	
	a.m.	p.m.
1	6.5	6.5
2	6.4	6.4
3	6.5	6.4
4	6.3	6.3
5	6.4	6.3
6	6.4	6.4

a.m. 12:00 a.m.–11:59 a.m. (PHT)
p.m. 12:00 p.m.–11:59 a.m. (PHT)

The plot where testing has been conducted was a paddy field with a fertile riverine alluvial soil, and however, due to irrigation, drainage and daily data of precipitation, its desirability to ideally grow food crops such as rice and other horticulture crops are paralyzed due to acidity [96]. Based on the testing, there have been no recorded entries for the mean of pH level. This implies that the pH level of soil/water is within the desirable soil pH range between 5.5 and 6.5 as illustrated in Table 2.

8.3 Dissolved Oxygen

In the wet rice farming and food crop production such as rice, water quality is also an important indicator. The water flows to the plot during rain or irrigation schedule. These farming activities are important to make sure that the semi-aquatic organisms can respire to survive. Oxygen dissolves in the surface water due to the process of aerating action of winds.

Dissolved oxygen (DO) is the amount of gaseous oxygen (O_2) that has been dissolved water expressed in milligrams per liter (mg/L). Food crop such as rice can grow in bunded fields or paddy ones. Both fields need continuous flooding or irrigation to make sure that there is sufficient water to avoid supersaturation that can cause gas bubble disease to aquatic and semi-aquatic organisms.

The very high concentration of DO may pose lethal effects to the aquatic and semi-aquatic organisms and might deliberately destroy the soil. Table 3 illustrates that there was a variation; however, these values are within the range of desirable values for growing food crops. It also shows that during the testing, the paddy field was conducive for growing wet rice as a food crop and other horticulture crops.

Table 3 Average dissolved oxygen

Test day	Dissolved oxygen (mg/L)	
	a.m.	p.m.
1	5.0	5.1
2	5.2	5.1
3	5.2	5.1
4	5.3	5.4
5	5.3	5.3
6	5.4	5.5

a.m. 12:00 a.m.–11:59 a.m. (PHT)
p.m. 12:00 p.m.–11:59 a.m. (PHT)

8.4 Water Level

In growing food crops, a recommended continuous flooding or scheduled irrigation should be best provided after transplanting to provide the best growth and conducive environment. A desirable level of 3 cm water level is being maintained and is increased according to environmental conditions, or an increase in plant height. Water level and continuous flooding are also important to make sure weeds and other pests will not flourish in the paddy field. Depending on the type of rice being grown, the water level should also be monitored to make sure that the system could support the rice growth stages of vegetative, reproductive, and ripening.

As illustrated in Table 4, during the tests, the water level was sent to the analytical reporting system by the sensors were higher during the first day. These water levels from two areas/plots were vouched by the farming activities since the first day of the testing was conducted a few hours after the scheduled weekly irrigation. During irrigation, the water level is maintained at 5 cm from the ground, and during drainage, 3 cm from the ground is maintained.

Table 4 Average water level

Test day	Plot 1		Plot 2	
	Water level (cm)		Water level (cm)	
	a.m.	p.m.	a.m.	p.m.
1	4.4	4.3	4.4	4.4
2	4.0	3.9	4.3	4.2
3	3.7	3.5	4.2	4.1
4	3.5	3.5	4.0	3.8
5	3.5	3.4	3.7	3.6
6	3.4	3.4	3.4	3.4

9 Conclusion and Recommendation

This chapter aimed at providing the rudimentary and ramifications of Industry 4.0 in the agricultural sector. It presented the issues and challenges and recent advancements in the agriculture during the Agriculture 4.0 era. This chapter also presented the design, development, and deployment of an application of Internet of things (IoT)-based precision agriculture technology using wireless sensor networks (WSN) in food and horticulture crop management by collecting and processing the Ambiental and environmental condition on the farm. The same application provides notifications and recommended courses of actions when needed.

In this chapter, it was elaborated in the design and a development phase that the IoT architecture was utilized by the author which includes the IoT device layer including the sensors and IoT gateway layer, and the IoT platform layer. In this precision agriculture technology, there were four (4) sensor nodes spatially deployed in the farm of food and horticulture crops. Each sensor unit has a temperature sensor, pH sensor, dissolved oxygen sensor, and water level sensors. These sensors will also gather sensory information from the paddy field to measure and monitor the value of the parameter specified for each sensor.

Also, an online crop management system (OCMS) was developed as part of the IoT platform layer. The application aimed at providing the farm environmental conditions including the temperature, water level, pH value, and dissolved oxygen measurements from the arable farm with food crops and horticulture crops being grown. The OCMS will help the farmers to the optimized farm operation and management, data management, file sharing, analyzing the farm Ambiental data, and for analytical reporting management on temperature in (°C), pH or water acidity in moles per liter, dissolved oxygen in milligrams per liter (mg/L), and water level measurements (cm).

Further studies could be done by analyzing the inter- and intra-field variabilities such as soil herbicide weed control, side dressing, late blight control, and haulm killing.

References

1. Matsuyama, K.: Agricultural productivity, comparative advantage, and economic growth. J. Econ. Theor. **58**(2), 317–334 (1992)
2. Kenessey, Z.: The primary, secondary, tertiary and quaternary sectors of the economy. Rev. Income Wealth **33**(4), 359–385 (1987)
3. Alexandratos, N.: World food and agriculture: outlook for the medium and longer term. Proc. Natl. Acad. Sci. **96**(11), 5908–5914 (1999)
4. Teixeira, E.I., Fischer, G., Van Velthuizen, H., Walter, C., Ewert, F.: Global hot-spots of heat stress on agricultural crops due to climate change. Agric. For. Meteorol. **170**, 206–215 (2013)
5. Matson, P.A., Parton, W.J., Power, A.G., Swift, M.J.: Agricultural intensification and ecosystem properties. Science **277**(5325), 504–509 (1997)
6. Rhoades, J.D., Loveday, J.: Salinity in irrigated agriculture. Agronomy **30**, 1089–1142 (1990)
7. Hallmann, J., Quadt-Hallmann, A., Mahaffee, W.F., Kloepper, J.W.: Bacterial endophytes in agricultural crops. Can. J. Microbiol. **43**(10), 895–914 (1997)

8. Teasdale, J.R.: Contribution of cover crops to weed management in sustainable agricultural systems. J. Prod. Agric. **9**(4), 475–479 (1996)
9. Bruinsma, J.: World agriculture: towards 2015/2030: an FAO study. Routledge (2017)
10. Adams, R.M., Hurd, B.H., Lenhart, S., Leary, N.: Effects of global climate change on agriculture: an interpretative review. Climate Res. **11**(1), 19–30 (1998)
11. Rosenzweig, C., Parry, M.L.: Potential impact of climate change on world food supply. Nature **367**(6459), 133 (1994)
12. Adams, R.M., Rosenzweig, C., Peart, R.M., Ritchie, J.T., McCarl, B.A., Glyer, J.D., Allen Jr., L.H.: Global climate change and US agriculture. Nature **345**(6272), 219 (1990)
13. Velten, S., Leventon, J., Jager, N., Newig, J.: What is sustainable agriculture? A systematic review. Sustainability **7**(6), 7833–7865 (2015)
14. Francis, C., Lieblein, G., Gliessman, S., Breland, T.A., Creamer, N., Harwood, R., Wiedenhoeft, M.: Agroecology: the ecology of food systems. J. Sustain. Agric. **22**(3), 99–118 (2003)
15. Wezel, A., Bellon, S., Doré, T., Francis, C., Vallod, D., David, C.: Agroecology as a science, a movement and a practice. A review. Agron. Sustain. Dev. **29**(4), 503–515 (2009)
16. De Schutter, O. Agroecology and the Right to Food. Report presented at the 16th session of the United Nations Human Rights Council [A/HRC/16/49], vol. 8 (2011)
17. Pretty, J.N.: Participatory learning for sustainable agriculture. World Dev. **23**(8), 1247–1263 (1995)
18. Frank, A.L., McKnight, R., Kirkhorn, S.R., Gunderson, P.: Issues of agricultural safety and health. Annu. Rev. Public Health **25**, 225–245 (2004)
19. About Water. Retrieved from https://www.unenvironment.org/explore-topics/water/about-water
20. Pimentel, D., Berger, B., Filiberto, D., Newton, M., Wolfe, B., Karabinakis, E., Nandagopal, S.: Water resources: agricultural and environmental issues. Bioscience **54**(10), 909–918 (2004)
21. Shoemaker, R.A., Harwood, J.L., Day-Rubenstein, K.A., Dunahay, T., Heisey, P.W., Hoffman, L.A., Klotz-Ingram, C., Lin, W.W., Mitchell, L., McBride W.D., Fernandez-Cornejo, J.: Economic Issues in Agricultural Biotechnology (No. 1474-2016-120826) (2001)
22. Dioxide, A.C.: Status and issues concerning agricultural emissions of greenhouse gases. Agric. Dimensions Glob. Clim. Change **229** (1993)
23. Fuhrer, J., Booker, F.: Ecological issues related to ozone: agricultural issues. Environ. Int. **29**(2–3), 141–154 (2003)
24. Giller, K.E., Witter, E., Corbeels, M., Tittonell, P.: Conservation agriculture and smallholder farming in Africa: the heretics' view. Field Crops research **114**(1), 23–34 (2009)
25. Hobbs, P.R., Sayre, K., Gupta, R.: The role of conservation agriculture in sustainable agriculture. Philos. Trans. R. Soc. B: Biol. Sci. **363**(1491), 543–555 (2007)
26. Tscharntke, T., Clough, Y., Wanger, T.C., Jackson, L., Motzke, I., Perfecto, I., Vandermeer, J., Whitbread, A.: Global food security, biodiversity conservation and the future of agricultural intensification. Biol. Conserv. **151**(1) (2012)
27. Pretty, J.N., Noble, A.D., Bossio, D., Dixon, J., Hine, R.E., Penning de Vries, F.W., Morison, J.I.: Resource-Conserving Agriculture Increases Yields in Developing Countries (2006)
28. Bennett, A.F., Radford, J.Q., Haslem, A.: Properties of land mosaics: implications for nature conservation in agricultural environments. Biol. Cons. **133**(2), 250–264 (2006)
29. Lipper, L., Thornton, P., Campbell, B.M., Baedeker, T., Braimoh, A., Bwalya, M., Hottle, R.: Climate-smart agriculture for food security. Nat. Clim. Change **4**(12), 1068 (2014)
30. Campbell, B.M., Thornton, P., Zougmoré, R., Van Asten, P., Lipper, L.: Sustainable intensification: what is its role in climate smart agriculture? Curr. Opin. Environ. Sustain. **8**, 39–43 (2014)
31. Neufeldt, H., Jahn, M., Campbell, B.M., Beddington, J.R., DeClerck, F., De Pinto, A., LeZaks, D.: Beyond climate-smart agriculture: toward safe operating spaces for global food systems. Agric. Food Secur. **2**(1), 12 (2013)
32. Harvey, C.A., Chacón, M., Donatti, C.I., Garen, E., Hannah, L., Andrade, A., Clement, C.: Climate-smart landscapes: opportunities and challenges for integrating adaptation and mitigation in tropical agriculture. Conserv. Lett. **7**(2), 77–90 (2014)

33. Steenwerth, K.L., Hodson, A.K., Bloom, A.J., Carter, M.R., Cattaneo, A., Chartres, C.J., Jenkins, B.M.: Climate-smart agriculture global research agenda: scientific basis for action. Agric. Food Secur. **3**(1), 11 (2014)
34. Rosset, P.M., Martínez-Torres, M.E.: Rural social movements and agroecology: context, theory, and process. Ecol. Soc. 17(3) (2012)
35. Altieri, M.A.: Agroecology: principles and strategies for designing sustainable farming systems. Agroecology in Action (2000)
36. Altieri, M.A.: Agroecology: The Scientific Basis of Alternative Agriculture (1983)
37. De Schutter, O.: Agroecology and the right to food. Report Presented at the 16th Session of the United Nations Human Rights Council [A/HRC/16/49], p. 8 (2011)
38. Robert, P.C.: Precision agriculture: a challenge for crop nutrition management. In Progress in Plant Nutrition: Plenary Lectures of the XIV International Plant Nutrition Colloquium, pp. 143–149. Springer, Dordrecht (2002)
39. Gebbers, R., Adamchuk, V.I.: Precision agriculture and food security. Science **327**(5967), 828–831 (2010)
40. Stafford, J.V.: Implementing precision agriculture in the 21st century. J. Agric. Eng. Res. **76**(3), 267–275 (2000)
41. Pierce, F.J., Nowak, P.: Aspects of precision agriculture. Adv. Agron. **67**, 1–85. Academic Press (1999)
42. Adamchuk, V.I., Hummel, J.W., Morgan, M.T., Upadhyaya, S.K.: On-the-go soil sensors for precision agriculture. Comput. Electron. Agric. **44**(1), 71–91 (2004)
43. Haboudane, D., Miller, J.R., Tremblay, N., Zarco-Tejada, P.J., Dextraze, L.: Integrated narrow-band vegetation indices for prediction of crop chlorophyll content for application to precision agriculture. Remote Sens. Environ. **81**(2–3), 416–426 (2002)
44. Mulla, D.J.: Twenty five years of remote sensing in precision agriculture: key advances and remaining knowledge gaps. Biosys. Eng. **114**(4), 358–371 (2013)
45. Basso, B., Ritchie, J.T., Pierce, F.J., Braga, R.P., Jones, J.W.: Spatial validation of crop models for precision agriculture. Agric. Syst. **68**(2), 97–112 (2001)
46. Lamb, D.W., Brown, R.B.: Pa—precision agriculture: remote-sensing and mapping of weeds in crops. J. Agric. Eng. Res. **78**(2), 117–125 (2001)
47. Schellberg, J., Hill, M.J., Gerhards, R., Rothmund, M., Braun, M.: Precision agriculture on grassland: applications, perspectives and constraints. Eur. J. Agron. **29**(2–3), 59–71 (2008)
48. Dombrowski, U., Wagner, T.: Mental strain as field of action in the 4th industrial revolution. Procedia Cirp **17**, 100–105 (2014)
49. Abramovici, M., Göbel, J.C., Neges, M.: Smart engineering as enabler for the 4th industrial revolution. In: Integrated Systems: Innovations and Applications, pp. 163–170. Springer, Cham (2015)
50. Falkenthal, M., Breitenbücher, U., Képes, K., Leymann, F., Zimmermann, M., Christ, M., Kempa-Liehr, A.W.: Opentosca for the 4th industrial revolution: automating the provisioning of analytics tools based on apache flink. In: Proceedings of the 6th International Conference on the Internet of Things, pp. 179–180. ACM (2016, November)
51. Dasig, Jr, D.: User experience of embedded system students on Arduino and field programmable gate array (FPGA). In: Proceeding of the Second International Conference on Advances in Applied Science and Environmental Engineering-ASEE, pp. 124–128 (2014, February)
52. Yahya, N.: Agricultural 4.0: its implementation toward future sustainability. In: Green Urea, pp. 125–145. Springer, Singapore (2018)
53. Weltzien, C.: Digital agriculture or why agriculture 4.0 still offers only modest returns. Landtechnik **71**(2), 66–68 (2016)
54. Hoogenboom, G., Jones, J.W., Wilkens, P.W., Porter, C.H., Batchelor, W.D., Hunt, L.A., Boote, K.J., Singh, U., Uryasev, O., Gijsman, A. J.: Decision support system for agrotechnology transfer version 4.0. University of Hawaii, Honolulu, HI (CD-ROM) (2004)
55. Falkenthal, M., Breitenbücher, U., Christ, M., Endres, C., Kempa-Liehr, A. W., Leymann, F., Zimmermann, M.: Towards function and data shipping in manufacturing environments: how cloud technologies leverage the 4th industrial revolution. In: Proceedings of the 10th Advanced Summer School on Service Oriented Computing, pp. 16–25 (2016)

56. Zysman, J.: The 4th service transformation: the algorithmic revolution. Berkeley Roundtable on the International Economy (2006)
57. Corallo, A., Latino, M.E., Menegoli, M.: From industry 4.0 to agriculture 4.0: a framework to manage product data in agri-food supply chain for voluntary traceability. Int. J. Nutr. Food Eng. **12**(5), 146–150 (2018)
58. Rose, D., Chilvers, J.: Agriculture 4.0: responsible innovation in an era of smart farming. Front. Sustain. Food Syst. **2**, 87 (2018)
59. De Clercq, M., Vats, A., Biel, A.: Agriculture 4.0: The Future of Farming technology. In: Proceedings of the World Government Summit, Dubai, UAE, pp. 11–13 (2018)
60. Kortuem, G., Kawsar, F., Sundramoorthy, V., Fitton, D.: Smart objects as building blocks for the internet of things. IEEE Internet Comput. **14**(1), 44–51 (2009)
61. Li, S., Da Xu, L., Zhao, S.: The internet of things: a survey. Inf. Syst. Front. **17**(2), 243–259 (2015)
62. Lee, I., Lee, K.: The Internet of Things (IoT): applications, investments, and challenges for enterprises. Bus. Horiz. **58**(4), 431–440 (2015)
63. Xia, F., Yang, L.T., Wang, L., Vinel, A.: Internet of things. Int. J. Commun Syst **25**(9), 1101 (2012)
64. Meyer, S., Ruppen, A., Magerkurth, C.: Internet of things-aware process modeling: integrating IoT devices as business process resources. In: International Conference on Advanced Information Systems Engineering, pp. 84–98. Springer, Berlin, Heidelberg (2013, June)
65. Wurm, J., Hoang, K., Arias, O., Sadeghi, A.R., Jin, Y.: Security analysis on consumer and industrial IoT devices. In: 2016 21st Asia and South Pacific Design Automation Conference (ASP-DAC), pp. 519–524. IEEE (2016, January)
66. Zanella, A., Bui, N., Castellani, A., Vangelista, L., Zorzi, M.: Internet of things for smart cities. IEEE Internet of Things J. **1**(1), 22–32 (2014)
67. Da Xu, L., He, W., Li, S.: Internet of things in industries: a survey. IEEE Trans. Industr. Inf. **10**(4), 2233–2243 (2014)
68. Miorandi, D., Sicari, S., De Pellegrini, F., Chlamtac, I.: Internet of things: vision, applications and research challenges. Ad Hoc Netw. **10**(7), 1497–1516 (2012)
69. Stankovic, J.A.: Research directions for the internet of things. IEEE Internet of Things J. **1**(1), 3–9 (2014)
70. Ronen, E., Shamir, A.: Extended functionality attacks on IoT devices: The case of smart lights. In: 2016 IEEE European Symposium on Security and Privacy (EuroS&P), pp. 3–12. IEEE (2016, March)
71. Blaauw, D., Sylvester, D., Dutta, P., Lee, Y., Lee, I., Bang, S., Kim, Y., Kim, G., Pannuto, P., Kuo, Y.-S., Yoon, D., Jung, W., Foo, Z., Chen, Y.-P., Oh, S., Jeong, S., Choi, M.: IoT design space challenges: Circuits and systems. In: 2014 Symposium on VLSI Technology (VLSI-Technology): Digest of Technical Papers, pp. 1–2. IEEE (2014, June)
72. Biswas, A.R., Giaffreda, R.: IoT and cloud convergence: opportunities and challenges. In: 2014 IEEE World Forum on Internet of Things (WF-IoT), pp. 375–376). IEEE (2014, March)
73. Tewari, A., Gupta, B.B.: Cryptanalysis of a novel ultra-lightweight mutual authentication protocol for IoT devices using RFID tags. J. Supercomputing **73**(3), 1085–1102 (2017)
74. Zhu, Q., Wang, R., Chen, Q., Liu, Y., Qin, W.: Iot gateway: Bridgingwireless sensor networks into internet of things. In: 2010 IEEE/IFIP International Conference on Embedded and Ubiquitous Computing, pp. 347–352. IEEE (2010, December)
75. Datta, S.K., Bonnet, C., Nikaein, N.: An IoT gateway centric architecture to provide novel M2M services. In: 2014 IEEE World Forum on Internet of Things (WF-IoT), pp. 514–519. IEEE (2014, March)
76. Dan, L.I.U., Xin, C., Chongwei, H., Liangliang, J.: Intelligent agriculture greenhouse environment monitoring system based on IOT technology. In: 2015 International Conference on Intelligent Transportation, Big Data and Smart City, pp. 487–490. IEEE (2015, December)
77. Guoqiang, S., Yanming, C., Chao, Z., Yanxu, Z.: Design and implementation of a smart IoT gateway. In: 2013 IEEE International Conference on Green Computing and Communications and IEEE Internet of Things and IEEE Cyber, Physical and Social Computing, pp. 720–723. IEEE (2013, August)

78. Vasisht, D., Kapetanovic, Z., Won, J., Jin, X., Chandra, R., Kapoor, A., Sinha, S., Sudarshan, M., Stratman, S.: FarmBeats: an IoT platform for data-driven agriculture. In: 14th {USENIX} Symposium on Networked Systems Design and Implementation ({NSDI} 17), pp. 515–529 (2017)
79. Bröring, A., Schmid, S., Schindhelm, C.K., Khelil, A., Käbisch, S., Kramer, D., Teniente, E.: Enabling IoT ecosystems through platform interoperability. IEEE Softw. **34**(1), 54–61 (2017)
80. Lea, R., Blackstock, M.: City hub: a cloud-based IoT platform for smart cities. In: 2014 IEEE 6th International Conference on Cloud Computing Technology and Science, pp. 799–804. IEEE (2014, December)
81. Huh, S., Cho, S., Kim, S.: Managing IoT devices using blockchain platform. In: 2017 19th International Conference on Advanced Communication Technology (ICACT), pp. 464–467. IEEE (2017, February)
82. Jayaraman, P.P., Palmer, D., Zaslavsky, A., Georgakopoulos, D.: Do-it-yourself digital agriculture applications with semantically enhanced IoT platform. In: 2015 IEEE Tenth International Conference on Intelligent Sensors, Sensor Networks and Information Processing (ISSNIP), pp. 1–6. IEEE (2015, April)
83. Yick, J., Mukherjee, B., Ghosal, D.: Wireless sensor network survey. Comput. Netw. **52**(12), 2292–2330 (2008)
84. Akyildiz, I.F., Su, W., Sankarasubramaniam, Y., Cayirci, E.: Wireless sensor networks: a survey. Comput. Netw. **38**(4), 393–422 (2002)
85. Mao, G., Fidan, B., Anderson, B.D.: Wireless sensor network localization techniques. Comput. Netw. **51**(10), 2529–2553 (2007)
86. Al-Karaki, J.N., Kamal, A.E.: Routing techniques in wireless sensor networks: a survey. IEEE Wirel. Commun. **11**(6), 6–28 (2004)
87. Van Dam, T., Langendoen, K.: An adaptive energy-efficient MAC protocol for wireless sensor networks. In: Proceedings of the 1st International Conference on Embedded Networked Sensor Systems, pp. 171–180. ACM (2003, November)
88. Xiangning, F., Yulin, S.: Improvement on LEACH protocol of wireless sensor network. In: 2007 International Conference on Sensor Technologies and Applications (SENSORCOMM 2007), pp. 260–264. IEEE (2007, October)
89. Doherty, L., El Ghaoui, L.: Convex position estimation in wireless sensor networks. In: Proceedings IEEE INFOCOM 2001. Conference on Computer Communications. Twentieth Annual Joint Conference of the IEEE Computer and Communications Society (Cat. No. 01CH37213), vol. 3, pp. 1655–1663. IEEE (2001, April)
90. Ye, W., Heidemann, J., Estrin, D.: Medium access control with coordinated adaptive sleeping for wireless sensor networks. IEEE/ACM Trans. Netw. (ToN) **12**(3), 493–506 (2004)
91. Beckwith, R., Teibel, D., Bowen, P.: Report from the field: results from an agricultural wireless sensor network. In: 29th Annual IEEE International Conference on Local Computer Networks, pp. 471–478. IEEE (2004, November)
92. Garcia-Sanchez, A.J., Garcia-Sanchez, F., Garcia-Haro, J.: Wireless sensor network deployment for integrating video-surveillance and data-monitoring in precision agriculture over distributed crops. Comput. Electron. Agric. **75**(2), 288–303 (2011)
93. Wark, T., Corke, P., Sikka, P., Klingbeil, L., Guo, Y., Crossman, C., Bishop-Hurley, G.: Transforming agriculture through pervasive wireless sensor networks. IEEE Pervasive Comput. **6**(2), 50–57 (2007)
94. Keshtgari, M., Deljoo, A.: A wireless sensor network solution for precision agriculture based on zigbee technology (2011)
95. Van Bodegom, P.M., Stams, A.J.M.: Effects of alternative electron acceptors and temperature on methanogenesis in rice paddy soils. Chemosphere **39**(2), 167–182 (1999)
96. Watanabe, H., Takagi, K., Vu, S.H.: Simulation of mefenacet concentrations in paddy fields by an improved PCPF-1 model. Pest Manag. Sci. Formerly Pestic. Sci. **62**(1), 20–29 (2006)

Bayesian Network for Development of Expert System in Pest Management

Niranjan Singh and Neha Gupta

Abstract Timely availability of decision support to the farmers on '*whether and what management option is required*' is imperative for effective pest management. In the absence of knowledge and expertise, farmers are over-dependent on pesticide dealers for pest management decision support in the country, which results in excessive, injudicious and irrational use of chemicals for the pest control and thus not only degrades the environment but also affects human health. For decades, the pest economic threshold level (ETL) has been the basis to select appropriate pest management option for decision support to the farmers, used in ICT-based expert systems for decision support in agriculture. This process requires quantitative information about pest activity, which needs to be scientifically observed in the farmers' fields. Moreover, in current pest management, decision-making depends upon a large range of agro-ecological information, besides pest activity. However, a large section of the farming community is not able to obtain this kind of information rather can observe qualitative information consisting of uncertainty. Neither there are the methods of selecting appropriate pest management option based on tentative pest relevant information, provided by farmers that can be used in expert systems of decision support in pest management. Bayesian Network (BN), the most suitable probabilistic reasoning approach for dealing with uncertain information has been effectively used in ecological decision-making nevertheless, very little efforts have been for its usage in pest management in the country. Application of Bayesian network may greatly assist in development of innovative methods of selecting appropriate management option based on uncertain or tentative information of identified agro-ecological factors besides crop condition that farmers provide. Application of the methods thus developed in information and communication technology (ICT) based expert systems (ESs) of agriculture would automate and speed up the process of providing real-time decision support on insect-pest management to the farmers.

N. Singh · N. Gupta (✉)
FCA, MRIIRS, Faridabad, India
e-mail: nehag2012@gmail.com

N. Singh
e-mail: attri.ns@gmail.com

P. K. Pattnaik et al. (eds.), *Internet of Things and Analytics for Agriculture, Volume 2*, Studies in Big Data 67,
https://doi.org/10.1007/978-981-15-0663-5_3

1 Introduction

In the absence of knowledge and expertise, farmers are over-dependant on pesticide dealers for pest management decision support, which results in excessive, injudicious and irrational use of chemicals for the pest control. This not only degrades the environment but also affects human health due to retention of pesticide residues in vegetables [1]. The major decision concern of farmers for pest management is '*whether and what pest management actions are required*'. Over the years, pest ETL has been the basis of pest management decision-making for selection and timing of pest management option if necessary, but this process requires quantitative information about pest activity, which needs to be scientifically observed from the farmers' fields. However, a large section of the farming community is not able to obtain this kind of information scientifically. Moreover, in current pest management, decision-making depends upon a large range of agro-ecological or pest relevant information [2], beside pest activity. Timely availability of correct advice on pest management decision-making to the farmers can result either in saving crop worth crores of rupees or in non-application of intervention saving the cost of intervention involved and thus saving the environment [1].

There have been various kinds of tools of decision support such as warning services, onsite devices and expert systems, used to provide support on decision-making in crop pest management. However, expert systems are the most significant and are widely used. Innovative methods or algorithms are important components of any dynamic expert system in pest management. There are no methods available to deal with pest relevant agro-ecological information consisting uncertainty that farmers provide for decision-making in management option. Bayesian network, a key computer technology dealing with probabilities in Artificial Intelligence, is one of the most effective and popular methods to deal with uncertain information and has been widely and effectively used in ecological decision-making. However, there have been little efforts to apply Bayesian network in development of innovative methods/techniques to deal with qualitative agro-ecological information consisting uncertainty for decision-making in crop pest management. Application of innovative algorithms/methods in expert systems of agriculture providing decision support systems on selection of appropriate pest management option, based on the tentative pest relevant information obtained by the farmers can facilitate and speed up the process of providing pest management decision support to the farmers.

2 Decision-Making in Pest Management

Decision-making is a mental process resulting in the selection of an action among several alternative solutions. Every decision-making process produces a final choice [3]. Decision-making starts with the identification of a problem, which requires the collection of all relevant information for critical analysis of the problem. This analysis leads to development of a set of available alternative courses of actions to solve the problem; only realistic solutions should be selected considering multiple criteria, e.g. effectiveness, benefits, costs and the constraints. Based on this analysis, the best solution is selected, and the decision is converted into an action [4]. Compared to other economic activities, decision-making in crop protection has received very little attention [5]. A better understanding of the decision-making process is necessary given the increased complexity of the decisions required in pest management.

The core of pest management framework is the decision-making process. Decision-making in pest management is a dynamic and complex process (Fig. 1) which requires much more knowledge and support than the conventional agriculture. The first decision concerns pest identification and second decision concerns '*whether and what pest management option is required*'. Whenever the decision-maker notices any pest activity while scouting the field, he needs to identify the pest and scientifically observe the pest and other relevant information necessary for pest management decision-making. After the analysing the information observed, first decision to be made is whether pest management action is required and if so, what is the most appropriate option; chemical, biological or physical; their doses, application methods, frequency, etc. In effective pest management, sustainable biological, physical and other non-chemical methods, which are safer to environment and human health, are preferred to chemical methods [4].

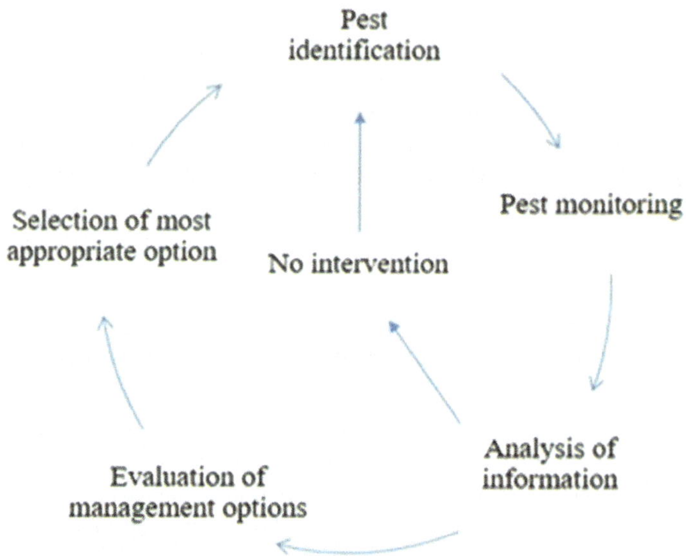

Fig. 1 Decision-making process in pest management

2.1 Basis of Decision Support in Pest Management

For several decades, pest economic threshold level (ETL) has been the basis for decision support on pest management to the farmers, which requires scientifically observed quantitative information about pest activity from farmers' fields. ETL is an important pest management concept which integrates biology and economics and uses pesticides, or other management actions, only when economic loss in anticipated [6]. ETL is defined as the lowest population density that will cause economic damage at which control measures should be initiated against an increasing pest population to prevent economic damage. It is expressed quantitatively and is calculated based on management cost per hectare, price of the farm produce/kilogram and expected damage (kilogram/pest). ETL is the best-known and widely used concept for pest management decision support [7], but there are many reasons for not to use this concept. Major problem of using ETL as the basis for pest management decision support is that it requires quantitative pest information scientifically observed from farmer's fields which large farming community is not able to provide. Moreover, it is based on parameters that are changing all the time, and that are often not known [8].

However, present-day pest management emphasises on agro-ecological situation of farmer field wherein decisions are based on large range of other pest relevant information [9] such as crop health, natural enemies and weather. Agro-ecological information-based pest management results in more reduction in chemical pesticide usage than ETL based and thus conserves the agroecosystems. Pest management decisions cannot be based on only a simple pest count but have to consider many other aspects such as crop health, natural enemies and weather condition. Therefore, agro-ecological information-based pest management decision support is more holistic approach than ETL. Even agro-ecological situation-based pest management decision support need quantitative information on the said aspects, but farmers can provide the qualitative information on these aspects, which consists of uncertainties.

2.2 Pest Management Expert Systems and Their Limitations

An agricultural expert system is a decision support system for agricultural extension agents who have to decide what advice to be offered to farmers and in turn who have to decide what action to take based on it. An important issue for pest management is to develop dynamic and interactive expert tools of decision support that can facilitate the process of providing timely and correct information on timing and selection of appropriate pest management intervention [10]. There are many pest management options available, and however, most appropriate one is selected based on more than one criterion. The timing of its application is imperative for the success of the intervention. Therefore, it becomes necessary to use knowledge-intensive decision-making for designing and developing expert systems to enhance the pest management programmes.

A properly designed expert system for decision support assists the managers or decision-makers in obtaining helpful information from raw data, documents, personal knowledge, and methods for identification and solution of the problem and making appropriate decision. In any process of decision-making, identification of decision-maker is the first step and very important, since all other process relevant elements depend upon it. Alternative decisions and outcomes are other essential elements of a decision-making process. Not all decision outcomes are appropriate; therefore, decision-maker has to assess them based on their advisability. A pest management expert system of decision support gives all the required information about decision-making criteria and management options to the user [11].

Generally, a pest management expert systems of agriculture provide all requisite information or expertise about pest identification; pest and pathogen life cycle; pest sampling methods and threshold calculators; decision-making criteria, available bio-control and chemical control methods; and their safety and environmental impact issues to the domain users. Many initiatives have been taken for the development of expert systems in the country by various public and private organisations. However, most of them are simple web-based static information systems; only few are the dynamic, providing real-time expert support for decision-making to the farmers based on pest situation in their fields. Major initiatives taken in the field of crop pest management in the country as well as world over have been summarised in Tables 1 and 2, respectively.

One of the main reasons for the poor application of expert systems of decision support in pest management in the country has been the failure of developers to target farmers as user, not pay attention to their level of knowledge, their need and requirements and furthermore to the environment in which they work. Most of the EXs developed in the country are either web-based databases or information systems targeted to researchers, KVKs, extension workers or State Department of Agriculture (SDA) personnel. Another lacuna of these systems is that they push information to the users rather than collecting real-time pest information from the farmer's field, process the same and then advice the appropriate pest management solution. For effective and successful pest management, farmer needs easy and timely decision support on selection of appropriate pest management option [34] based on the pest situation in their fields.

Most of the available pest management EXs in the country are either web-based information systems or web-based databases that address only specific aspect of pest management, i.e. pest identification and their management strategies, whereas the most important pest management decision concern of the farmers is to have decision support on '*whether and what pest management actions are required*'. Few expert systems provide ETL-based real-time decision support to the farmers by capturing quantitative pest information scientifically obtained by trained pest scouts from farmer fields. These systems produce ETL-based pest reports showing whether pest incidence has reached near ETL or crossed ETL' to the experts to select

Table 1 Brief description of major expert systems of pest management developed in India

Reference	Name of the system	Area of application	Decision support on
Sharma et al. [12]	OPMAS	Pest management	ETL-based real-time decision support for cotton pest management
Cpsskerala.in [13]	CPSS	Pest management	Pest identification and field pest condition-based crop pest management
Iffcokisan.com [14]	IFFCO Kisan	Agriculture	Market prices, weather forecast, agro-advisory, animal husbandry, horticulture
Padma et al. [15]	IDSS for Apple	Pest management	Integrated pest management of apple orchard
Dath and Balakrishnan [16]	Expert System on Coconut Pest Management	Pest management	Pest management in coconut crop
Prabhuraj [17]	Electronic solutions for agri. pests (e-SAP)	Pest management	ETL-based real-time decision support for pest management in selected crops
Hasan et al. [18]	CaneDES	Nutritional disorders and pest management	Diagnosis of various kinds of disorders in sugarcane
Kumbhar and Singh [19]	e-Sagu	Crop production	Advice on crop pest management
Katkamwar et al. [20]	AGREX	Crop production and protection	Fertilizer application and disease diagnosis in rice
Sharma et al. [21]	E-Pest Reporting and Alert system	Pest management	ETL-based real-time decision support for pest management in pulses
Marwaha [22]	AgriDaksh	Crop production	Pest identification and variety selection
Islam et al. [23]	EXOWHEM	Crop production	Wheat agrimony, varieties and pest management

Table 2 Brief description of major expert systems of pest management developed world over

Reference	DSS name	Area of application	Decision support on
Martin and Goebel [24]	DECIPESTS	Pest management	Pest identification and their level based pest management in sugarcane
Asi.ucdavic.edu [25]	UC IPM	Pest management	Pest identification and location specific pest management decision-making
Pest.cabweb.org [26]	Crop protection compendium	Pest management	Pest identification and management
Carhuapoma et al. [27]	Insect life cycle modelling (ILCYM)	Pest management	Temporal temperature scenario-based pest distribution and risk maps.
Damos and Karabatakis [28]	Real agro-net topology and Insect smart model application (RANTISMA)	Crop health	Real-time moth phenology forecasts
Orellana et al. [29]	SAIFA	Pest management	Integrated production for the olive crop
Vincent et al. [30]	HDSS	Pest management	Weather-based pest forecasting and their management
Samietz et al. [31]	SOPRA	Pest management	Insect-pest monitoring and their management in fruit orchards
Bourgeois et al. [32]	CIPRA (Computer centre for agricultural pest forecasting)	Pest management	Weather-based prediction of pest, crop phonological and postharvest disorders affecting storage of apple in Canada
Magarey et al. [33]	NAPPFAST North Carolina State University (NCSU)	Pest management	Plant pest forecasting system

appropriate pest management option to be advised to the farmers. However, scientific observation of such quantitative pest information from farmers' fields requires large number of qualified and trained workforce, which is a difficult and costly proposition for state agencies. Therefore, can the farmers themselves do the field pest scouting? Yes, they can. However, large farming community cannot record the quantitative pest observations as per pest-specific scientific methods rather can observe the qualitative information consisting of uncertainty. Moreover, in the present-day pest management decision-making, other relevant factors such as the presence of natural enemies, weather and crop condition affecting the level of pest activity are to be considered necessarily for determining the appropriate pest management option. However, there are not methods available to deal with this kind of qualitative pest relevant information consisting uncertainty for selection of pest management options, which can be used in pest management expert systems.

3 Bayesian Network and Its Application in Decision-Making

Bayesian network, a key computer technology dealing with probabilities in Artificial Intelligence, is one of the most effective and popular methods of modelling uncertain knowledge expression and reasoning such as environmental management [35, 36]. BNs emerged from artificial intelligence research wherein originally they emerged as formal modes to analyse decision approaches under uncertain conditions [37]. Bayesian network can be used to estimate the effectiveness of alternate management decisions or policies because of being causal in nature. BNs have the ability to sustain clarity through explicit causal assumptions and so are generally used to model when the relationships to be described are not easily expressed using mathematical notation. New computational methods and techniques keep increasing BNs' abilities and range of practical applications [38].

In contrast to the other approaches or techniques used in environment studies, Bayesian networks use probabilistic expressions rather than deterministic to express the relationships among variables of the system. Bayesian network accounts the lack of knowledge in the network by the usage of Bayesian probability theory, which allows the subjective estimation of the probability of occurrence of a particular outcome that is to be combined with more objective data quantifying the frequency of occurrence in finding the conditional probabilistic relationships. Bayesian networks are very appropriate technique to deal with systems where uncertainty is inherent due to it being accounted in the model, as this is an important issue in ecological systems [39].

Bayesian networks have many other attractive properties, which make them particularly applicable in data analysis and decision-making. BNs have simple causal and graphical structure and can be easily extended and altered. BNs can integrate missing data readily by using Bayes' theorem. They can easily be understood without much mathematical background and knowledge. They have been found to possess good accuracy of prediction with small sample size. BNs can also be applied for predicting the probable values of states of a system for different future scenarios. Bayesian networks are also useful for participatory processes. They can assist in evaluating the alternative decisions to optimise the desired outcome. BNs can also help in processes of social learnings. BNs express the system as network of reciprocal actions among system variables from main cause to outcome; subject to all assumptions of cause-effect are clearly made [39]. Reasoning approaches are very useful in dealing with uncertain information [40].

A Bayesian network is a combination (G; P) [41, 42] where

- $G = (V; E)$ is a DAG where set of nodes $V = \{X1, X2, \ldots, Xn\}$ represents the variables of the system and E, a set of arcs represents direct conditional dependencies between the variables (nodes);
- P represents the set of conditional probability distributions comparing conditional probability distribution P (Xi/pa (Xi)) for each variable X given the set of parents pa (Xi) in the graph.

The joint probability distribution over V can be recovered from the set P of conditional probability distributions by the application of following chain rule:

$$P(X1,\ X2, \ldots, Xn) = \prod_{i}^{n} P\,(Xi/\text{pa}\,(Xi))$$

Conditional probabilities of BNs can be computed based on the experimental or trial data, results produced by the models and domain expert's knowledge elicitation [43]. Uncertainty is clearly represented by way of presentation of the probabilities [44].

3.1 Bayesian Network Theory

BNs are probabilistic graphical model consisting of variables of interest of a system and their conditional relationships [45]. A BN consists of 'nodes' that represent the set of variables of interest and 'links' represent conditional interdependencies (cause-effect relationship) among nodes via a directed acyclic graph (DAG) [46]. Conditional probability distribution captures the quantum of dependence between system variables and thus explains the relationships between the nodes. For instance, if node A of a system affects another node B, then nodes A and B be linked to each other by an arc having direction from node A to node B. Herein, node A is called as parent of node B, and node B is referred to as child of node A. If node B further affects another node C, then node B be the parent of node C. Figure 2 shows the conditional or causal relationship among nodes A, B and C of a basic Bayesian network.

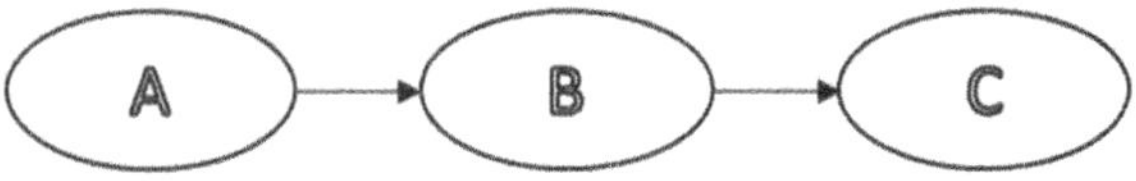

Fig. 2 Conditional or causal relationship among BN nodes

A BN can comprise different kinds of nodes, i.e. nature nodes, decision nodes and utility nodes. Nature nodes are the variables, which can be controlled by actions of decision-maker and represent the empirical or calculated parameters and the probabilities of occurrence of various states. Input nodes are the nodes without parents and can be expressed in as either constant or categorical states with associated marginal probability distributions. The variables or events, which can be directly implemented by the decision-maker, are represented by a decision node. Decision node represents two or more alternative decision options that a manager can choose from. A decision node does not have probabilities associated with it. Utility nodes [47] should always accompany decision nodes. A utility node represents the value of an outcome or a decision. Utility node can be directly linked to the decision node [48]. BNs representing and solving decision problems under uncertainty are also known as Bayesian decision networks [49].

Bayesian network uses Bayes' probability theorem to propagate information/evidences among the nodes. Bayes' theorem defines how observed evidence updates the prior information/knowledge about a hypothesis. In Bayes' theorem, prior probability represents the possibility of an input parameter being in a specific state whereas conditional probability estimates the possibility of a parameter state given the states of input parameters affecting it. Estimate of likelihood of a parameter to be in particular state is the posterior probability, given the input parameters, conditional probabilities and the rules controlling how the probabilities combine. The network is solved using Bayes' rule:

$$P\,(\mathrm{A}|\mathrm{B}) = P\,(\mathrm{B}|\mathrm{A})\,P\,(\mathrm{A})/\,P\,(\mathrm{B})$$

Where P (A) is the prior probability of parameter A; P (A|B) is the posterior probability, the probability of A given new data B; and P (B|A) the conditional probability of parameter B given existing data A [39].

3.2 Major Instances of Bayesian Network in Agriculture

Bayesian networks (BNs) are widely being used for environmental modelling to integrate multiple issues and components of the system; for utilising information from various sources; and to handle the missing and uncertain data [47]. BNs have been adequately used in ecological decision-making [50]. They have also been used in agriculture nonetheless; there are very few instances of their usage in pest management [51].

Tari [52] illustrated the possible applications of BNs for assessing the yield response of winter wheat to fungicide programmes. Therein, experimental data from ADAS trials has been used with some assumptions in order to complete the data.

Gu et al. [53] applied Bayesian network to handle the uncertainty in climate prediction information for studying risk and benefits to crop production in Scotland. The BN establishes relationship between model of fababean growth and a model of weather generator and thus is able to respond to the queries on fababean production under different climate predictions.

Borsuk et al. [54] developed a Bayesian network to know about possible reasons for reduction in the status of health of brown trout population in Switzerland. Further investigation was done by Borsuk et al. [55], wherein BN was applied to determine the importance of historical causes of anthropogenic changes and also for predicting the effects of management actions proposed.

Bromley et al. [44] developed Bayesian network for integrated management of four basins from Europe while focussing on management of water. The project aim was to devise an active and functional method to help the managers in several types of objective decision-making and simultaneously making sure that various stakeholders actively participate in the process of decision-making.

Smith et al. [56] also devised a Bayesian network, which interacted with expert knowledge and geospatial data about the favoured habitats to map the suitability of Julia Creek dunnart (*Sminthopsis douglasi*) habitat in north-west Queensland. These were earlier thought to be extinct [39].

Chen and Yu [57] presented application of Bayesian network for development of a disease identification system for maize crop. Noisy-or model and conversion from certainty to probability were applied in the development of Bayesian network. Thus, a BN-based disease identification system for maize crop was developed. The practice proves Bayesian network as an effective tool for maize disease diagnosis.

Steventon and Daust [58] developed a BN for modelling the epidemic of mountain pine beetle in north-central British Columbia, Canada. Subsequently, this was applied for examining the response of habitat and population size of pine beetle to management scenarios with built in parameter uncertainty.

Meysam and Mahdi [59] studied the BN usage for locust pest management. However, it was used to find the effect of various trap plants on population. This was an instance of using Bayesian network for examining the effects of trap plants on locust population.

Douglas and Newton [60] presented about usage of BNs for supporting protected area management by way of developing habitat suitability models to address the conservation of eight species. The BNs were developed using scientific literature and expert knowledge and were evaluated with the field survey results. Subsequently, they were applied to see the effect of various management practices such as tree cutting, grazing and burning on suitability of habitat of species.

Abbal et al. [61] developed a DSS tool for vine-growers to examine quality of vineyard to be planted. Herein, vineyard quality is determined by the probability of wine sale profitability. Bayesian network being widely used to deal with uncertainty in management of natural resource was constructed using the relationship between environment and vineyard status parameters. Information for Bayesian network construction was obtained from vine growing experts. The system facilitates the assessment of possible impact of vine grower's decisions on the quality of new vineyards.

4 Conceptual Bayesian Network for Decision-Making in Pest Management

Construction of conceptual Bayesian network for pest management system comprises the following components:

- Framework of pest management system
- Structure of Bayesian network for decision-making in pest management
- Parameterization of the Bayesian network

Construction of BN structure for pest management decision support required identifying the important variables, establishment of links between variables and computation of conditional probabilities. Crop protection system includes field, plants, pests and their natural enemies.

4.1 Framework of Pest Management System

In crop protection, all the decisions related to pest management are made by the farmers at field level. Regular crop monitoring is the cornerstone of pest management. Figure 3 shows the components of field monitoring in crop pest management.

During field monitoring, farmer has to observe all the pest relevant agro-ecological information and crop condition to estimate the pest risk to the crop. Crop condition is measured in terms of crop phenology. Weather and the presence of natural fauna are the most important factors affecting pest activity. These observations are recorded weekly. Based on the estimate of overall field agro-ecological situation, he has to timely decide '*whether and what pest management action is required*' for effective and successful pest management.

4.2 Structure of Conceptual Bayesian Network

Bayesian networks are strictly one-way because they represent a conditional probability distribution that describes the relative likelihood of each value of the 'child' node (end of the arrow) conditional on every possible combination of values of the 'parent' nodes (start of each arrow). For instance, if a link is directed from node E towards node H as in Fig. 4, then each value of node E conditionally depends upon

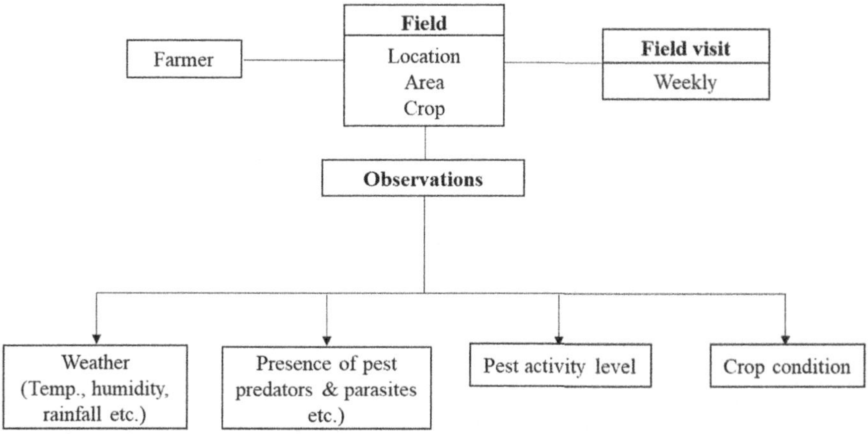

Fig. 3 Components of field monitoring for crop pest management

Fig. 4 Basic structure of a Bayesian network

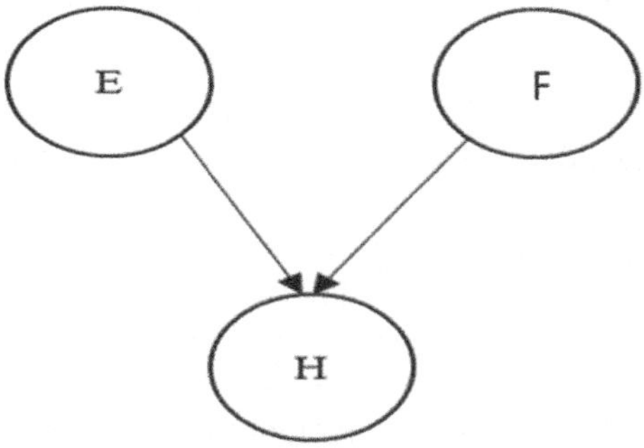

the possible combination of values of node H since E and F parent nodes are having causal relationship with child node H.

BN structure is usually made of 'box and arrow' diagram. A BN can be constructed using two approaches, using expert knowledge elicitation and based on data learnings. Data learning-based approach is useful to construct BN for complex systems that are difficult to be understood. Pest management decision-making processes are complex in nature wherein quality and comprehensive historical data is rarely available. Hence, it becomes difficult to determine accurate structure based on data learnings only. So, the outcome of the data-based structural learnings can be enhanced by combining it with expert knowledge elicitation together. While developing the conceptual BN structure, all variables (nodes) of the system influencing the outcome of decision-making process are considered.

Identifying all the variables of the system and establishing links between them results in structure of conceptual Bayesian network representing the system under consideration. Description of pest management framework and knowledge elicited from experts and published literature revealed prominent variables influencing

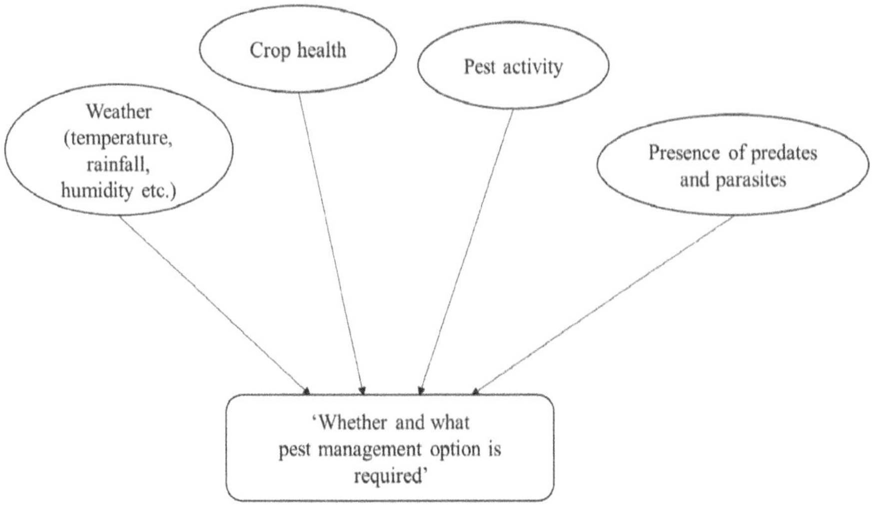

Fig. 5 Structure of conceptual BN for decision-making in pest management

decision-making in pest management system are weather (temperature, rainfall and humidity), pest activity, the presence of pest natural enemies (parasite and predators) and crop condition.

Weather comprising temperature, rainfall, humidity; the presence of pest natural enemies comprising predators and parasites; and pest activity and crop condition are the broad factors to be considered for determining '*whether and what pest management intervention is required*'. Using these variables, structure of conceptual BN for decision-making in pest management was established as shown in Fig. 5.

Once structure of the conceptual BN established, we can start with it to have refined version of the network. Identification of significant variables ('nodes') of the system and exclusion of non-significant variables are usually done based on review of the literature, data analysis, experts' opinion and discussion with the stakeholders. Once the nodes have been selected, links are revised based on the relationship among them. If a variable is not at all related to the decision-making in the system, it may be excluded because inclusion of insignificant variables can make the BN structure more complex. Since inclusion of insignificant variables costs extra time and efforts but it does not add any value to the BN.

Significance of these factors depends upon the location and cropping season. A factor may be significantly affecting pest activity in a particular location and cropping season, while other may not be having any significant effect. Thus, using data records obtained from organisations working in the field of pest management and the knowledge elicited from domain experts and literature, key variables (nodes) of the system relevant to the process of decision-making in pest management and their relationship with each other are identified. The variables, which are not significant, are excluded, and thus, the structure of conceptual network is refined. Mutually exclusive states of each variable (node) are also determined through the analysis of data records and domain expert inputs.

4.3 Parameterisation of the Bayesian Network

After identification of nodes, they are assigned the states to be taken by each of them. Each BN node represents either an observable or measurable system variable. Potential values or conditions that can be assumed by a node are represented by its states. Type and quality of available data determines the types of the state and their number. The initial values for each node can be determined using available data records or based on the knowledge elicited from the literature or through expert consultation. The states of a node may be categorical, e.g. large, medium, small, Boolean, e.g. yes or no, discrete such as integers or continuous variables, e.g. $\leq$10, >10 and $\leq$20 and >20. However, the decision about the division of range into sub-ranges should not be arbitrary rather it should be done based on data analysis such as descriptive, plotting, classification or percentile analysis. Discretisation can be useful wherever a variable has specific breakpoint that is significant to management. Domain expert consultation may also be combined with data analysis. Sometimes, defining fewer state of a variable result in information loss and assigning too many states to a variable may complicate the BN. For Bayesian networks of decision-making, states representing the important thresholds can be used as alternatives.

As stated above, weather conditions, the presence of pest natural enemies, level of pest activity and crop health are the prominent factors broadly considered for pest management decision-making. Weather conditions are described by a set of variables, i.e. maximum temperature, minimum temperature, rainfall, humidity; presence of natural enemies by population of predators and population of parasites, as the variables, whereas pest activity and crop health are not represented as set of variables. Set of possible mutually exclusive states of these variables or nodes in conceptual BN for decision-making in pest management may be categorised as shown in Table 3.

After defining the state type and its numbers, conditional probabilities for the state values of each child node are estimated based on the state values of its parent

Table 3 Possible state values of prominent variables of decision-making in pest management

Variable (s)	Possible state values
Temperature	Extremely hot, hot, warm, cold, frigid
Rainfall	High, moderate or low
Humidity	High, moderate or low
Presence of predators	Abundant, moderate, low, scarce
Presence of parasites	Abundant, moderate, low, scarce
Pest activity	High, medium, low
Crop health	Good, average, poor

nodes. Probability of priory expecting a node to be into a definite state can be determined from known frequencies or can assume a uniform distribution representing uncertainty [48].

The relationship between parent and child node is defined by the conditional probability table (CPT). The CPT presents the probability value of a node being in a state provided combination of probability values of states of parent nodes. CPT dimension of each node is the multiple of numbers of child node states and its parent nodes states. If a node has no parents, it can be described probabilistically by a marginal probability distribution. Figure 6 shows the CPT of a basic Bayesian network consisting of three nodes. Node H will be in its high, medium and low states, given the states of node E and F in the network. The nodes E and F are root nodes and so are defined by marginal probabilities. H is the child node of E and F nodes, and thus, probabilities of H node states are conditional given the combination of states of E and F nodes. Parent nodes E and F are the causing factors of child node.

The CPT values associated with each state of a node can be estimated using expert knowledge elicitation, observed data, probabilistic or empirical equations and results of simulation models or their combination [62]. Probability distribution quantifies the uncertainties associated with each relationship. It is very important to record probability computation method including any assumptions and limitations. Construction of Bayesian network is complete once the probabilities of all nodes are computed. Once completed, BN is used for scenario analysis. Different scenarios be it set of management actions or system observations, can easily be analysed using the BN. Bayesian networks are simple means of scenario analysis that allows the users to propagate evidences as node input using well-defined node distribution. The effect of a scenario is tested through its effect on other nodes of the network by propagating the probabilities. Fast propagation of evidence through the network is one of the main

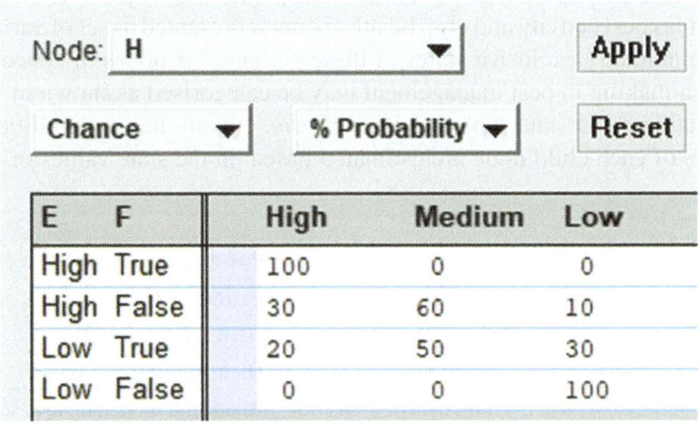

E	F	High	Medium	Low
High	True	100	0	0
High	False	30	60	10
Low	True	20	50	30
Low	False	0	0	100

Fig. 6 CPT of node in basic structure of a BN

advantages of BNs. Therefore, BNs are used to instantly find the effect of observed conditions or decisions at one node in the whole system. Subsequently, the network is evaluated or validated.

5 Evaluation and Validation of the Bayesian Network

Evaluation and validation of a developed Bayesian network assist in determining whether its results are feasible and satisfactory. The criteria to evaluate or validate BN's performance depend upon its objective. Initially, the interactions of different components of the network can be examined by application of various input scenarios and determine whether the results are logical and reasonable. Expected results can include the network showing realistic performance. Preferably accuracy of the network performance should be examined with empirical data; however, many a times data is either not available or is partially available. Data other than that used to develop the Bayesian network should be used for evaluation and validation of the model. Many BN softwares have the functionalities to allow data to be checked against network predictions. These softwares revise the probabilities of all the samples within the case, except the unobserved nodes and then produce the beliefs of each unobserved node. These values are compared with actual node value. Another method of BN evaluation is sensitivity analysis, which is done to find sensitive parameters. This analysis ranks the variables according to the importance relative to the variables of the interest. This can be used to confirm whether response of network is correct and matches to the expectations. Sensitivity analysis also assists in identification of the variables having significant influence on the results, and subsequently [62]. There are other forms of evaluation such as critique of assumptions, ability of the model to perform under different scenarios [63], whereas validation of the Bayesian decision network can be done by cross-examining the outcome of the BN against the results of existing methods of decision-making.

In case of limited availability or complete unavailability, qualitative forms of evaluation and validation for instance peer review are valuable. By application of different input values, outcome of the network in the form of resulted probabilities can be examined by the reviewers whether it is consistent and logical. Aguilera et al. [64] did a review of over hundred BN applications in environmental management and found that 38% did not do any validation. This may be because Bayesian network is frequently applied in cases of scarce data availability. Network evaluation should be done even with limited data. In case, evaluation or validation results are not logical and reasonable then network structure and its assumptions need to be reassessed that includes readjustment of network structure, refinement of conditional probabilities or variable states.

Development of BN is an ongoing process. Major advantage of BN is that is can be easily updated. The BNs are very useful in cases of limited data and knowledge availability and thus can be built with scare or incomplete information.

Environmental and crop pest management are the systems complex in nature, where data availability is ether scarce or limited. Therefore, evaluation and validation of BNs developed for these systems can be done using combination of data and exert knowledge elicitation.

6 Conclusion

Many expert systems have been developed to provide expert decision support in pest management by private and public organizations world over. However, most of them are web-based database or information, which provides information on pest identification and management. There have been very little efforts to develop interactive and dynamic expert systems providing real-time decision support to the farmer based on the uncertain agro-ecological situation in their fields. There are no methods available to deal with uncertain agro-ecological information provided by farmers for selection of appropriate pest management option to be advised to them. Application of Bayesian network, a probabilistic reasoning approach, can help in development of innovative method or technique for selecting appropriate pest management option based on uncertain agro-ecological information that farmers provide. Application of method or technique thus developed speed up and automate the process of providing real-time decision support to the farmers based on agro-ecological situation in the their fields.

References

1. Singh, N., Sardana, H.R., Kumar, S.S.: Pest management information system (PMIS) for tomato: a database tool for IPM. Indian J. Entomol. **74**(2), 202–204 (2012)
2. Satyagopal, K., Sushil, S.N., Jeyakumar, P., Shankar, G., Sharma, O.P., Sain, S.K., Saha, S.: AESA based IPM package for tomato. National Institute of Plant Health Management, Hyderabad, AP (2014)
3. March, J.G.: Primer on decision-making: how decisions happen, p. 289. The Free Press, New York, NY, USA (1994)
4. Rossi, V., Caffi, T., Salinari, F.: Helping farmers face the increasing complexity of decision-making for crop protection. Phytopathologia Mediterranea **51**(3), 457–479 (2012)
5. Eom, S., Kim, E.: A survey of decision support system applications (1995–2001). J. Oper. Res. Soc. **57**(11), 1264–1278 (2006)
6. Damos, P.: Modular structure of web-based decision support systems for integrated pest management. A review. Agron. Sustain. Dev. **35**(1), 1347–1372 (2015)
7. Uniyal, M.: Integrated Pest Management: SATNET Asia Training Manual. Retrieved from http://www.satnetasia.org (2015)
8. Jeykumar, P.: AESA Based IPM. Retrieved form http://vikaspedia.in/agriculture/crop-production/integrated-pest-managment/aesa-based-ipmdn (2016)

9. Prakash, A., Bantur, J.S., Prasad, M.S., Tanwar, R.K., Sharma, O.P., Bhagat, S., … Jeykumar, P.: Integrated Pest Management for Rice. National Centre for Integrated Pest Management, New Delhi (2014)
10. Damos, P., Savopoulou-Soultani, M.: Temperature-driven models for insect development and vital thermal requirements. Psyche **2012**(1), 1–13 (2012)
11. Damos, P., Agron, S.D.: Modular structure of web-based decision support systems for integrated pest management. Agron. Sustain. Dev. **35**(4), 1347–1372 (2015)
12. Sharma, R., Kumar, A., Tanwar, R.K., Singh, N.: Web-enabled data extraction system for cotton pests. Indian J. Agric. Sci. **88**(6), 841–844 (2018)
13. Cpsskerala.in.: Crop-Pest Surveillance System. Retrieved from http://www.cpsskerala.in (2017)
14. Iffcokisan.com: AgroICT Services. Retrieved from http://www.iffcokisan.com (2016)
15. Padma, T., Mir, S.A., Shantharajah, S.P.: Intelligent decision support system for an integrated pest management in apple orchard. In: Sangaiah, A., Abraham, A., Siarry, P., Sheng, M. (eds.) Intelligent Decision Support Systems for Sustainable Computing, pp. 225–245. Springer, NY, USA (2017)
16. Dath, A., Balakrishnan, M.: Expert system on coconut pest management. Int. J. Adv. Res. Comput. Sci. Softw. Eng. **6**(3), 442–446 (2016)
17. Prabhuraj, A.: Electronic Solutions Against Agricultural Pests (e-SAP). Retrieved from http://www.mmp.cips.org.in/documents/Workshops/2015/30jul-1aug/eSAP_A_Prabhuraj.pdf (2015)
18. Hasan, S.S., Solomon, Baitha A.S., Singh, M.R., Sah, A.K., Kumar, R., Sudhir, S.: CaneDES, a web-based expert system for disorder diagnosis in sugarcane. Sugar Tech **17**(4), 419–427 (2015)
19. Kumbhar, V., Singh, T.P.: A comprehensive study of application of decision support system in agriculture in Indian context. Int. J. Comput. Appl. **63**(14), 6–11 (2013)
20. Katkamwar, N., Kumar, B., Kalaskar, K.: An expert system approach for improvement of agriculture decision. Singaporean J. Sci. Res. **5**(1), 8–12 (2013)
21. Sharma, O.P., Singh, N., Bhardwaj, A., Vennila, S., Bhagat, S., Mehta, N., Kumar, K.: ICT based pest management system for sustainable pulse production, a case study. In: Maumbe, B., Patrikakis, C. (eds.) E-Agriculture and Rural Development, Global Innovations and Future Prospects, pp. 169–183. IGI Global, Hershey, PA, USA (2013)
22. Marwaha, S.: AgriDaksh—a tool for developing online expert system. 3rd National Conference on Agro-Informatics & Precision Agriculture (AIPA) proceedings, New Delhi, India, 1–3 August 2012, pp. 18–23 (2012)
23. Islam, S.N., Kundu, S., Sabir, N., Sharma, K., Farooqi, M., Poswal, R., Agarwal, H.O., Chaturvedi, K., Sharma, R.K., Sharma, A.: Selection of wheat *Triticum aestivum* variety through expert system. Indian J. Agric. Sci. **82**(1), 43–47 (2012)
24. Martin, P., Goebel, F.R.: DECIPESTS, a decision-support system for pest management in sugarcan. Int. Sugar J. **11**(9), 978–982 (2017)
25. Asi.ucdavic.edu: UC IPM portal. Retrieved from http://asi.ucdavis.edu/programs/sarep/what-is-sustainable-griculture/practices/ipm (2017)
26. Pest.cabweb.org: CABI-Crop Protection Compendium. Retrieved form http://pest.cabweb.org/cpc/cpchp.htm (2017)
27. Carhuapoma, P., Sporleder, M., Mujica, N., Kreuze, J., Simon, R., Kroschel, J.: Insect life cycle modelling ILCYM—an open source tool for entomologists. In: Presented in International Conference on Statistics and Big Data Bioinformatics in Agricultural Research, CRISAT, Hyderabad, 21 November 2016 (2016)
28. Damos, P., Karabatakis, S.: Real-time pest modelling through the World Wide Web, decision-making from theory to praxis. Integr. Protect. Fruit Crops **91**(1), 253–258 (2013)
29. Orellana, F.J., del Sagrado, J., del Águila, I.M.: SAIFA, A web-based system for integrated production of olive cultivation. Comput. Electron. Agric. **78**(2), 231–237 (2011)
30. Vincent, P.J., Jay, F.B., Gary, G.G., Brad, P., Gerald, V.T., Wendy, E.J.: A web-based decision support system to enhance IPM programs in Washington tree fruit. Pest Manag. Sci. **66**(6), 587–595 (2010)

31. Samietz, J., Graf, B., Höhn, H., Höpli, H.U., Schaub, L.: SOPRA, phenology modelling of major orchard pests—from biological basics to decision support. Acta Hortic. **803**(1), 35–42 (2008)
32. Bourgeois, G., Plouffe, D., Chouinard, G., De Ell, J.: The apple CIPRA network in Canada, using real-time weather information to forecast apple phenology, insects, diseases and physiological disorders. Acta Hort. **803**, 29–34 (2008)
33. Magarey, D.R., Fowler, G., Borchert, D.M., Sutton, T.B., Colunga-Garcia, M., Simpson, J.A.: NAPPFAST, An internet system for the weather-based mapping of plant pathogens. Plant Dis. **91**(4), 336–345 (2007)
34. Bajwa, W.I., Leonard, Coop, Kogan, M.: Integrated pest management (IPM) and internet-based information delivery systems. Neotropical Entomol. **32**(3), 373–383 (2003)
35. Bi, C., Chen, G.: Bayesian networks modeling for crop diseases. In: Li, D., Liu, Y., Chen, Y. (eds.) Computer and Computing Technologies in Agriculture IV, CCTA 2010: 4th IFIP TC 12 Conference proceedings, Nanchang, China, 22–25 October 2010, pp. 312–320 (2011)
36. Uusitalo, L.: Advantages and challenges of Bayesian networks in environmental modelling. Ecol. Model. **203**(3–4), 312–318 (2007)
37. Varis, O.: Bayesian decision analysis for environmental and resource management. Environ. Model Softw. **12**(2–3), 177–185 (1997)
38. Mead, R., Paxton, J., Sojda, R.: Applications of Bayesian networks in ecological modelling. In: The Second IASTED International Conference on Environmental Modelling and Simulation. St. Thomas, Virgin Islands (2006)
39. Pollino, C.A., Henderson, C.: Bayesian Networks, A Guide for Their Application in Natural Resource Management & Policy. Retrieved from http://www.utas.edu.au/data/assets/pdf_file/588474/TR_14_BNs_a_resource_guide.pdf (2010)
40. Pan, H., McMichael, D.: Fuzzy Causal Probabilistic Networks—A New Ideal and Practical Inference Engine. Retrieved from http://citeseerx.ist.psu.edu/viewdoc/download? (1998)
41. Jensen, F.V., Nielsen, T.D.: Bayesian Networks and Decision Graphs, 2nd edn. Springer-Verlag, NY, USA (2007)
42. Kjrulff, U.B., Madsen, A.L.: Bayesian Networks and Influence Diagrams. Springer-Verlag, NY, USA (2008)
43. Borsuk, M.E., Reichert, P., Peter, A., Schager, E., Burkhardt-Holm, P.: Assessing the decline of brown trout (*Salmotrutta*) in Swiss rivers using a Bayesian probability network. Ecol. Model. **191**(1–2), 224–244 (2006)
44. Bromley, J., Jackson, N.A., Clymer, O.J., Giacomello, A.M., Jensen, F.V.: The use of Hugin to develop Bayesian networks as an aid to integrated water resource planning. Environ. Model Softw. **20**(2), 231–242 (2005)
45. del Águila Isabel, M., del Sagrado, J.: Metamodeling of Bayesian networks for decision-support systems development. In: 8th Workshop on Knowledge Engineering and Software Engineering (KESE), Montpelier, USA, 949, 12–19 (2012)
46. Pearl, J.: Probabilistic Reasoning in Intelligent Systems: Networks of Plausible Inference. Morgan Kaufmann Publishers Inc., San Francisco, USA (1988)
47. Singh, N., Gupta, N.: Bayesian network for decision support on pest management of tomato fruit borer, *H. armigera*. Int. J. Eng. Technol. **6**(4), 168–170 (2017b)
48. Kragt, M.E.: A Beginners Guide to Bayesian Network Modelling for Integrated Catchment Management (Technical Report No. 9, Landscape Logic). Department of Environment, Water Heritage and the Arts, Australia (2009)
49. Singh, N., Gupta, N.: Decision-making in integrated pest management & Bayesian network. Int. J. Comput. Sci. Inf. Technol. **9**(2), 31–36 (2017a)
50. Rivot, E., Prevost, E., Parent, E., Bagliniere, J.L.: A Bayesian state-space modelling framework for fitting a salmon stage-structured population dynamic model to multiple time series of field data. Ecol. Model. **179**(4), 463–485 (2004)
51. Fabre, F., Pierre, J.S., Dedryver, C.A., Plantegenest, M.: Barley yellow low dwarf disease risk assessment based on Bayesian modelling of aphid population dynamics. Ecol. Model. **193**(3–4), 457–466 (2006)

52. Tari, F.: A Bayesian network for predicting yield response of winter wheat to fungicide programmes. Comput. Electron. Agric. **15**(2), 111–121 (1996)
53. Gu, Y., Crawford, J.W., Peiris, D.R., Grashoff, C., McNicol, J.W., Marshall, B.: Modelling fababean production in an uncertain future climate. Agric. For. Meteorol. **79**(4), 289–300 (1996)
54. Borsuk, M.E., Stow, A.C., Reckhow, K.H.: Ecological prediction using causal Bayesian networks: a case study of eutrophication management in the Neuse River estuary. Duke University, North Carolina, USA (2002)
55. Borsuk, M.E., Stow, A.C., Reckhow, K.H.: A Bayesian network of eutrophication models for synthesis, prediction and uncertainty analysis. Ecol. Model. **173**(2–3), 219–239 (2004)
56. Smith, C.S., Howes, A.L., Price, B., McAlpine, C.A.: Using a Bayesian belief network to predict suitable habitat of an endangered mammal—The Julia Creek *dunnart Sminthopsis douglasi*. Biol. Cons. **139**(3–4), 333–347 (2007)
57. Chen, G., Yu, H.: Bayesian network and its application in maize diseases diagnosis. In: Li, D. (eds.) Computer and Computing Technologies in Agriculture-II proceedings. Wuyishan, China, 18–20 August 2007, pp. 917–992 (2008)
58. Steventon, J., Daust, D.K.: Management strategies for a large-scale mountain pine beetle outbreak, Modelling impacts on American martens. For. Ecol. Manage. **25**(7), 1976–1985 (2009)
59. Meysam, M., Mahdi, S.: Bayesian network and pest management: a case study of trap plants on locust population: Adv. Environ. Biol. **4**(2), 147–151 (2010)
60. Douglas, S.J., Newton, A.C.: Evaluation of Bayesian networks for modelling habitat suitability and management of a protected area. J. Nat. Conserv. **22**(3), 1–12 (2014)
61. Abbal, P., Sablayrolles, J., Matzner-Lober, E., Boursiquot, J., Baudrit, C., Carbonneau, A.: A decision support system for vine growers based on a Bayesian network. J. Agric. Biol. Environ. Stat. **21**(1), 131–151 (2015)
62. Pollino, C.A., White, K.A., Hart, B.T.: Examination of conflicts and improved strategies for the management of an endangered Eucalypt species using Bayesian networks. Ecol. Model. **201**(1), 37–59 (2007)
63. Jakeman, A.J., Letcher, R.A., Norton, J.P.: Ten iterative steps in development and evaluation of environmental models. Environ. Model Softw. **21**(5), 602–614 (2006)
64. Aguilera, P.A., Fernández, R., Fernández, A., Rumi, R., Salmerón, A.: Bayesian networks in environmental modelling. Environ. Model Softw. **26**(12), 1376–1388 (2011)

An IoT and Wireless Sensor Network-Based Technology for a Low-Cost Precision Apiculture

D. D. Dasig Jr. and J. M. Mendez

Abstract The current environmental ramifications due to industrial agriculture, climate change, and pathogens have destructed the habitat and loss of biodiversity. These ecological modifications have threatened the apiculture including the bees and wild pollinators which plays a vital role and are indispensable in the terrestrial ecosystems. This chapter presents the foundational space of apiculture, challenges, and emerging techniques of the apiarists in beekeeping including precision apiculture, and a low-cost IoT and remote sensing technology deployed for data gathering and monitoring the vigor and productivity of beehive colonies. The system has redundancy components and utilized the three-level hierarchical model including wireless node, local information server, and the cloud data server. The system also consists of sensors deployed across distinct locations to capture the beehives' temperature, relative humidity, and loads of the hives and honeycombs, and stored to the local and cloud server. This system will complement the urgent need to countervail the honeybee colonies' collapse by helping the apiarists, apiculture subsector, and agriculture sector to embark emerging precision apiculture technologies.

Keywords Precision apiculture · Internet of things · Wireless network sensors · Beekeeping · Bees

D. D. Dasig Jr. (✉)
Graduate Studies, College of Science and Computer Studies, De La Salle University-Dasmariñas, DBB-B, West Ave, 4115 Dasmariñas, Cavite, Philippines
e-mail: dhuniel2006@yahoo.com

J. M. Mendez
College of Computer Studies and Engineering, Lorma Colleges, San Fernando La Union 2500, Philippines
e-mail: janelli.mendez@lorma.edu

P. K. Pattnaik et al. (eds.), *Internet of Things and Analytics for Agriculture, Volume 2*, Studies in Big Data 67,
https://doi.org/10.1007/978-981-15-0663-5_4

1 Introduction

Internet of things (IoT) has allowed countless sectors and companies to boost effectiveness through the implementation of centralized surveillance solutions for the most significant procedures over the previous few years. In apiculture or beekeeping, one such instance can be discovered. Honeybees are creepy crawlies or the insects of significant of critical monetary worth and by pollinating plants provide such a crucial provider to agriculture, to the apiculture particularly. Honeybees face an astonishing population decrease without a clear comprehension of the cause. This situation brought and causes serious and genuine natural, money, or economical related and environmental mishaps, also known as the disorder from the colony. Inadequate information collection is one shortcoming in investigating and handling colonies of honeybees. Sensor systems that track in-hive features continually could enable fresh methods for research and apiary leadership.

The pollinators of insects are vital to forest ecosystems and many farmer and agricultural enterprises. They guarantee that certain environmental procedures, including plant reproduction, are maintained. Bees are the most advanced insect pollinators because they can effectively carry and store pollen [1]. Honeybees are essential not only for their manufacturing of honey but also for the equilibrium of the ecosystem since they are crucial for the pollination of plant flowers [2].

Honeybees are environmental insects and live in big, well-structured family communities. Honeybees are commonly known as social insects. Superorganism as referred to by entomologists or the social insects are heavily engaged and developed insects that perform many complex tasks that are not performed by many isolation insects. Interaction, complicated nest design, environmental monitoring, labor defense, and partition are just a few of the behaviors established in the social colonies. These astonishing behaviors are one of the most interesting beings in the world, social insects generally and honeybees.

Typically, three types of adult bees form a honeybee colony. A colony is composed of a queen, the drones, and the workers. In nest construction, several thousand working bees collaboratively work together to build their nest including those activities of gathering their foods and brood rearing. Each worker has a specific task to perform in connection with their adult age. But survival and reproduction take the colony as a whole combined effort [3].

There is only one queen in each colony, except during and after swarming or substitute activities. The fact is that the queen is the only sexually developed bee from the women; her primary role in the colony is reproduction. The queen spawns pollinated and unfertilized eggs. In the springtime and early summertime, the queens in the beehives lay the highest number of eggs including fertilized and the unfertilized ones. The queens can lay eggs of approximately up to 1500 eggs daily.

In early October, they gradually stop laying eggs and generate limited number of eggs or no eggs at all by early following spring (January). A queen lays up to 250000 eggs per year, and over a million lives are being produced by queen.

The most important bees in the colony are drones (male bees). They are usually only present in late summer and spring. The head of the drone is far bigger than the queen and the worker's compound eyes tally at the uppermost of the dome. Drones do not have a stinger, wax glands, or pollen baskets. The drones mainly fertilize the virgin king when she sees her, but this function is performed only by a tiny number of drones. About a week after emergence, drones mature sexually and die immediately after maturation. While drones do not perform such good job for the colony, its presence is considered essential for the ordinary functioning of the colony.

Worker bees are the lowest corporate adults, and most bees are present in the group or colony. They are women who are sexually developed and do not lay spawns under ordinary hives. Workers are structured with speciality facilities like brood glands for food, perfume glands, waxing glands, and pollen baskets that enable them to carry out the entire hive's work. They are purifying and polishing cells, feeding larvae, taking care of the queen, collecting waste, handling incoming nectar or a sap, constructing peas, watching the entrance and air condition, and ventilating the peat in the original weeks as adolescents. They drill for the water, sap, pollens, and propolis (plant sap) later as held bees.

At the point when a colony winds up to a point that it becomes queenless, the ovaries of a few specialists or worker bees create and laborers start to lay unfertilized eggs. Similarly, the development of the laborers' ovaries is accepted to be hindered by the nearness of brood and the queen and her synthetic substances or chemicals. The manifestation of laying laborers or worker bees in a colony describes and implies that the colony has been queenless for at least one to several weeks already. Notwithstanding, laying laborers or worker bees may additionally be found in ordinary "queenright" colonies throughout the swarming period and as soon as the colony is being ruled and controlled by a pitiable queen.

Every one of the three sorts of grown-up bees goes through three formative stages before developing as grown-ups: egg, hatchling or larvae, and pupa. The three phases are all things considered brood. While the formative stages are comparable, they do vary in terms. Unfertilized eggs become rambles or the drones, while those fertilized ones become laborers or worker bees, and or the queens in the colonies. The colonies' health and nutrition are primarily important, and significantly, it impacts in development of female honeybees; hatchlings or larvae bound to end up laborers or worker bees, and get less illustrious jelly and more a blend of nectar and pollens contrasted with the bounteous measures of imperial or royal jelly that the colony ruler or queen larvae would receive.

Physical variables, including weight, temperature, humidity, air gases, vibrating, sound, and forager traffic, are continuously monitored in most scientists as the price and size of electronic detectors reduce as their accuracy and capability rise. To obtain data about the behavior and phenology of colonies, researchers adopted various methods to collate the results.

Temperature affects the behavior of bees. Rarely do they operate when it is below 57 °F or above 100 °F. If the temperature is less than 55 °F, they cannot fly. Bees gather outside shaded hives during very warm days and do not operate. But in the summer, too much shade is irritating for bees. Honeybees keep the brood nest temperature between 32 and 35 °C dynamically so that the brood usually delves. Once the nest temperature is too high, the bees ventilate the warm water out of the nest by fanning or use processes for evaporation.

Using accuracy apiculture, it enables apiarists to control beehives for many potential purposes as mentioned in several studies, data analytics and best practices on beekeepers' real-time monitoring of bees [4, 5], and learning how to reduce the resources and time allocated in beekeeping activities without compromising the quality and productivity of honeys.

This paper's general objective is twofold: to develop a wireless sensor system for Internet surveillance of beehive microclimate variables, including temperature, relative humidity, and weight that could enhance techniques of apicultural research and development. The second objective of the study is to utilize IoT architecture and wireless sensor network in the design and development of low-cost in-hive models for precision apiculture.

The users of the study project are La Union's beekeepers in the Philippines. The project aimed to assist them to take care of the bees and verify their returns regularly. Beekeepers of La Union often have the ability and skills to be a beekeeper of the world class. Not only did this project create things simpler for beekeepers, but it also made beekeeping a company or career more in demand. The unit has been implemented on one of La Union's beekeepers, making the bee farm more efficient and exceptional.

Table 1 exemplifies the honey-gathering calendar and the flow of honey. The season of honey supply is the busiest moment in the year of a beekeeper. The best way for the beekeeper to assess the arrival and termination of the honey flow is to control adjustments in the hives' weight. The honey flow month is between February and March and finishes in May. The beekeeper also extracts surplus honey from the hive during the honey flow season. June to October month is the season in which bee has low honey owing to the season's shortage of food. Therefore, beekeepers feed them sugar for them to survive in the rainy season.

For the beekeepers to monitor the beehive, a mobile application was developed for the project. The mobile application displays the current temperature and humidity, the current weight of the hive, and the bee counter which indicates the bees that go inside and outside the hive [6, 7]. The mobile application would require Internet

Table 1 Honey calendar

Honey flow activities	Month											
	1	2	3	4	5	6	7	8	9	10	11	12
Pre-honey flow season	x											
Honey flow season		x	x	x								
Harvesting time		x	x	x								
Post-honey flow					x							
Critical period for feeding bees						x	x	x	x	x		

1 January, *2* February, *3* March, *4* April, *5* May, *6* June, *7* July, *8* August, *9* September, *10* October, *11* November, *12* December connectivity for the user to monitor the beehive wherever they are. The application also sends weekly reports of the status of the hive; this serves as an overview of the beehive in the past days, and it would make the beekeeper to maintain the beehive. A database would also be developed for storage of data that comes from the hive, and this would make data easier to access. The beehive must also have connectivity to the Internet so data that will be gathered inside would be sent to the database.

This project explored the possibilities of Internet of things in beekeeping for enhancing or substituting manual inspection of hives from a user-centered perspective and at the same time provides an automated cooling mechanism to assist the colony. Using IoT-enabled devices for monitoring beehives made the work of collecting data more efficient. Since the observed beehives might be in distant places, it is preferred to use technology to record and to transfer data [4, 6].

The research project has the following constraints: The project is purely for monitoring and data-gathering purposes only; it cannot act if the honey is ready to be harvested. The beekeeper is still the one responsible for maintaining the good status of the honeybees and their beehives. The weight of the bees and the honeycomb will also be included in the measurement of the weight of the beehive; hence, the accuracy of measuring the weight is reduced.

2 Related Literature

Though the honeybee is frequently associated with honey manufacturing, honeybees' function in agriculture is predominantly for pollination, allowing for fruit development. Pollination is the fertilization phase of the spermatophyte or fruit-bearing plant development process. Male gametophyte cells are contained in pollen. The male pollen grains germinate when transmitted to the stigma of a flower and reach the female gamete cells, and this method allows for the growth of reproduction and fruit. The wind acts as pollen vectors along with birds, insects, and other livestock.

In the first years of the twentieth century, interest in constant monitoring of honeybee colonies started. In 1914, a manually gathered data was released related to beehive temperature in 1907 for several days [8]. In 1926, thermocouples were used to control the temperature in a beehive [9]. The sensor technology development and electronic data acquisition systems have enabled the measurement procedures to be improved. Different kinds of tracking techniques were used in this manner, from easy observation of information in the hive [10] to systems that can analyze those data [11], while a summary of some of the continuous monitoring methods was concurred by specifying the parameters (temperature, O_2, CO_2, relative moisture, weight, vibration, etc.) and the duration of each technique.

Apidictor [12] was among the first electronic bee monitoring devices. This scheme comprised of a low-pass frequency filter to detect sound changes within the hive up to two to three weeks before swarming. An analysis of sound captured on a computer with a sound card and controlled temperature and relative humidity through a pyrometer was also conducted [13]. Three omnidirectional microphones captured the noise. Researchers also defined an imaging system at the entrance to the in-hive [14] to find out how many times a bee entered and left the hive. To recognize the bee under research, they used a tag connected to the body of the bee.

In another study, supervised bees' wintering and measuring the beehive temperature were investigated. The information was sent to a processing computer to determine the hive status by applying an algorithm [15], while RFID scheme was utilized to determine the frequency and duration of honeybee queens' nuptial journeys [16]. A 12-bit resolution data logger was used to evaluate the evolution of a beehive's weight [16] by looking at the hourly weight of two hives in a solar panel operated system [17].

The improvement of the performance and size of the microcontrollers has allowed the development of low-cost beehive monitoring systems based on Arduino®, Make®, Sparkfun®, and Adafruit® [18]. An example of an Arduino-based monitoring system is proposed by a group of researchers [19]. The system used an Excel database to store the temperature and relative humidity information in a microSD memory card. To be able to process this information later, the beekeeper had to go to the beehive colony and download the SD memory card content for each hive to his/her computer.

Honeybee hive surveillance systems based on WSNs have been used in the past few years. These devices have the benefit of being able to monitor the environmental circumstances of the hive remotely in a non-intrusive manner. Technological advancements paved the way toward better apiculture with another study using a primary unit scheme that communicated wirelessly with each node in the hive, acting as an Internet gateway at the same time [20]. Using an embedded SHT15 sensor, the system tracked the temperature and relative humidity. The information gathered was sent to a remote database server by the primary unit. Another development on apiculture was the Raspberry Pi-based temperature monitoring scheme as an Internet gateway [21]. The temperature sensors situated through a one-wire network in the hives linked to the Raspberry Pi. If only one Raspberry Pi was used for all beehives, the writers noted the low price of the solution.

In the advent of wireless sensor networks in precision agriculture, a wireless sensor network (WSN) was used to monitor a hive colony and gather the most important information about the beehive activity/environment and its surroundings [22]. Each beehive has been monitored by two nodes, and each node has been implemented by Libelium (Zaragoza, Spain) through a Waspmote®. Several heterogeneous sensors (temperature, CO_2, pollutants, NO_2, etc.) were used to monitor the honeybees in the hive. Two commercial extension boards have been used, one for each Waspmote because of the large number of sensors. At a frequency of six samples/day, each sensor was read. The gathered specimens were stored for backup on an SD card and transmitted to the base station once every 24 h on a Zigbee network. The base station was a bridge, sending the information over a 3G/GSM link to the remote server. Also, a surveillance scheme was used to suggest multiple algorithms in order to detect changes in the hive and warn the apiculturist of prospective threats to the colony [23]. A weather algorithm has been suggested based on the CO_2 concentrations of the hive.

Similarly, researchers have designed and developed a scheme for measuring the temperature of the beehive using an LM35 sensor using an Arduino platform [24]. The information has been transmitted wirelessly to an Internet-connected desktop base station via an XBee-PRO module. The records were processed to discover patterns to detect future absconding circumstances in the heat stress in the shed. The information was gathered.

In another study, it described an environmental control system to monitor bee activity frequency [25]. The system monitors the frequency of bee entering and exiting from the hive, as well as indoors and outdoors, the temperature and moisture. Accordingly, the precision of the frequency was based on hive activity, a 10% average mistake and 20% maximum.

The beam also adheres to Poisson's ratio, where a beam, when in tensile straining, will extend in length in the direction of the force, but its cross-sectional area (CSA) will contract. When a beam experiences bending, its top surface experiences a tensile strain and its bottom surface experiences a compressive strain (depending on the direction of bending). If the beam is behaving elastically, then the strain is directly proportional to stress, and hence to the applied force. These changes in the beam can be used to measure the force applied precisely, by means of strain gauges [26].

There have been recent advances in the development of load cells, pressure sensors, and strain gauges which made these technologies more attractive for use in weight measuring applications. Load cells have become a standard in beehive weighing systems. In the application of a single, robust load cell used to weigh an entire beehive [14], the maximum weight the load cell can accommodate is 200 kg, which satisfies the requirements of an average-sized beehive. The jig used to translate the weight of the beehive to the load cell allows for uniform weight distribution and can cope should there exist an imbalance of weight occur in the beehive.

A prototype was developed to collect images and audio within a beehive [27]. The platform utilizes a Libelium Waspmote, and Raspberry Pi in order to process and store the data. Additionally, microphones, accelerometers, thermal and infrared cameras were used, along with emergency notifications to the user in case an undesirable event occurs to the beehive. The goal was to utilize the said equipment's to collect data in a discreet manner.

A proposed framework to classify system-planning methodologies was also made available for selecting a proper methodology to develop an information system that is a key factor for its success [28]. The framework aimed at identifying appropriate methodology utilized to implement myBee and was concurred successful based on the obtained observations. In addition, the authors mentioned that several methodologies for implementing and planning information systems are available; thus, myBee's architecture is flexible to be utilized in various fields, which simplifies and streamlines the development process.

3 Methodology

The researchers employed descriptive-developmental research design. A descriptive secondary data analysis was used by looking into the manual records, active observation, documentation, databases of the apiculturists as inputs to the project planning. During the development phase, prototyping was employed to allow the researchers to improve the system quality based on the requirements' specifications. It also facilitated user involvement in the development making way for reduced time and cost. The researchers utilized the IoT-architectured embedded system and wireless sensor networks in designing the low-cost precision apiculture. The system was tested and piloted in the province of La Union, Philippines, where apiculture is one of the viable economic means.

3.1 The IoT Architecture of Beehive

In the design framework, the IoT architecture with three domains or layers was considered in the system including the IoT device, IoT gateway, and IoT platform layer as illustrated in Fig. 1. The IoT device layer is the IoT architectural component responsible for the observation and data gathering. The input domain includes the sensing instrumentation tools, counter, temperature sensor, moisture sensor, and the weight sensor of the hives.

In this architecture, the IoT gateway layer is responsible for the connectivity, the protocol, data aggregation, filtering and processing, data security, data management.

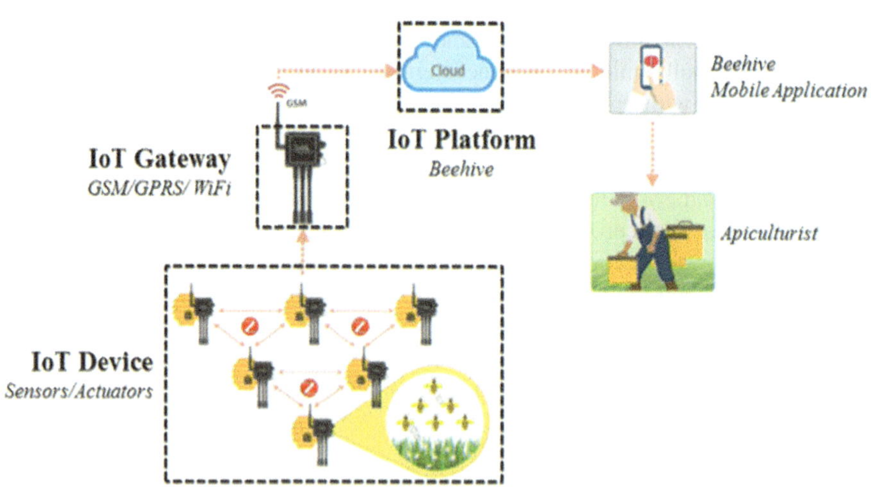

Fig. 1 IoT architecture of beehive

A wireless sensor network (WSN) was developed using a low-energy adaptive clustering hierarchy (LEACH) while Zigbee, GSM/GPS, and Wi-Fi protocols were used in the system.

The third layer of the architecture is the IoT platform. This layer affords the cloud computing services in the system to provide public access and data sharing capabilities for the end-users. This layer allows system provisioning through the "Beehive" online apiculture management system with a mobile application. The IoT platform layer allows management as well as the automation of connected devices in this IoT-architectured low-cost precision apiculture.

3.2 *Beehive Hardware Components*

The study envisioned to improve productivity in beekeeping of the apiculturists without modifying the internal components and configurations of beehives commonly known as a man-made nest of any bee colony. In these modern days, Langstroth hives are commonly used by apiculturists which is a modular beehive that is vertically oriented with entrance for the bees. The hives have also boxes and frames for brood and honey. In this study, the hives developed by the researchers are illustrated in Fig. 2 with the hardware requirements for bees counting and hive status. The embedded system of this project is enclosed in this hive.

Beehive embedded system includes the IoT devices such as sensors and actuators, microcontroller, the Arduino compatible modules for IoT gateways, and was programmed and configured according to the desired system functions.

Fig. 2 Langstroth beehive type

Fig. 3 Arduino ATmega328P

3.2.1 Beehive Microcontroller (Arduino ATmega328P)

A microcontroller used in the system is the Arduino ATmega328 as illustrated in Fig. 3. It consists of both a physical programmable circuit board with and IDE. ATmega328 is an input and output processing low-voltage microcontroller. The program that comes from the Arduino IDE also was stored. ATmega328 is used as an Arduino off-board and will work the way it works. The latter is one of the most important parts of this de-appliance because it is necessary for all the device's major parts DHT22, HX711, and ESP8266.

3.2.2 Temperature and Humidity Sensor

In this IoT-architectured precision apiculture, one of the most important data-gathering sensors in the IoT device layer is the temperature and moisture sensor. In precision apiculture, the efficiency and effectivity of measuring the conditions of the hive can be improved even without human intervention, and it is very crucial that these ambiental conditions such temperature and moisture can be determined. The sensor DTH22 is compatible with the development board Arduino ATmega328P that was used in the system. The sensor utilizes thermistor and a moisture resistor to evaluate temperature. The element is linked to the digital pin ATmega328 and placed inside the machine to monitor the beehive's temperature and humidity (Fig. 4).

3.2.3 Load Cells (HX711) Information

Apiculturists need to remove the hexagonal prismatic wax cells or the honeycomb from the hives and make observations to know the amount of honey in the hives. These removals of honeycomb activities interfere and lower the production in the apiculture

Fig. 4 Temperature and humidity sensor

industry due to interfering factors to bee colony activities of honey production. In this system, apiculturists need not remove the honeycomb from the hives.

The design utilized a semi-bridged weight sensor and an HX711 Load Cell-Amplifier as illustrated in Fig. 5 to measure the study quantity of honey in the hives. HX711 is a precision 24-bit analog-to-digital converter (ADC) developed for weighing scales purposes that can be used in small scale to industrial application by having it interfaced with a bridge sensor. In this design, a 200 kg maximum weight can be measured in the scheme.

Further, to improve the system's ability to measure the honey in the hives, multiple sensors were used simultaneously. In this context, modifications on the configuration were performed in HX711 Load Cell-Amplifier at the level of the millivolt; it is used to make these modifications to the level that can be read by the Arduino development board. The load cells are connected to the HX711 sensor, and the HX711 sensor was connected to the Arduino development module so that Arduino can evaluate the incoming information and transmit to the server. In this research, there is a

Fig. 5 Load sensor (HX711)

Fig. 6 Bee counter (QRE1113)

proportional placement of four load cells on the plywood tray and a plywood shelf and to be connected to a mobile module.

3.2.4 Bee Counter (QRE1113) Information

Apiculturists need to learn and monitor the honeybee colony's reproduction or swarming situation. This can be done by monitoring how many bees left the original hive location with the old queen. A prime swarming happens when the queen bee left the colony with 60% or a large group of workers, and these activities should be tracked by the system using counter. The QRE1113 was used for this purpose to afford important advantages for the apiculturists such that swarming is a good indication of healthy hives (Fig. 6).

3.2.5 Wi-Fi Module

In this IoT-based precision apiculture architecture, the ESP8266 Wi-Fi module was utilized as an IoT gateway layer. In this layer, the Wi-Fi gateway serves the connectivity services between the IoT devices and the IoT platform. ESP8266 is a low-cost SOC with TCP/IP protocol stack which allowed the Arduino module to access the Wi-Fi network and provide services and connection points between the temperature and humidity sensors, and load sensors to the Beehive online and mobile application.

Fig. 7 Wi-Fi module (ESP8266)

This module also supports additional security in the IoT network as well as in data security and transports (Fig. 7).

3.2.6 Solar Charger

A solar charger as illustrated in Fig. 8 is a voltage or current regulator that is connected to the solar panel. It prevents the battery from overcharging. The solar panel absorbs sunlight and will generate direct current DC energy from it. It will then be converted into an alternating current AC to make it usable. A battery bank will be used to store energy for long-lasting and non-stop consumption. The battery bank will give life to the whole device and will serve as the main source of power.

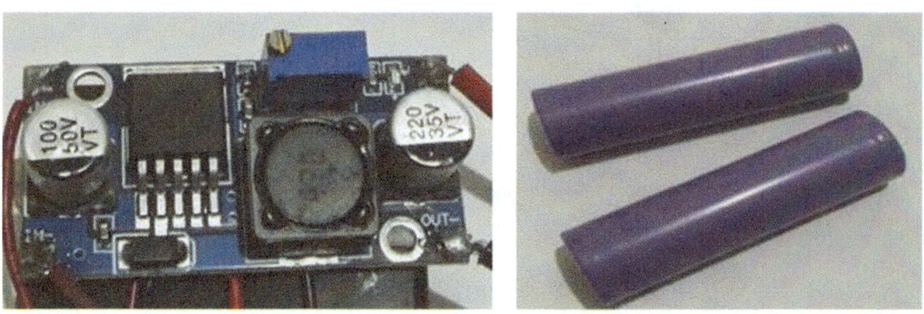

Fig. 8 Battery charger

3.3 Beehive System Architecture

In this developed precision apiculture technology, beehive IoT devices were expected to fulfill functional criteria: (i) Use a minimally invasive technique, (ii) operate in distant fields for lengthy periods, and (iii) enable real-time surveillance and reporting. It was highlighted and defined that various automatic surveillance system architectures for real-time beehive surveillance system afforded numerous advantages [26].

This system employed the IoT architecture with three domains or layers including the IoT device layer, IoT gateway layer, and IoT platform layer. The system utilized five stages of IoT architecture including the sensors and actuators; IoT gateway, protocols, and data acquisition; edge computing, datacenter management, and cloud computing; and reporting. Temperature and humidity sensors were utilized to gather and communicate Ambiental parameters of the hives including temperature and humidity sensors, load sensor through wireless links and communicate to the cloud. For the system data aggregation and Internet gateways, a wireless sensor network (WSN) was developed with a planned spatially deployed sensor nodes in La Union, while GSM/GPS and Wi-Fi TCP/IP protocols were used in the IoT gateway.

Cloud computing services were embedded to provide public access and data sharing of Beehive application with a mobile application for the end-users (beekeepers and professional apiculturists). On the last stage of the architecture, a system dashboard is available for system administrators, and vectoral or numerical forms sent to the users (Fig. 9).

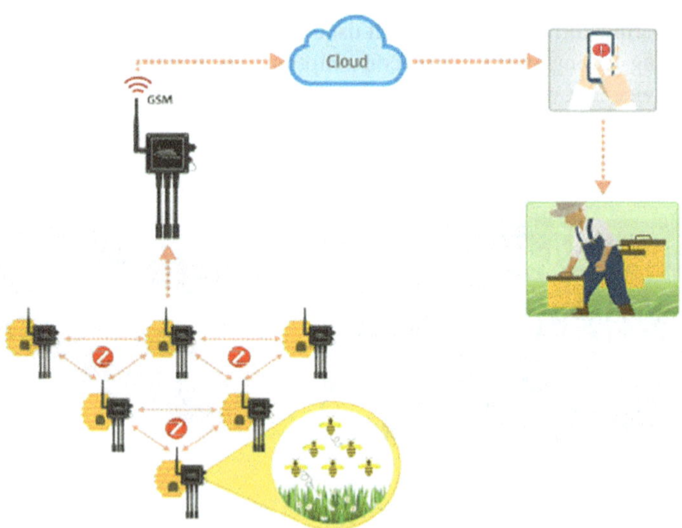

Fig. 9 System architecture

4 A Detailed Look of the System

A schematic diagram illustrated in Fig. 10 has been specifically designed to allow configurations on specific functionalities (components) in developing flexible and agile maintenance system architecture. In this diagram, all sensors are linked to the ATmega328 together with the Wi-Fi module. The ESP8266 works as it connects from ATmega and transmits data over the Internet.

The DHT22 receives its data from the ATmega328 and measures the temperature and humidity inside the device. The HX711 module has four load cells linked to it and mounted under the beehive. The load cells put a balanced measurement on the four corners of the under the beehive.

The schematic diagram illustrated in Fig. 11 is the connections of the beehive solar panel, battery, and CPU fan components. The solar panel absorbs energy from the sun and transforms it into electricity. The electricity from the solar panel will flow straight into the solar charge controller and its rechargeable batteries. The CPU fan is connected to the ATmega328 and was installed on the top of the hive to prevent too much heat, and help the water flow inside the beehive.

The schematic diagram illustrated in Fig. 12 is about the bee counter components, devices, and their connectedness. To achieve the counting function of bees leaving and coming into the hives, QRE1113 was utilized with two reflectance detectors per

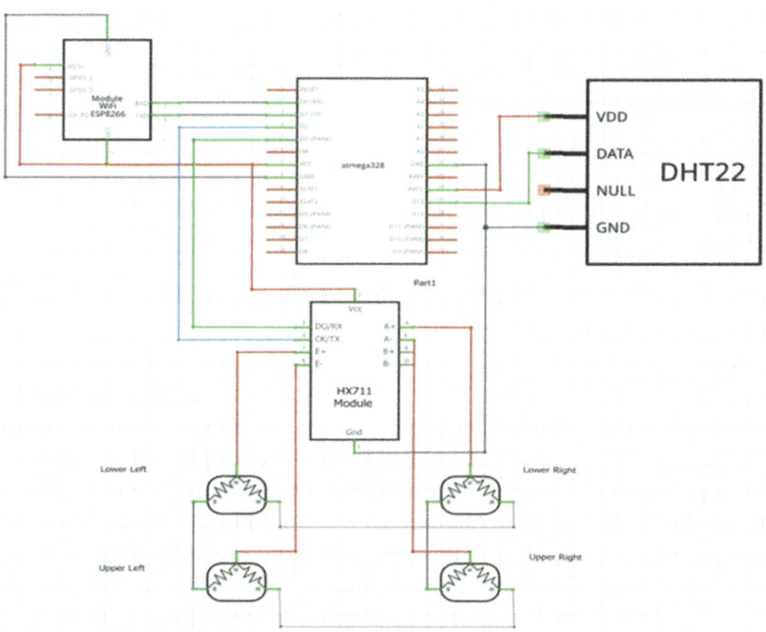

Fig. 10 Schematic diagram of the ATmega328 sensors and ESP8266

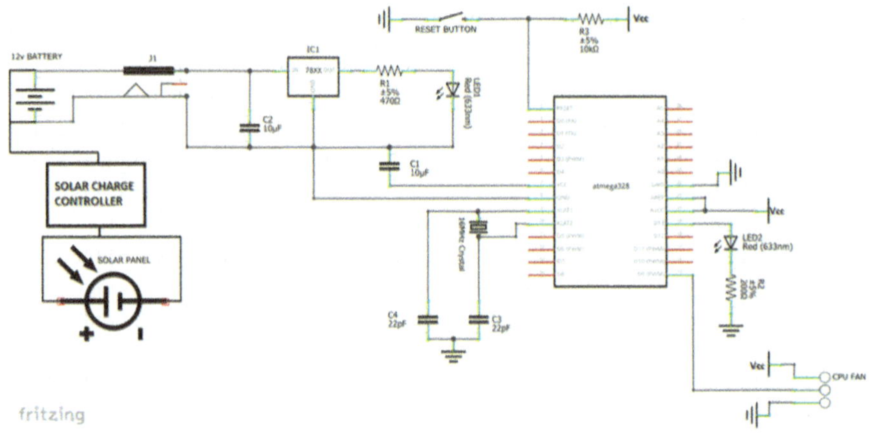

Fig. 11 Schematic diagram of solar panel, battery, CPU fan

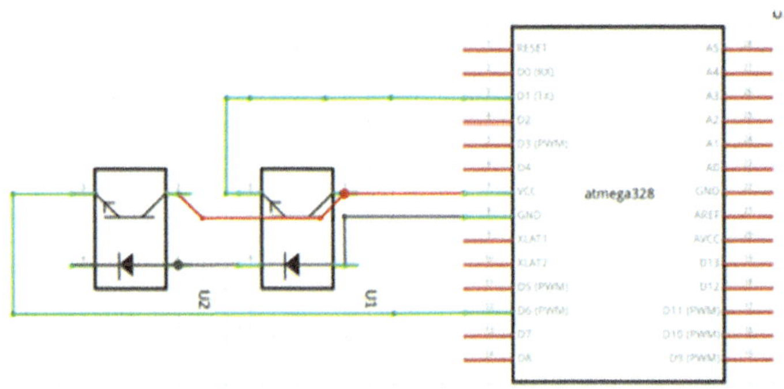

Fig. 12 Schematic diagram of the QRE1113

gateway; one sensor was used to detect the outgoing bees and the other one to detect the incoming bees.

A beekeeper or apiarist performs numerous hive management activities to make sure that apiculture production is high. Before the application of technologies in apiculture, apiarists were doing manual beekeeping. Figure 13 illustrates the flowchart of manual beekeeping. The apiculturist monitors the hive's ambiental conditions. Hives' ideal temperatures should be between 60 and 100 °F (15–38 °C) if its too hot bees do not work because they cluster in unshaded hives. In traditional beekeeping, an apiarist checks the weather. IF it is too hot, THEN, go to the beehive and cool it down, check the frames, IF there is enough honey, THEN harvest. Now, IF the weather is not hot or between 15 and 38 °C, THEN check the frames, IF there is enough honey, THEN harvest.

In this developed precision apiculture technology, the apiarists can either use the online application using a web browser or launch a mobile application in a

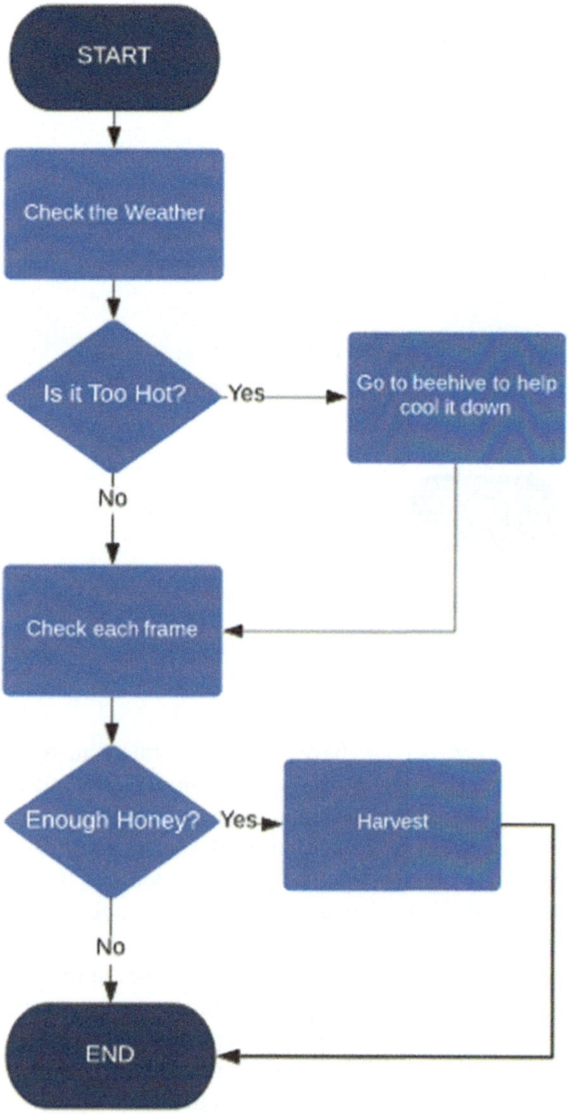

Fig. 13 Flowchart of manual beekeeping

smartphone where the mobile app was installed. Apiarists will be able to select which hives conditions they would like to check and monitor. Apiarists are also informed via notification or dashboard as to whether it is good to harvest already based on the load or weight of the honey from beehive's frame. The mobile app is designed with icons representing (temperature, humidity, fan, and weight) that a user can select from to improve user experience. Figure 14 illustrates the flowchart of the developed precision apiculture technology.

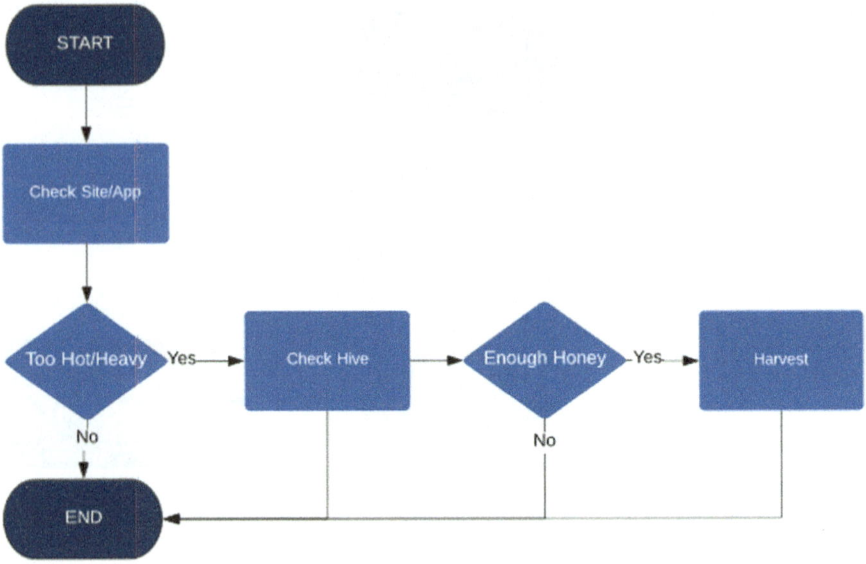

Fig. 14 Flowchart of the developed precision apiculture technology

4.1 Device Prototype

The two sensor nodes have been incorporated into the roof of a hive where the samples can effectively gather the hive conditions. This also resulted in effective operation without interfering with the beekeeper's activity. The data outlined in the next subsections of this paper was gathered using this prototype to conduct the data gathering and system evaluation. The data collected was used to evaluate the effectiveness of classification algorithms and recommended features of the hardware's future design (Fig. 15).

Figure 16 illustrates the actual beehives in the bee farm of Bacnotan, La Union. The actual hive is a modular beehive oriented vertically with frames inside. It has an entrance for the bees and several boxes inside the enclosure intended for the bee brood and honey. It is also top capped to protect the beehives from weather and water. There are three nodes deployed in this location.

Bacnotan is a first-class municipality in the province of La Union where apiculture is one of the livelihoods and viable means for economic development. It has the La Union Honeybee Center where visitors can have the first-hand experience on the lifecycle of bees and honeys from bees are manufactured through a live demonstration.

The power consumption measurements of the nodes were carried out on the wireless sensor networks installed for testing the precision apiculture system. There were three nodes and a local database server in the network. The measurements were carried out with the active node which continually acquires information and sends it to

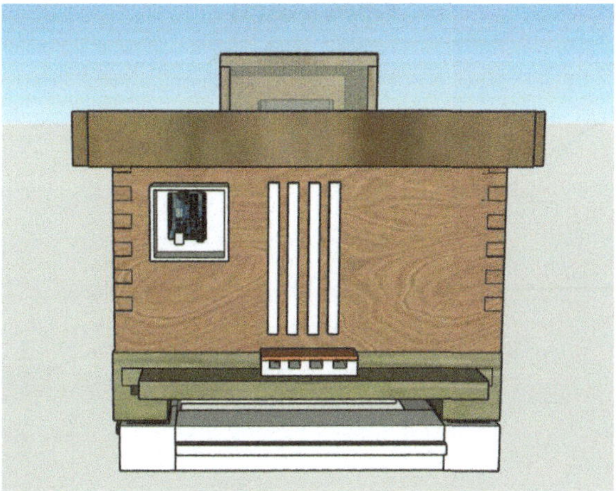

Fig. 15 Device prototype

Fig. 16 Actual deployed devices (sensor nodes)

the local server. In the first two cases, a Wi-Fi module was set in receiving mode, and the acquired data was not transmitted via the node. In the latter case, the nodes continually transferred the obtained information.

The battery-compensation capacitors work when an electricity supply failure occurs in one of the components of the system and sensor networks installed in beehives. The primary energy of the nodes installed within the ring was intended for 75 h to assess the independence of the nodes and the local database server was shut down. During that time, the sequence of operations is like the one described above, except that the Wi-Fi module does not transmit data and the samples are stored on the

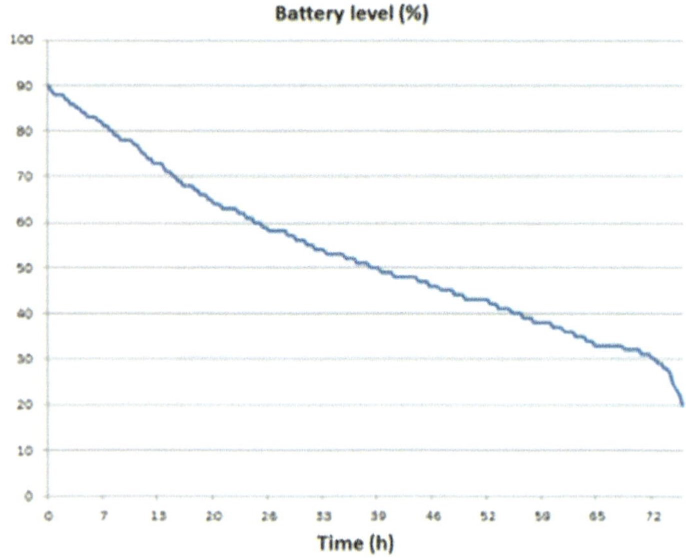

Fig. 17 Sensor nodes' power consumption

microSD card. With a rechargeable lithium-ion battery of 3.7 V/2300 mAh, the battery level of the sensor nodes that were analyzed fell an average of 70%, as shown in Fig. 17. This time of autonomy is enough to solve the problem since the global server of the developed system generates an alarm when losing communication. Therefore, it was not necessary to install a battery with higher capacity.

In this IoT-architectured precision apiculture, a mobile application was developed to support the users in the IoT platform layer. The mobile application allows the end-users (beekeeper, apiculture professionals, and researchers) to check and monitor the measurements of different sensors in the beehive such as temperature, humidity, fan, and weight of the beehives. The app is also used for hive inspection, management, and maintenance by using smartphones and tablets PCs. The "Bee Watch" as illustrated in Fig. 18 was developed using Android Studio and APIs.

Figure 18 shows the user interface (UI) of the Bee Watch mobile application. The app allows the users to check and monitor the measurements transmitted by the IoT devices to the IoT platforms including the temperature, humidity, and weight of the beehive. The fan is also included in the design to allow users to activate fan remotely using the mobile app when the weather or temperature reading indicates it is higher than the heat threshold of the application. Similarly, the mobile app will also inform the users if the fan is working or not with an indicator (Fig. 19).

The illustration presented in Fig. 20 is the user interface for the Bee Watch weight feature. This feature provides the weight of the bees and the weight of the honey. This feature informs the user on the rate of change in weight as well as the health of the colony. It will let the beekeeper know if swarming conditions are likely to happen

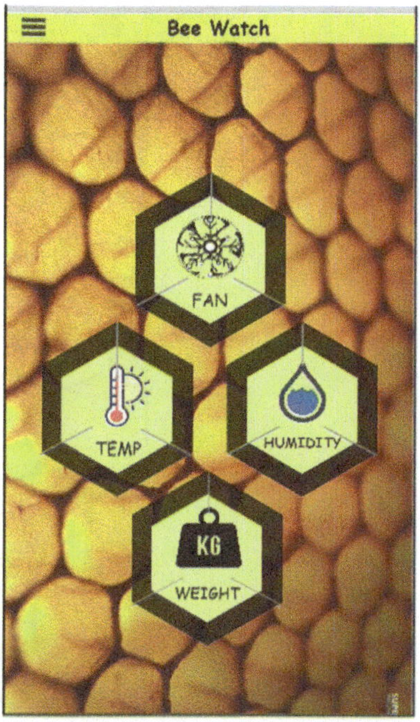

Fig. 18 Beehive mobile application “Bee Watch”

Fig. 19 Bee Watch user interface for fan status

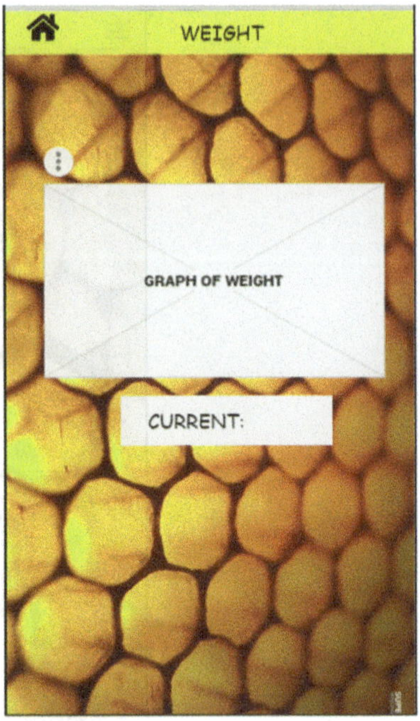

Fig. 20 Bee Watch user interface for weight

or have happened already as shown in the current weight of the beehive base on the time stamp.

In Fig. 21, the average weight of the hives is shown over 32 days as an average development. It describes that from day 26, the weight of the beehives can be observed as stable. The changes in weight were associated with honey production depending on the quality of incoming pollen resources and swarming conditions. Based on the illustration in Fig. 21, by the end of the period of bloom, the weight of beehive was

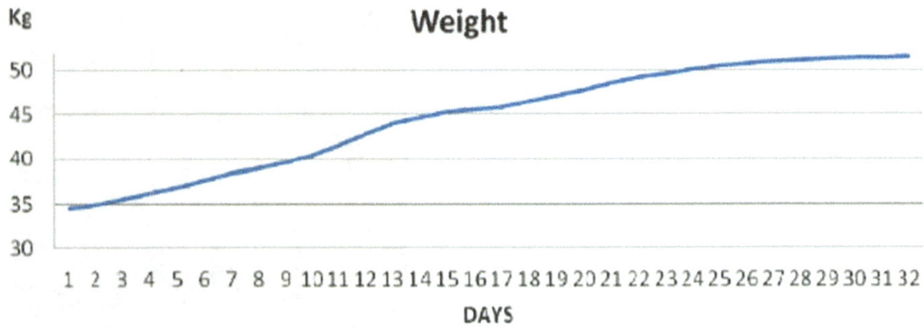

Fig. 21 Average weight of the hives

stabilized which implied that the beekeeper can now remove the honey from the hives.

When the weather is warmer than the device threshold, the outfitted fan activates help serve as a cooling system and maintain an appropriate temperature in the hives. Figure 22 illustrates the average temperature measurements in Celsius recorded during the experimentation for 32 days. The information indicates the ambient temperature outside the beehives and the temperatures in various fields in the beehives: mid-brood; region in the periphery of the same brood comb with honey/pollen reserves, and honeycombs separated from brood combs. The figure corresponds to the four hives' average daily temperature.

Figure 23 illustrates the environmental moisture and humidity of the beehives in various regions: mid of brood; the area in the periphery of the same brood comb with the honey/pollen reserves; and honeycombs separated from the brood combs.

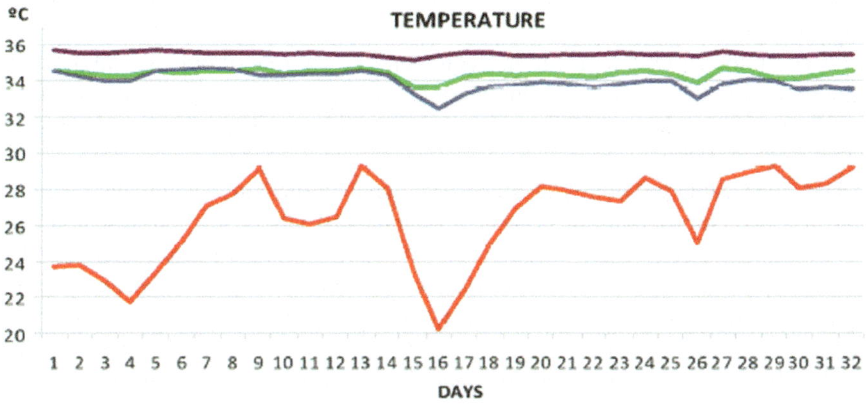

Fig. 22 Average temperature of the hives

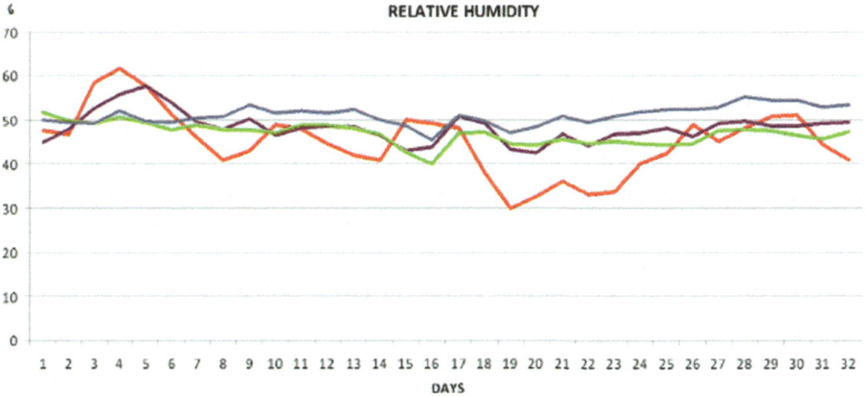

Fig. 23 Average relative humidity

The information matched the average moisture content of four hives per day. The humidity inside of the beehive stayed more stable than beehives' outside humidity.

5 Conclusion and Recommendation

In this IoT-architectured precision apiculture technology, the researchers aimed at assisting the beekeepers on apiary inspection, monitoring, management, and maintenance. The descriptive-developmental methodology was explored using the prototyping techniques to facilitate user involvement in the project development stages. The IoT architecture and wireless sensor networks (WSNs) were used in the design and implementation of the system. The three layers of the IoT architecture were configured based on the results of system analysis and design. IoT architecture layers were considered in the system including the IoT device, IoT gateway, and IoT platform layers.

Based on the system deployed in Bacnotan, La Union, Philippines, and the data gathered during the 32-day experimentation, the researchers concur that IoT architecture can be complemented with a wireless sensor network approach in the design of IoT gateway layer.

Also, the system provides real-time surveillance of the defined ambiental parameters of a bee colony that has become a popular instrument for both research and practical beekeeping, as the tiny, precise, and robust sensors make it easy for the electronic parts to deploy and retain. Hive weight monitoring allows beekeepers to monitor hive health and can record quality control during pollination. Precision beekeeping is still evolving, but multiple weight surveillance systems for bee colonies are already accessible. The developed and deployed project acquired samples and dataset from all hives of apiary synchronously, with a real-time hive weight measurement that is an innovation in the monitoring of wireless beehive systems.

Similarly, wireless beehive system allows apiarists to monitor the temperature, relative humidity, and weight monitoring non-intrusively, hence, it promotes beehives to become healthy and increase honey's productivity, and effectively perform the overall aspects of apiary management using the Bee Watch mobile application.

The embedded system of this project was powered by a solar panel and compensation capacitors in the rechargeable battery. However, redundancy components are made available should power interruption will deter the transmission of ambiental data to the servers. Aside from the analytical tools and reports the system provides, it is also a reliable low-cost IoT-architectured beehive monitoring system based on a WSN.

The researchers recommend that empirical investigation should be conducted based on the data gathered compared with other "top of the line apiary technologies" available in other countries. Future researchers may consider on swarm intelligence to address the decreasing populations of bees and applying machine learning techniques to improve beehives' conditions and increase honeybee production and manufacturing.

References

1. Patrício-Roberto, G.B., Campos, M.J.O.: Aspects of landscape and pollinators—what is important to bee conservation? Diversity **6**, 158–175 (2014)
2. Goulson, D., Nicholls, E., Botías, C., Rotheray, E.L.: Bee declines driven by combined stress from parasites, pesticides, and lack of flowers. Science **347**, 1255957 (2015)
3. Matthias, A.B., Hanno, H., Charlotte, K.H., Robin, F.A.M.: Brood temperature, task division and colony survival in honeybees: a model. Ecol. Model. **221**(5), 769–776 (2010)
4. Evans, H.: Remote bee hive monitoring. Arnia [Online]. Available: http://www.arnia.co.uk/. Accessed on 22 Sept 2014 (2014)
5. Dongyun W., Kai W., Mingcong D.: The application study of intelligent PID algorithm for the internal combustion engine control system. In: International Conference on Mechatronics and Automation, China, pp. 923–927 (2010)
6. Arnia Remote Hive Monitoring. Arnia [Online]. Available: http://www.arnia.co.uk/. Accessed on 22 Sept 2014 (2014)
7. Southwick, E.E.: Metabolic energy of intact honeybee colonies. Comparative. Biochem Physiol **71**, 277–281 (1982)
8. Gates, B.N.: The Temperature of the Bee Colony; Bulletin of the U.S. Department of Agriculture; U.S. Department of Agriculture: Washington, DC, USA (1914)
9. Dunham, W.: Hive temperatures for each hour of a day. Ohio J. Sci. **31**, 181–188 (1931)
10. Odoux, J.F., Aupinel, P., Gateff, S., Requier, F., Henry, M., Bretagnolle, V.: Ecobee: a tool for long-term honey bee colony monitoring at the landscape in West European intensive agroecosystems. J. Apic. Res. **53**, 57–66 (2014)
11. Kviesis, A., Zacepins, A., Riders, G.: Honey bee colony monitoring with implemented decision support system. In: Proceedings of the 14th International Scientific Conference Engineering for Rural Development (ERDev), Jelgava, Latvia, 20–22 May 2015, pp. 446–451 (2015)
12. Woods, E.F.: Means for detecting and indicating the activities of bees and conditions in beehives. US Patent 2,806,082, 10 Sept 1957
13. Ferrari, S., Silva, M., Guarino, M., Berckmans, D.: Monitoring of swarming sounds in bee hives for early detection of the swarming period. Comput. Electron. Agric. **64**, 72–77 (2008)
14. Chen, C., Yang, E., Jiang, J., Lin, T.: An imaging system for monitoring the in-and-out activity of honey bees. Comput. Electron. Agric. **89**, 100–109 (2012)
15. Zacepins, A., Stalidzans, E.: Architecture of automatized control system for honey bee indoor wintering process monitoring and control. In: Proceedings of the 13th International Carpathian Control Conference (ICCC 2012), Podbanske, Slovakia, 28–31 May 2012, pp. 772–775 (2012)
16. Heidinger, I.M.M., Meixner, M.D., Berg, S., Büchler, R.: Observation of the mating behavior of honey bee (*Apis mellifera* L.) queens using radio-frequency identification (RFID): factors influencing the duration and frequency of nuptial flights. Insects **5**, 513–527 (2014)
17. Meikle, W.G., Holst, N.: Application of continuous monitoring of honeybee colonies. Apidologie **46**, 10–22 (2015)
18. Meikle, W., Holst, N., Mercadier, G., Derouané, F., James, R.: Using balances linked to linked to dataloggers to monitor honey bee colonies. J. Apic. Res. Bee World **45**, 39–41 (2006)
19. Bromenshenk, J.J., Henderson, C.B., Seccomb, R.A., Welch, P.M., Debnam, S.E., Firth, D.R.: Bees as biosensors: chemosensory ability, honey bee monitoring systems, and emergent sensor technologies derived from the pollinator syndrome. Biosensors **5**, 678–711 (2015)
20. Sánchez, V., Gil, S., Flores, J.M., Quiles, F.J., Ortiz, M.A., Luna, J.: Implementation of an electronic system to monitor the thermoregulatory capacity of honeybee colonies in hives with open-screened bottom boards. Comput. Electron. Agric. **119**, 209–216 (2015)
21. Kviesis, A., Zacepins, A., Durgun, M., Tekin, S.: Application of wireless sensor networks in precision apiculture. In: Proceedings of the 14th International Scientific Conference Engineering for Rural Development (ERDev), Jelgava, Latvia, 20–22 May 2015, pp. 440–445 (2015)
22. Zacepins, A., Kviesis, A., Ahrendt, P., Richter, U., Tekin, S., Durgun, M.: Beekeeping in the future—smart apiary management. In: Proceedings of the 17th International Carpathian

Control Conference (ICCC), TatranskaLomnica, Slovakia, 29 May–1 June 2016, pp. 808–812 (2016)

23. Murphy, E., Popovici, E., Whelan, P., Magno, M.: Development of a heterogeneous wireless sensor network for instrumentation and analysis of beehives. In: Proceedings of the IEEE International Instrumentation and Measurement Technology Conference, Pisa, Italy, 11–14 May 2015 (2015)
24. Murphy, E., Magno, M., Whelan, P., Popovici, E.: b+WSN: Smart beehive for agriculture, environmental, and honey bee health monitoring—preliminary results and analysis. In: Proceedings of the 10th IEEE Sensors Applications Symposium (SAS), Zadar, Croatia, 13–15 April 2015, pp. 1–6 (2015)
25. Kridi, D.S., Carvalho, C.G.N., Gomes, D.G.: Application of wireless sensor networks for beehive monitoring and in-hive thermal patterns detection. Comput. Electron. Agric. **127**, 221–235 (2016)
26. Jiang, J.-A., Wang, C.-H., Liao, M.-S., Su, Y.-L., Chen, W.-S., Huang, C.-P., Chen, C.-H., Yang, E.-C., Chuang, C.-L.: A WSN-based automatic monitoring system for the foraging behaviorof honey bees and environmental factors of beehives. Comput. Electron. Agric. **123**, 304–318 (2016)
27. Kviesis, A., Zacepins, A.: System architectures for real-time bee colony temperature monitoring. Procedia Comput. Sci. **43**, 86–94 (2015)
28. Libelium Communications. Waspmote Catalogue. Available online: http://www.libelium.com/products/waspmote. Accessed on 10 Oct 2016
29. Benchoff, B.: Building a digital scale from scratch. Hackaday, 12 June 2013. [Online]. Available: http://hackaday.com/2013/06/12/building-a-digital-scale-from-scratch/. Accessed on 6 Jan 2015

Adaptive Encryption Technique for Collaborative Cloud Environments

Y. Harold Robinson, E. Golden Julie and A. Vegi Fernando

Abstract In recent times, the data storage in collaborative cloud environment has distorted a lot when compared with prior times. Security flattering is one of the major imperative challenges in database storage and classification. Data stored in databases are susceptible and thus focusing the safety only on the susceptible data that minimizes the delays or troubles in the classification. This paper describes an extremely innovative and novel methodology for securing numeric data in databases. It presents a realistic solution to the problem where numeric data are transformed into alphanumeric types and hence there is minimal probability of storing encrypted data in the existing numeric field. The proposed algorithm allows translucent verification intensity encryption that does not modify the data field category or the fixed time span.

Keywords Encryption · Decryption · Database · Security · Authentication

1 Introduction

The most excellent technique of securing the data is to limit the access rights to the information which can be achieved by the procedure of authentication and authorization [1] in collaborative cloud environment. A consumer should be checked for authentication before being allowed to access the information and must only be allowed to attain the operations for which access rights are obtainable [2, 3]. If the

Y. Harold Robinson (✉) · A. Vegi Fernando
Department of Computer Science and Engineering, SCAD College of Engineering and Technology, Cheranmadevi, India
e-mail: yhrobinphd@gmail.com

A. Vegi Fernando
e-mail: a.vegifernando@gmail.com

E. Golden Julie
Department of Computer Science and Engineering, Anna University Regional Campus, Tirunelveli, India
e-mail: goldenjuliephd@gmail.com

P. K. Pattnaik et al. (eds.), *Internet of Things and Analytics for Agriculture, Volume 2*, Studies in Big Data 67,
https://doi.org/10.1007/978-981-15-0663-5_5

data to be accessed is on a local machine, applying access control is easy, but if data is accessible from a remote client, user credentials and data needs to be secured over the network [4]. In such situations, security protocols should be used. In case that a malicious user somehow breaches the above security provisions and gets access to data; the only solution is scrambling the data [5]. Security also requires that data should be kept safe from unauthorized access [6, 7]. And the best line of defence is physical security (placing the machine to be protected behind physical walls) [8, 9]. Some of the algorithms used for achieving security are Diffie-Hellman [10], DES [11], TDES [12], IDEA [13] and AES [14]. Usage of these existing cryptographic techniques is meant to secure the data, which is not suitable for protecting the sensitive data of database [15]. Balaji et al. [16], the result is that most conventional database systems have leaks that attacker can use to penetrate the database [17]. No matter what degree of security is put in place, sensitive data in database are still vulnerable to attack [18]. A remedy, therefore, is to turn to cryptographic means of storing data [19]. Encrypted data stored in a database can prevent their disclosure to attackers even if they manage to circumvent the access control mechanism [20, 21]. Thus, cryptographic techniques can ensure excellent security for databases, by reducing the whole security process down to the protection of only few cryptographic keys [22]. Nonetheless, the time cost involved in encrypting and decrypting data items can greatly degrade the performance of the database system [23]. A compromise solution must be found between performance and security, by encrypting only sensitive data in the database [24]. The database is usually assumed to be trustworthy [25]. Under this assumption, the goal is to achieve security against external attacks and probably moreover alongside users trying to acquire data afar their constitutional rights, for instance, by various forms of statistical conjecture [26]. Encryption is a healthy established knowledge for defensive responsive information. Nevertheless, formerly encrypted, information can no longer be easily queried sideways from accurate matches [27]. Encryption can be used to provide an extra level of security [28]. Unfortunately, the use of standard encryption techniques for this purpose results in degradation in the performance of the database system [29, 30]. The main source of the problem is that the standard techniques do not preserve order and therefore the database indices such as B-tree can no longer be used for answering range queries [31–33].

The database is shared by multiple parties such as internal users, partners, contractors and other sensitive data stored in database could be a target to attackers [34]. The attackers are not only external but also from within the organization [35]. Adding the database encryption, valuable information in database becomes more secure since the encrypted data ensure the confidentiality of the data [36] in collaborative cloud environment.

2 Proposed Scheme

The proposed research is to secure the numeric data in the databases. It presents a practical solution to the problem that exists in the standard algorithms where numeric data is converted to binary or alphanumeric type and hence encrypted data is not possible to be stored in the existing numeric column. The 3KDEC algorithm allows record-level encryption that encrypts numeric data and gives numeric encrypted value; hence eliminating the need to change the data type of column. 3KDEC is a symmetric key block encipher algorithm. 3KDEC aims to provide a simple and efficient method for database encryption, overcoming the shortcomings of the prior database encryption methods with an objective to suggest how to encrypt the entire contents of the database without changing its structure. It provides a method of decrypting only the data of interest, wherein queries are not changed because of the encryption, ensuring that existing applications can use the encrypted database without the need for any changes in the application software.

2.1 3KDEC Algorithm

3KDEC algorithm is a symmetric key block cipher algorithm. Symmetric key algorithm is one which uses single secret keys for both encryption and decryption. Encryption/decryption can be considered as an electronic lock where sender puts the message in a box and locks the box using the shared secret key; receiver unlocks the box with the same key and takes out the message. The original message is called the plaintext and the message sent through the channel after encryption is called the cipher text. To create cipher text from the plaintext, sender uses an encryption algorithm and a shared secret key. To create plaintext from the cipher text, receiver uses a decryption algorithm and the same secret key.

The key can be visualized as a set of values/numbers that the cipher as an algorithm operates. In symmetric key encipherment, the encryption and decryption algorithms are inverses of each other. If P is the plaintext, C is the cipher text and K is the key, the encryption algorithm EK(x) creates the cipher text from; the decryption algorithm DK(x) creates the plaintext from the cipher text. Encryption algorithm EK(x) and the decryption algorithm DK(x) are inverses of each other and they cancel the effect of each other when applied one after the other on the same input.

Encryption: C = EK (P)

Decryption: P = DK (C)

That is,

$$P = DK\ (C) = DK\ (EK\ (x)) = EK\ (DK\ (x)) = x$$

Block ciphers mean a group of plaintext symbols of size *m* (where *m* >1) are encrypted together creating a group of cipher text of the same size.

2.2 *Key Components of 3KDEC Algorithm*

3KDEC algorithm uses:

- Numeric data to be encrypted.
- Three keys (stored as three 3 × 3 matrix)
- Substitution Box (commonly referred as S-Box) and an inverse S-box
- Variable number of Rounds (which can be 3, 6 or 9).

Algorithm operates on following steps in sequence for set number of rounds.

- Key Expansion
- Find and Substitute
- Row Shift
- Add set Key.

2.2.1 Key Expansion

In this step, the single key of the three keys is expanded to as many as three keys summing up the total of nine keys to be used in the maximum nine rounds. Key_1 is expanded as Key_{10}, Key_{11} and Key_{12} and similarly Key_{20}, Key_{21}, Key_{22}, Key_{30}, Key_{31} and Key_{32}. The complexity of key expansion is directly depending on the number of rounds. In case of three rounds, there will be algorithm complexity of one; while in case of six rounds, there will be a complexity of two and similarly three in case of nine rounds. The process of expansion of Key_1 into its constituent Key_{10}, Key_{11} and Key_{12} is:

- Key_{10} is same as Key_1
- Key_{11} is shifting the row 1 zero time, row 2 elements one time and row 3 elements two times with respect to original key_1
- Key_{12} is shifting the row 1 one time, row 2 elements two times and row 3 elements zero time with respect to the original.

With variable number of rounds and varying key expansions complexity of the algorithm increased.

2.2.2 Find and Substitute

In this step, the byte to be encrypted is found and substituted independently to provide the confusion effect. There is no fixed mechanism or any mathematical correlation in the formation of S-box. The entries of S-box can be different in different encryption processes.

So, the simple structure and variable entries of substitution box make the algorithm eligible to be used as a personal encryption algorithm where different S-Box variants can be used in encryption process depending on the party we are dealing with and the varying complexity levels can be set according to our needs.

2.2.3 Rowshift

This transformation step shifts towards the left. The number of shifts depends on the row number of the matrix. This means the first row of matrix is shifted zero time, second row of the matrix is shifted one time and the third row is shifted two times towards the left. During the decryption process, the inverse row shift process is carried out and the shifting is done towards the right. The number of shifts is same as the row number.

where,
Row 0: no shift
Row 1: one shift
Row 2: two shifts

2.2.4 Add Setkey

With each round, the matrix is added using XOR operation with the above-expanded keys. This means during first round, encryption Key_{10} is used. In the next round Key_{11} and then Key_{12} and so on.

2.2.5 Steps of 3KDEC Algorithm

Step 1: Encryption Step
Consider the sample input as follows (Figs. 1, 2, 3, 4 and 5):

Input M = 1234
Key 1 = 520255270
Key 2 = 104425204
Key 3 = 459105254

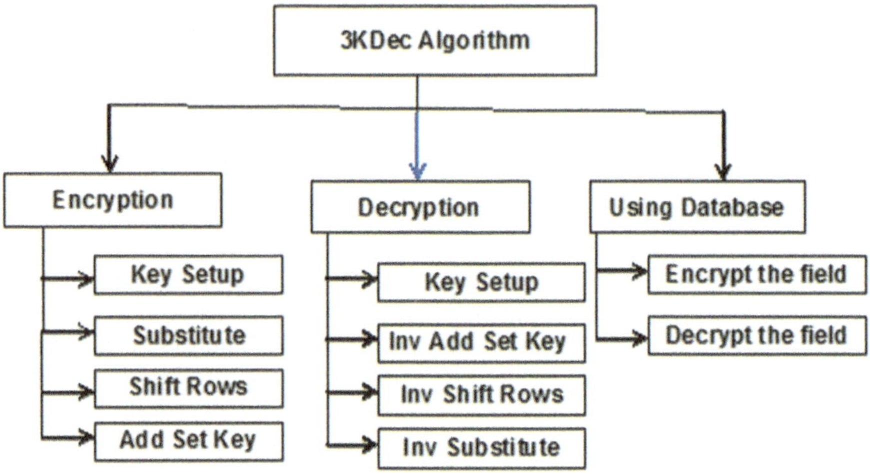

Fig. 1 Architecture of 3KDEC algorithm

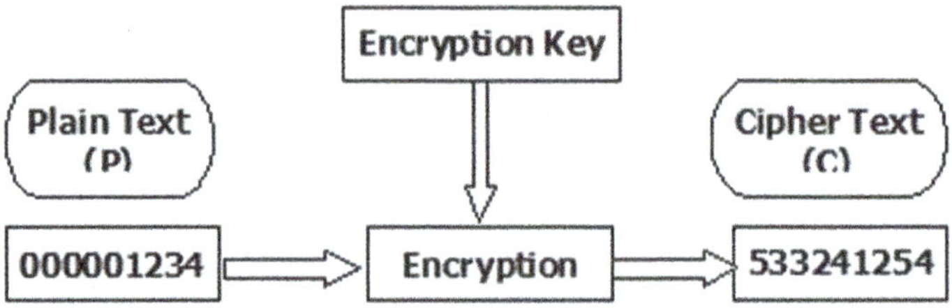

Fig. 2 Block diagram for encryption

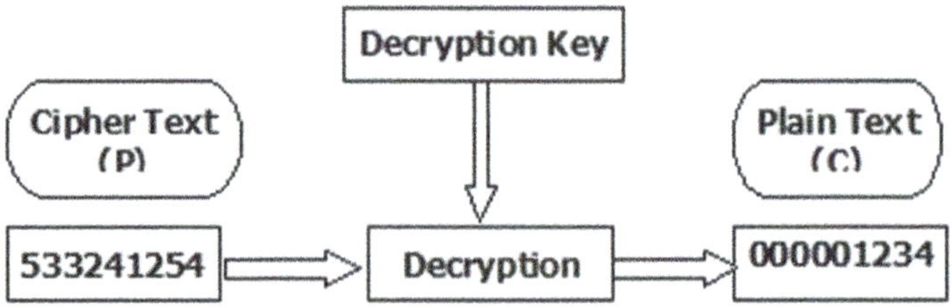

Fig. 3 Block diagram for decryption

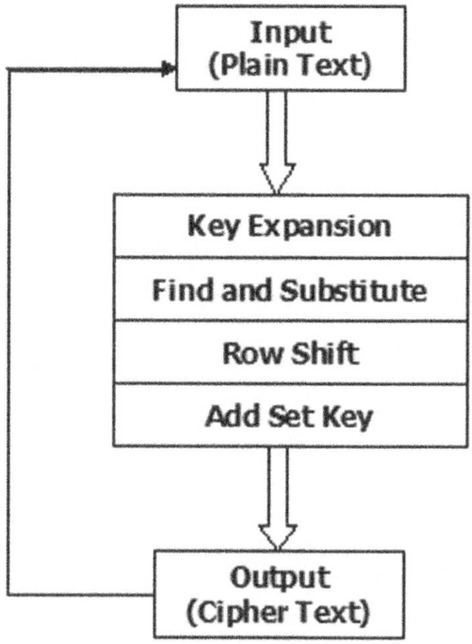

Fig. 4 Flow diagram for encryption

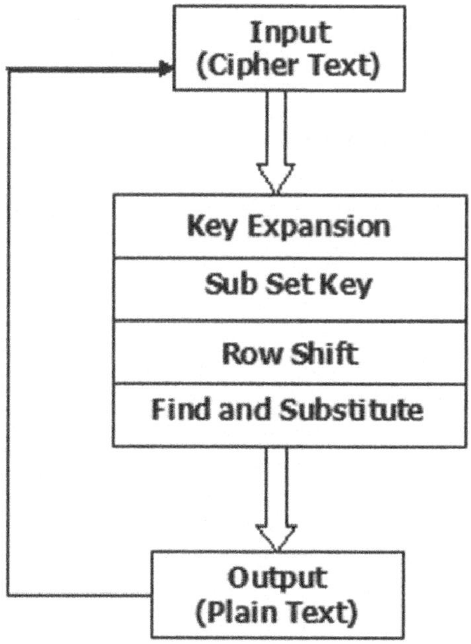

Fig. 5 Flow diagram for decryption

Step 2: Key Generation Steps

Input Key 1

5	2	0
2	5	5
2	7	0

The Generated Key 10

5	2	0
2	5	5
2	7	0

The Generated Key 11

5	2	0
5	5	2
0	2	7

The Generated Key 12

2	0	5
5	2	5
2	7	0

Input Key 2

1	0	4
4	2	5
2	0	4

The Generated Key 20

1	0	4
4	2	5
2	0	4

The Generated Key 21

<table>
<tr><td>1</td><td>0</td><td>4</td></tr>
<tr><td>2</td><td>5</td><td>4</td></tr>
<tr><td>4</td><td>2</td><td>0</td></tr>
</table>

The Generated Key 22

<table>
<tr><td>0</td><td>4</td><td>1</td></tr>
<tr><td>5</td><td>4</td><td>2</td></tr>
<tr><td>2</td><td>0</td><td>4</td></tr>
</table>

Input Key 3

<table>
<tr><td>4</td><td>5</td><td>9</td></tr>
<tr><td>1</td><td>0</td><td>5</td></tr>
<tr><td>2</td><td>5</td><td>4</td></tr>
</table>

The Generated Key 30

<table>
<tr><td>4</td><td>5</td><td>9</td></tr>
<tr><td>1</td><td>0</td><td>5</td></tr>
<tr><td>2</td><td>5</td><td>4</td></tr>
</table>

The Generated Key 31

<table>
<tr><td>4</td><td>5</td><td>9</td></tr>
<tr><td>0</td><td>5</td><td>1</td></tr>
<tr><td>4</td><td>2</td><td>5</td></tr>
</table>

The Generated Key 32

<table>
<tr><td>5</td><td>9</td><td>4</td></tr>
<tr><td>5</td><td>1</td><td>0</td></tr>
<tr><td>2</td><td>5</td><td>4</td></tr>
</table>

Step 3: SBOX

0	1	3	3	4	5	6	7	8	9
7	3	6	1	0	9	2	5	4	8

Step 4: Round Operations
The Encryption Steps
The Input data given:

0	0	0
0	0	1
2	3	4

The Round Steps
The Round 1
Substitution Step

7	7	7
7	7	3
6	1	0

Row shift Step

7	7	7
7	3	7
0	6	1

Add set Key Step
The key used is:

7	7	7
7	2	7
0	6	1

+

5	2	0
2	5	5
2	7	0

=

2	9	7
9	8	2
3	3	1

Repeating the same steps up to 9 rounds

The Round 9
S-Box Step

0	4	9
1	7	3
0	0	0

Row shift Step

0	4	9
7	3	1
0	0	0

Add Set Key Step:

0	4	9		5	9	4		5	3	3
7	3	1	+	5	1	0	=	2	4	1
0	0	0		2	5	4		2	5	4

The Final Output

5	3	3
2	4	1
2	5	4

For the Input Data M = 1234, the cipher data obtained is: 533241254. The time taken to perform these nine rounds is (in Milliseconds): 15

Step 5: Decryption Step
Consider the Input cipher text given is

Input CP = 533241254
Key 1 = 520255270
Key 2 = 104425204
Key 3 = 459105254

Step 6: Key Generation Steps

The Key generation steps in decryption are same as that of encryption because both the encryption and decryption of this algorithm take only same secrete key.

Step 7: Inverse SBOX

0	1	2	3	4	5	6	7	8	9
4	3	6	1	8	7	2	0	9	5

Step 8: Round Operations

In decryption, the round operation is performed by taking last key of encryption process as the first key, i.e., in the reverse order. Here, the cipher text is taken as the input and the original output is obtained by performing the round operations in the reverse order, i.e., Round 9 of the encryption process is the Round 1 of decryption process.

Step 9: The Decryption Steps

The Cipher Text given:

5	3	3
2	4	1
2	5	4

This is the input to the next round.

Step 10: The Round Steps

The Round 1

Inverse add set key Step:

5	3	3		5	9	4		0	4	9
2	4	1	+	5	1	0	=	7	3	1
2	5	4		2	5	4		0	0	0

Row shift Step

0	4	9
1	7	3
0	0	0

S-Box Step

4	8	5
3	0	1
4	4	4

The above operation is repeated
The Round 9
Inverse add Set Key Step:

7	7	7		5	2	0		7	7	7
7	3	7	-	2	5	5	=	7	3	7
0	6	1		2	7	0		0	6	1

Row shift Step

7	7	7
7	7	3
6	1	0

S-Box Step

0	0	0
0	0	1
2	3	4

At final the Original data is extracted 1234 (Figs. 6, 7, 8 and 9)

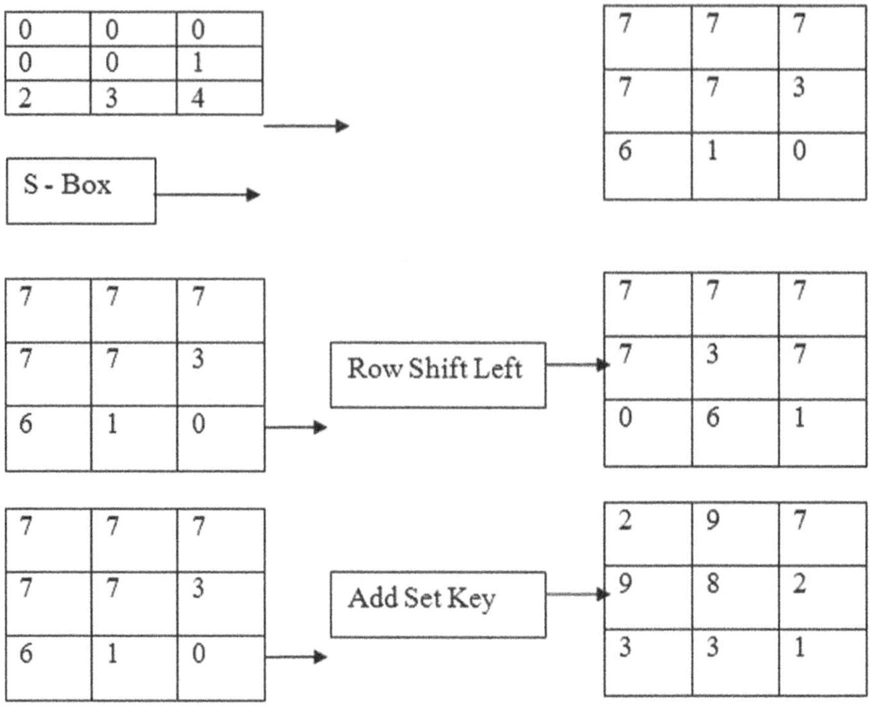

Fig. 6 Unit testing for encryption process

3 Performance Evaluation

As the encryption is from numeric to numeric, one cannot know that the information is encrypted. It appears content as if the encrypted data is itself the original content. As the number of keys is more and the key combination increases to 10^{27} which makes guessing of keys are very harder. As S-Box has simple structure and variable entries so the 3KDEC algorithm can be used as Personal Encryption Algorithm where different encryption processes can be carried out with varying degree complexity depending on the user requirements. Since 3KDEC algorithm encrypts numeric to numeric data, encrypted data is possible to be stored in the existing numeric field, thereby algorithm does not change the data field type and set fixed data length.

A. *Analysis of 3KDEC Algorithm*

To analyse the computational complexity of a cryptographic algorithm, we need an encryption algorithm to have low level of complexity and we need an algorithm used by the cryptanalyst who is trying to break the code to have high level of complexity. This means we need to perform encryption and decryption in short span of time, but at the same time, we want the intruder to run her computer infinitely if trying to break the code.

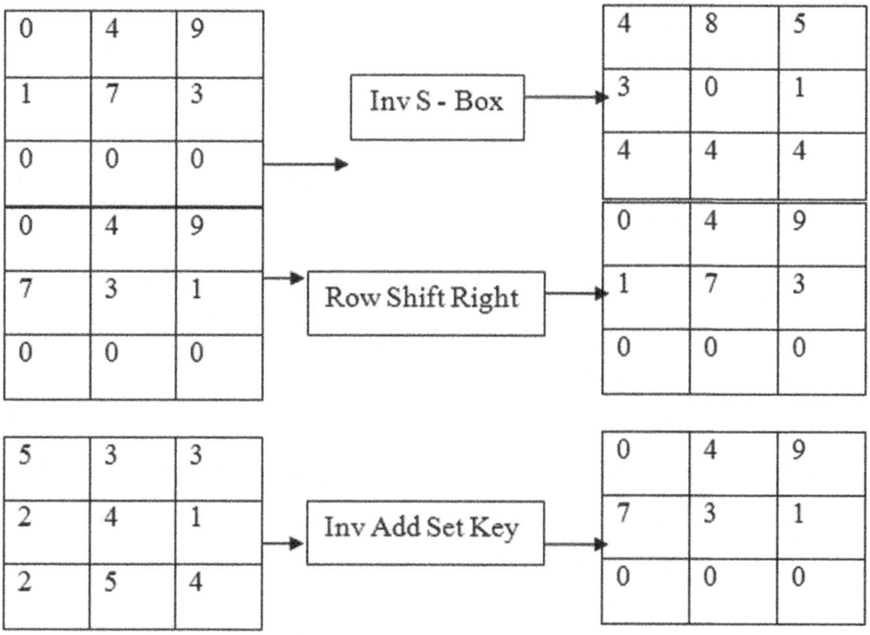

Fig. 7 Unit testing for decryption process

Fig. 8 Integration testing for encryption process

Fig. 9 Integration testing for decryption process

Complexity of program is based on two types of resources: space complexity of algorithm which refers to amount of memory needed to store the algorithm and the data and the time complexity which refers to amount of time needed to run the algorithm and get the results.

The time complexity of 3KDEC algorithm can be measured (independent of the computer on which it is running) by defining the bit-operation complexity which counts the number of bit operations the computer needs to perform to create the output from the input.

Bit operation can be defined as the time required for the computer to add, subtract, multiply or divide two single bits or to perform one single bit shift. For example: bit-operation complexity of function that adds two integers each having, d decimal digits can be calculated as: $d \log_2 10$.

B. *Security of 3KDEC Algorithm*

The main strength of 3KDEC algorithm is that it converts the numeric data input to numeric data output which makes it harder to see and guess whether the input is even encrypted or not.

(1) Brute-force attack

Also known as exhaustive key search method that tries to use all the possible keys. In 3KDEC algorithm, the three keys are expanded based on the number of rounds; there will be 10^{27} different key combinations. Lack of weak keys is another advantage of 3KDEC algorithm.

(2) Statistical attacks

The strong confusion and diffusion provided by the substitution box lookup and row shift transformations remove any frequency pattern in the plaintext.

(3) Differential and linear attack

The algorithm was designed to be resilient to linear and differential attacks. Differential cryptanalysis/chosen plaintext attack needs the analysis of encryption algorithm to collect information about plaintext–cipher text relationships; with the goal of finding the cipher key (Figs. 10, 11, 12, 13, 14, 15, 16, 17, 18, 19, 20, 21 and 22).

The probabilistic relationship can be created from the information about the S-box input/output table. In case of 3KDEC algorithm, the S-box values are not fixed and do not depend on any algebraic structure. The sender may use different S-box values each time attempting to encrypt the input. So, the attacker has to even guess what were the values used for the S-Box in encrypting that data.

Linear Cryptanalysis/Known-Plaintext Attack where the attacker needs to have information about plaintext/ciphertext pairs in addition to the intercepted cipher text that needs to be broken along with the assumption that key has not changed. Relationship between previous pair is used to analyse the current cipher text. But the strong point in the algorithm is that every time the three keys as well as S-box entries can be changed which makes it difficult to find the appropriate relationships among the plaintext and the cipher text pair.

Fig. 10 Encryption

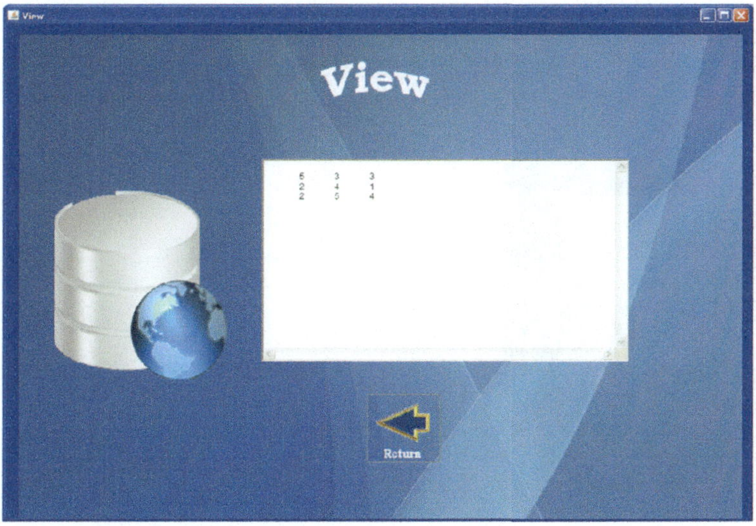

Fig. 11 Encrypted data

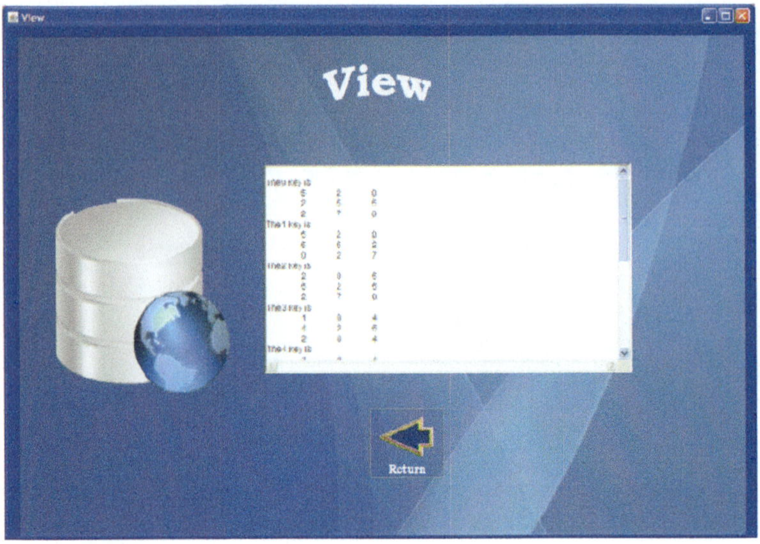

Fig. 12 Key generation

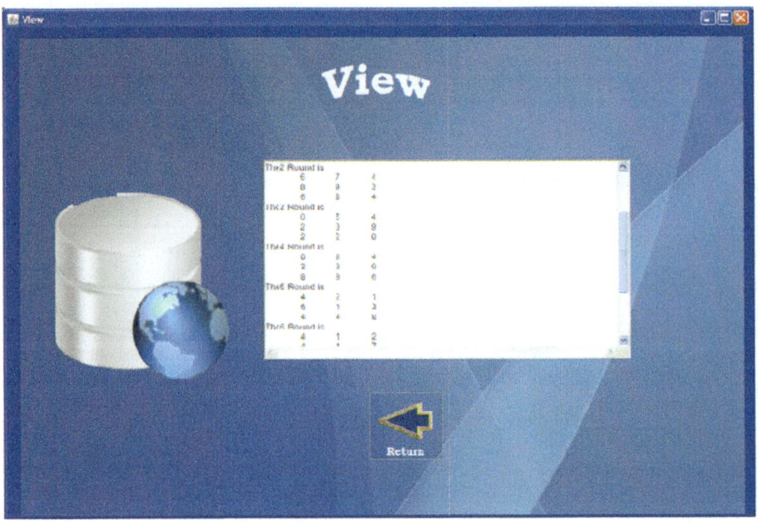

Fig. 13 Rounds

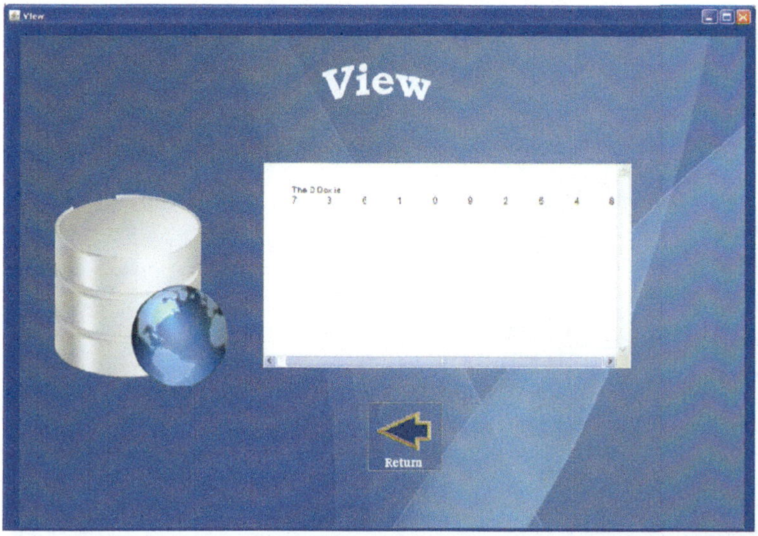

Fig. 14 S-Box

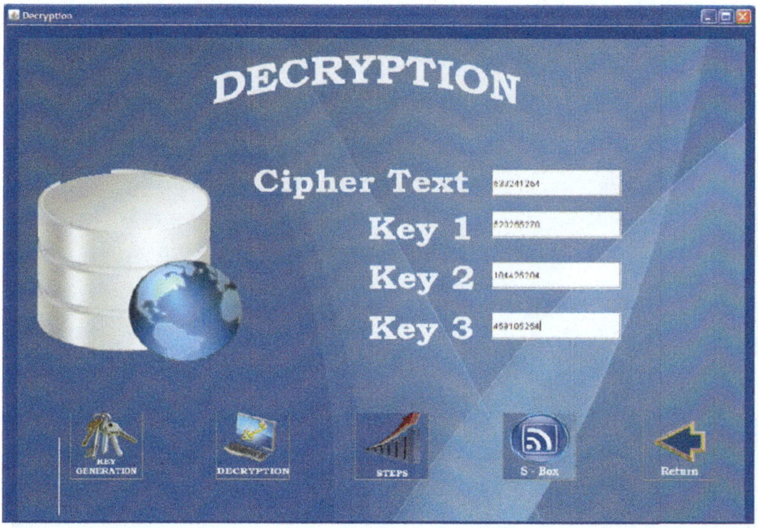

Fig. 15 Decryption

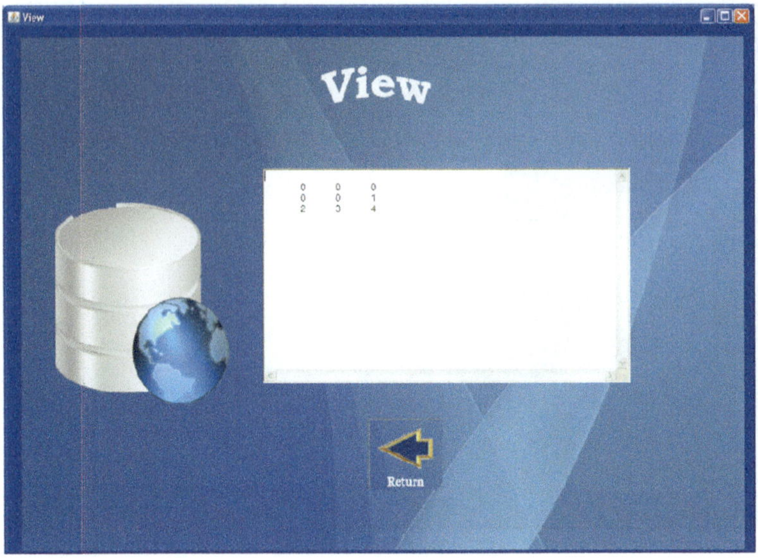

Fig. 16 Decrypted data

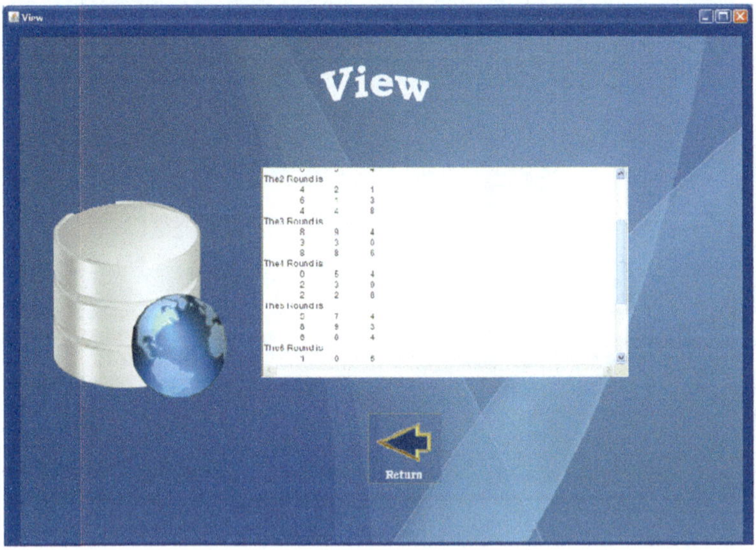

Fig. 17 Rounds

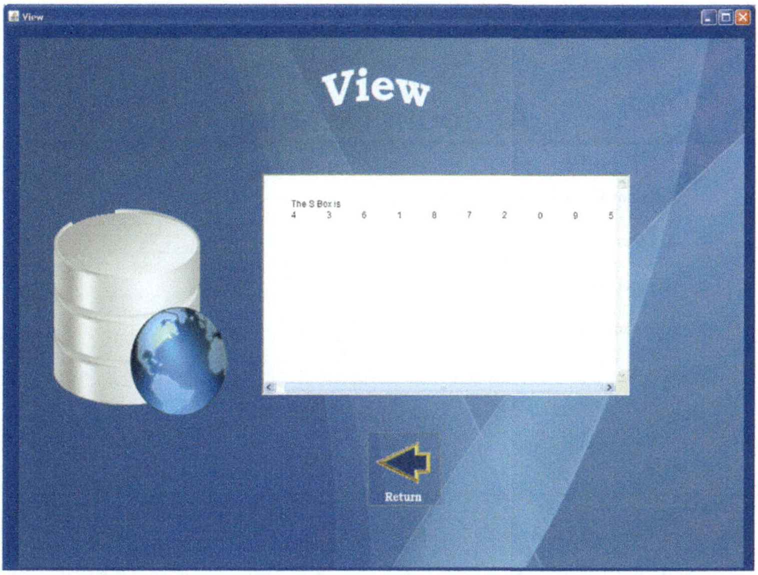

Fig. 18 Inverse S-Box

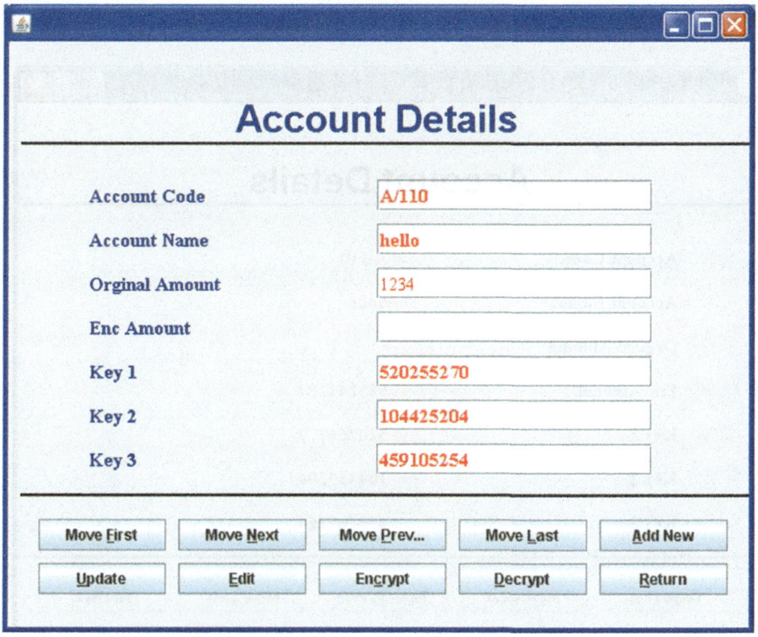

Fig. 19 Original amount before encryption

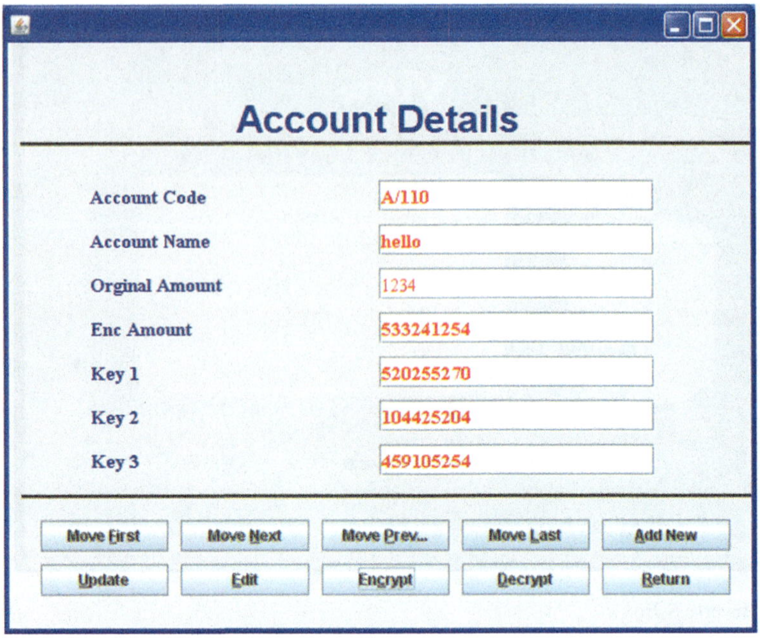

Fig. 20 Original amount after encryption

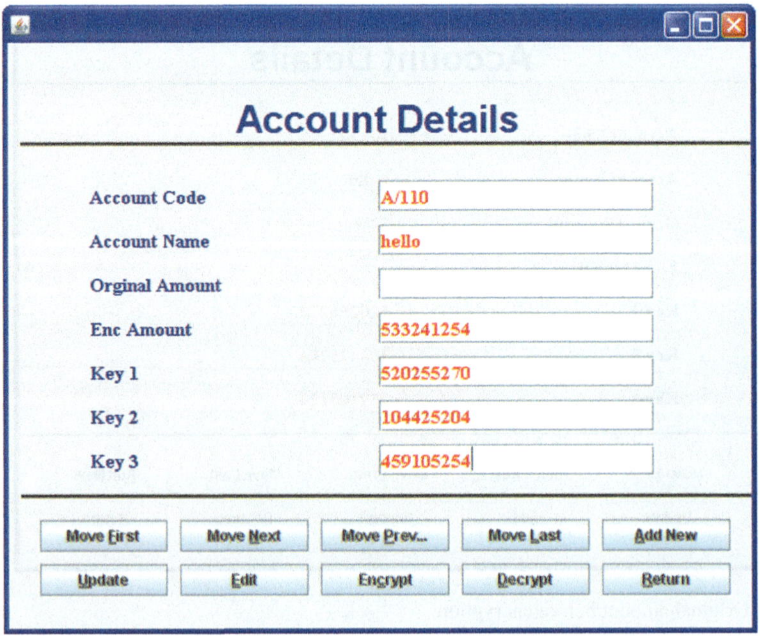

Fig. 21 Original amount before decryption

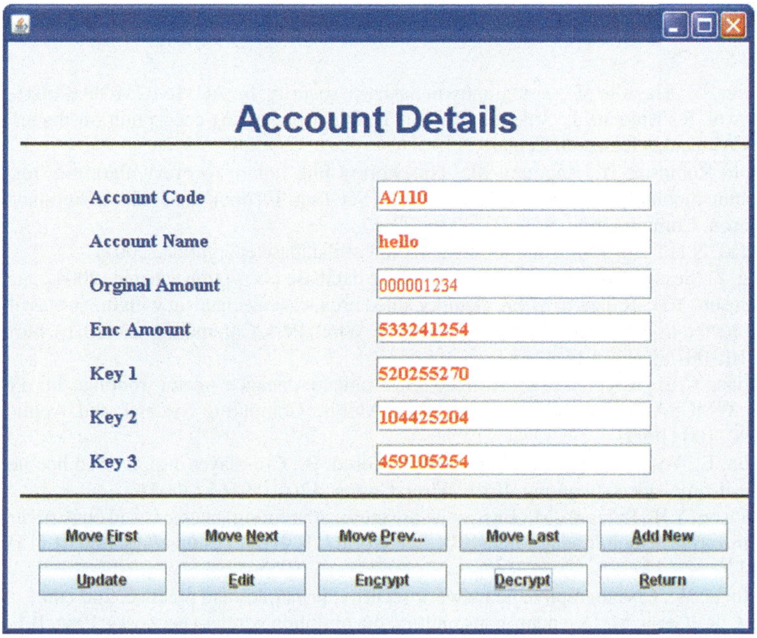

Fig. 22 Original amount after decryption

4 Conclusion

Understanding the need to secure your data is the first step towards securing it. In today's age, every detail—personal to corporate secrets—is present in form of data. For computers and networks which store and transfer this data, it is just numbers. It is for us to realize that the damage caused to this data may fall into the hands of an unscrupulous person. Whether the data is on your laptop, desktop or on an organization's storage network, it must be secured and should not come in the hand of an unauthorized entity. Proper access control mechanism should be enforced for securing the data. While in motion, data should be well protected. It is advisable to encrypt the data before putting it on a network even if it passes through a secure channel. The algorithm can be implemented for securing any corporate-related accounting information to data of personals use. This algorithm does not take into consideration decimal point numeric data which opens up another area of research and improvements. In 3KDEC algorithm, the numeric data is converted to numeric so that the data field length and the data field type will not change. In future, this work can be modified and used in mail services and wireless. The S-box can be made dynamic so that it will be more secure. It is focused only on numeric values, but we can use this with slight modification for floating point numbers.

References

1. Maurer, U.: The role of cryptography in database security. In: ACM SIGMOD (2004)
2. Agrawal, R., Kiernan, J., Srikanth, R., Xu, Y.: Order preserving encryption for numeric data. IBM Almandan Research Center
3. Harold Robinson, Y., Rajaram, M.: Trustworthy link failure recovery algorithm for highly dynamic mobile adhoc networks. World Acad. Sci. Eng. Technol. Int. J. Electr. Comput. Energ. Electron. Commun. Eng. **9**(2), 233–236 (2015)
4. Arshad, N.H.: The design and implementation of database encryption (2007)
5. Yang, Z., Sesay, S., Chen, J., Xu, D.: A secure database encryption scheme (2004)
6. Robinson, Y.H., Rajaram, M.: A memory aided broadcast mechanism with fuzzy classification on a device-to-device mobile Ad Hoc network. Wirel. Pers. Commun. 1–23 (2016). https://doi.org/10.1007/s11277-016-3213-0
7. Perkins, C.E., Royer, E.M.: Ad-Hoc On-Demand distance vector routing. In: Proceedings WMCSA'99 Second IEEE Workshop Mobile Computing Systems and Applications, pp. 90–100 (1999)
8. Setton, E., Yoo, T., Zhu, X., Goldsmith, A., Girod, B.: Crosslayer design of ad hoc networks for real-time video streaming. IEEE Wirel. Comm. **12**(4), 59–65 (2005)
9. Robinson, Y.H., Rajaram, M.: Energy-aware multipath routing scheme based on particle swarm optimization in mobile ad hoc networks. Sci. World J. 1–9 (2015). https://doi.org/10.1155/2015/284276
10. Stallings, W.: Cryptography and network security: principles and practice, 2nd edn
11. Lin, C.R., Gerla, M.: Asynchronous multimedia multihop wireless networks. Proc. IEEE Infocom **1**, 118–125 (1997)
12. Biswas, S., Morris, R.: Opportunistic routing in multi-hop wireless networks. ACM SIGCOMM Comput. Commun. Rev. **34**(1), 69–74 (2004)
13. Rozner, E., Seshadri, J., Mehta, Y., Qiu, L.: Soar: simple opportunistic adaptive routing protocol for wireless mesh networks. MobileComput. IEEE Trans. on **8**(12), 1622–1635 (2009)
14. Kaur, K., Dhindsa, K.S., Singh, G.: Numeric to numeric encryption of databases: using 3Kdec algorithm. In: IEEE International Conference IACC 2009, pp. 1501–1505 (2009)
15. Harold Robinson, Y., Rajaram, M.: Establishing pairwise keys using key predistribution schemes for sensor networks. World Acad. Sci. Eng. Technol. Int. J. Comput. Electr. Autom. Control Info. Eng. **9**(2), 608–612 (2015)
16. Balaji, S., Robinson, Y.H., Rajaram, M.: SCSBE: secured cluster and sleep based energy-efficient sensory data collection with mobile sinks. Circ. Syst. **7**, 1992–2001 (2016). https://doi.org/10.4236/cs.2016.78173
17. Chang, C., Chan, C.W.: A database record encryption scheme using RSA public key cryptosystem and its master keys. In: The International Conference on Computer networks and Mobile Computing (ICCNMC), pp. 312–315 (2013)
18. Robinson, Y.H., Balaji, S., Rajaram, M.: ECBK: enhanced cluster based key management scheme for achieving quality of service. Circ. Syst. **7**, 2014–2024 (2016). https://doi.org/10.4236/cs.2016.78175
19. Li, Y., Chen, W., Zhang, Z.L.: Optimal forwarder list selection in opportunistic routing. In: 2009 IEEE 6th International Conference on Mobile Adhoc and Sensor Systems MASS'09, pp. 670–675. IEEE (2009)
20. Füßler, H., Widmer, J., Käsemann, M., Mauve, M., Hartenstein, H.: Contention-based forwarding for mobile ad hoc networks. Ad Hoc Netw. **1**(4), 351–369 (2003)
21. Yang, S., Zhong, F., Yeo, C.K., Lee, B.S., Boleng, J.: Position based opportunistic routing for robust data delivery in MANETs. In: GLOBECOM 2009-2009 IEEE Global Telecommunications Conference, pp 1–6. IEEE (2009)
22. Harold Robinson, Y., Rajaram, M., Golden Julie, E., Balaji, S.: Dominating set algorithm and trust evaluation scheme for secured cluster formation and data transferring. World Acad. Sci. Eng. Technol. Int. J. Comput. Electr. Autom. Control Info. Eng. **10**, 388–393 (2016)

23. Darehshoorzadeh, A., Cerda-Alabern, L. Distance progress based opportunistic routing for wireless mesh networks. In: 2012 8th International Wireless Communications and Mobile Computing Conference (IWCMC), pp. 179–184. IEEE (2012)
24. Gupta, S., Dhurandher, S.K., Woungang, I., Kumar, A., Obaidat, M.S.: Trust-based security protocol against blackhole attacks in opportunistic networks. In: 2013 IEEE 9th International Conference on Wireless and Mobile Computing, Networking and Communications (WiMob), pp. 724–729. IEEE (2013)
25. Harold Robinson, Y., Rajaram, M., Golden Julie, E., Balaji, S.: Tree based data fusion clustering routing algorithm for illimitable network administration in wireless sensor network. World Acad. Sci. Eng. Technol. Int. J. Comput. Electr. Autom. Control Inf. Eng. **10**(6), 1123–1130 (2016)
26. Mahmoud, M.M., Lin, X., Shen, X., et al.: Secure and reliable routing protocols for heterogeneous multihop wireless networks. IEEE Trans. Parallel Distrib. Syst. **26**(4), 1140–1153 (2015)
27. Robinson, Y.H., Rajaram, M., Julie, E.G., Balaji, S.: TBOR: tree based opportunistic routing for mobile Ad Hoc networks. world Acad. Sci. Eng. Technol. Int. J. Comput. Electr. Autom. Control Inf. Eng. **10**(6), 1115–1122 (2016)
28. Cho, J.-H., Swami, A., Chen, R.: A survey on trust management for mobile ad hoc networks. Commun. Surv. Tutorials IEEE **13**(4), 562–583 (2011)
29. Yu, H., Shen, Z., Miao, C., Leung, C., Niyato, D.: A survey of trust and reputation management systems in wireless communications. Proc. IEEE **98**(10), 1755–1772 (2010)
30. Michiardi, P., Molva, R.: Core: a collaborative reputation mechanism to enforce node cooperation in mobile ad hoc networks. In: Advanced Communications and Multimedia Security, pp. 107–121. Springer (2002)
31. Ramalakshmi, S., Robinson, Y.H.: ATMPH: Attaining optimal throughput capacity of MANET with power control in heterogeneous network. Program. Device Circ. Syst. **6**(4), 111–115 (2014)
32. Ayyasamy, A., Venkatachalapathy, K.: Context aware adaptive service based dynamic channel allocation approach for providing an optimal QoS over MANET. Int. J. Eng. Technol. (IJET). **6**(3), 1465–1479 (2014). ISSN: 0975-4024
33. Harold Robinson, Y., Rajaram, M.: A novel approach to allocate channels dynamically in wireless mesh networks. World Acad. Sci. Eng. Technol. Int. J. Comput. Electr. Autom. Control Inf. Eng. **8**(10), 1865–1868 (2014)
34. Han, G., Jiang, J., Shu, L., Niu, J., Chao, H.-C.: Management and applications of trust in wireless sensor networks: a survey. J. Comput. Syst. Sci. **80**(3), 602–617 (2014)
35. Robinson, Y.H., Rajeswari, S.R.: Energy-based dynamic encryption for wireless sensor networks. Wirel. Commun. **3**(9), 661–663 (2011)
36. Buchegger, S., Le Boudec, J.Y.: Performance analysis of the confidant protocol. In: Proceedings of the 3rd ACM international symposium on Mobile ad hoc networking & computing, pp. 226–236. ACM (2002)

Role of Internet of Things and IT in Health Care

Rohit Sharma and Rajendra Prasad Mahapatra

Abstract The aim of this chapter is to introduce; however, IoT is remodeling health care and also the role of Information Technology in health care (Lee et al. in 2015 EEE International Conference on Industrial Engineering and Engineering Management (IEEM). IEEE, pp. 1855–1859, 2015 [1]). The IoT applications are obscurity essential in remodeling lives of individuals than in health care. Internet of Things indicates to physical devices, like a measuring system, weight scale and patients very important watching devices (glucose, force per unit area, vital sign, and activity watching, etc.) connect with the web and transforms data from the physical to the digital world. In step with Gartner, there will be nearly twenty-six billion devices on the web of things by 2020. This data provides insight into the supplements and health actions to boost the health, while not the hindrance of the daily routine. During this chapter, we will explore in depth the importance of the Internet of Things devices in health care and also the role of Information Technology in managing the massive volume of high-security patient's information of medical. Expected key exclude from this chapter is current challenges, trends, case study, and a project management expertise for a true world. This chapter proposes few IoT applications in rural health care and ways to boost primary health desires of the developing nations (Rizwan et al. in International Conference on IEEE Emerging Technological Trends (ICETT). pp. 1–7, 2016 [2].

Keywords Health care · Cloud · Internet of Things · Data analytics

1 Introduction

Internet of Things (IoT) might be a promising model to incorporate numerous correspondence arrangements and innovations. The Internet of Things characterizes wherever each physical articles to be associated with any place and whenever abuse net

R. Sharma (✉) · R. P. Mahapatra
SRM Institute of Science and Technology, Ghaziabad, India
e-mail: rohitapece@gmail.com

R. P. Mahapatra
e-mail: hod.cse@ncr.srmuniv.ac.in

P. K. Pattnaik et al. (eds.), *Internet of Things and Analytics for Agriculture, Volume 2*, Studies in Big Data 67,
https://doi.org/10.1007/978-981-15-0663-5_6

and be prepared to build up themselves to various gadgets. Lately, there are essential advances inside the field of IoT. It will deliver modern upheaval in an exceptionally sizable measure of utilizations like tending frameworks, sensible living, setting recognition, and sensible homes. These square measure a few IoT applications, and among those tending frameworks thought of one among the chief crucial difficulties inside the blessing scene. Well-being and eudemonia are one among the preeminent promising application territories of IoT innovation [3].

Distinctive utilize cases encapsulate rising a patient's consistence to treatment and prescription in healing facilities, centers, and diverse consideration offices. Medicinal gadgets like individual home-utilize analytic gadgets or low-end symptomatic and imaging gadgets that square measure used by portable specialists square measure one among the key innovation parts. A modernized tending framework should give higher tending administrations to people whenever and from wherever in amicable way. Well-being is one among the world difficulties for humankind. Solid individuals will downsize weight on the as of now overwhelmed healing facilities and centers. Hence to remain individuals sound, a great in vogue tending framework is pivotal. IoT has a few advantages in tending, through the usage of sensors, locators and instrumentation, and so forth. These empower the ID and patient after online, the areas of the specialist, and monitor the report of the patient [4].

The term IoT was first authored by Kevin Ashton. The RFID group characterizes web of things in light of the fact that the overall system of interconnected questions unambiguously accessible upheld ordinary correspondence conventions [5]. It consolidates antiquated fields like administration systems, wireless gadget networks, embedded systems, and automation to encourage device-to-device (D2D) correspondence through the net. The musing was first utilized at the Auto-Id focus at MIT. Radio-frequency identification (RFID) was viewed as a pre-imperative for executing frameworks that were delegated IoTs.

The tending advancement is predicated on rising the human life quality and remote sensors square measure put on the body of the patients to watch their well-being state. Continuous patient recognition frameworks square measure upheld for the point of early identification of clinical crisis. They are acclimated endlessly track the imperative indications of the patient, for instance: pulse oximetry, respiration rate, temperature, heart rate, rate inconstancy, circulatory strain, skin temperature and electrical wonder, blood liquor focus, and so forth. The sensors square measure feasible to catch changed data with respect to the patients, in regards to their physical, mental, and action state. Distinguished data is delivered to a restorative focal unit conveyed by any authorized individual [3, 6].

The restorative focal unit investigates the gotten data and demonstrates the outcomes to medicinal experts to get a handle on patient's well-being state. Remote body sensors can give the probability of nonstop recognition, early identification, and quick association in crisis conditions. It causes tending experts to stay away from move from one place to an alternate, and conjointly, it serves to patients to maintain a strategic distance from substitute lines. Sensors can encourage recognition for the senior citizens and individuals all square measure having perpetual illnesses like polygenic issue, malignant growth, asthma, memory decay, and manifestation heart condition, and so forth, they will maintain a strategic distance from movement to clinics from one place to various and time to see their well-being state.

At the moment, the term IoT is frequently utilized as an offering move. This move is named "vaporware"; promoting an item that is not offered in any case or does not exist. Firms are utilizing this move to establish the connection. IoT innovation are now offered to have an expedient appropriation once they have completed their advancement.

As every one of them square measure in unsteady condition which will create to crisis at any minute, for this condition, the sensors can caution the medicinal laborers remotely in order to intervene them. The patients can even utilize the remote sensors themselves for following their well-being every day and even remotely through the net [7].

2 About (IoT) Internet of Things

The term IoT was first authored by Kevin Ashton. The RFID group characterizes web of things in light of the fact that the overall system of interconnected questions unambiguously accessible upheld ordinary correspondence conventions [5]. It consolidates antiquated fields like administration systems, wireless gadget networks, embedded systems, and automation to encourage D2D correspondence through the net. The musing was first utilized at the Auto-Id focus at MIT. RFID was viewed as a pre-imperative for executing frameworks that were delegated IoTs. Today, it's applications for every close to home also as business clients.

From the edge of individual clients, social insurance, e-learning, domotics are the key fields while from business clients' point of view, computerization, providing, and mechanical creating are the vital spaces [8].

A. **Inescapable Computing**

The headways and union of miniaturized scale electromechanical frameworks (MEMS) innovation, remote interchanges, and computerized physical science have LED to the occasion of smaller than usual gadgets. These gadgets have the adaptability to detect, process, and impart remotely [9].

These smaller than expected gadgets alluded as hubs interconnect to make a remote gadget systems (WSN).

B. **Applications**

In the Internet of Things (IoT), its capability to assemble, record and break down new data streams faster and a ton of precisely by making gadgets accumulate and share information specifically with each other and furthermore the cloud. The Internet of Things can affect numerous application areas. The applications might be grouped upheld the sort of system openness, inclusion, scale, heterogeneousness, repeatability, client association, and effect [10].

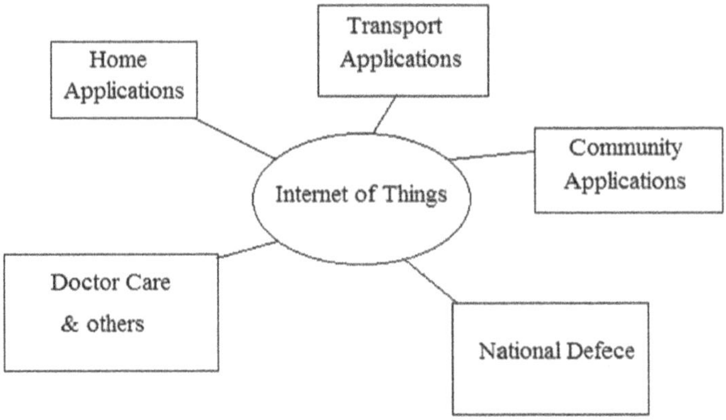

Fig. 1 Figure showing the Internet of Things application areas

These applications might be ordered into four areas: Personal and help, venture, utilities, and versatile. This can be portrayed in Fig. 1, that speaks to personal and help IoT at the measure of an individual or home, endeavor IoT at the extent of a network, utility IoT at a national or local scale and portable IoT that is some of the time unfurl crosswise over elective spaces chiefly because of the character of property and scale [11].

3 The IoT in Health Care

The point of this section is to a great degree general and does not state an incredible arrangement concerning its motivation. In this manner first, some foundation information is imperative to offer a motivation why this subject brought about an extremely postulation [12].

The Internet of Things might be an exceedingly respected point nowadays and is generally seen as each sort of articles that impart misuse the net. For some clients, this can be detectable once looking for transmission gadgets. Its intense at present to search out a television or radio that is not associated with the net. This pattern can proceed inside the years to return by interfacing extra questions. This may prompt huge systems and increase the monstrous data slant that began inside a similar way on the grounds that the snare of Things [13].

To comprehend this thought, the essential advance is to appear at anyway it works while not delving as well such a great amount into the specialized points of interest. Next, a definition is required to have a common comprehension. This can be tended to inside the first a piece of the writing survey.

A fascinating point is to imagine anyway firms can utilize this mechanical idea to make worth for his or her clients. An accommodating method for this reason for existing is gazing at the plans of action of those organizations. This can be the second a piece of the writing survey, plans of action for firms that utilization web of things applications in their valuable suggestion. The instructional exercise writing regarding this matter dressed to be rare since it is such an ongoing idea [14].

Seeing plans of action of firms that utilization the net of things might be an expansive subject for a section of a year. Constraining this subject to a specific division would slash this down and gives the opportunity to appear at it in extra detail. Inside the writing, there is typically an outline of different segments with the most essential prospects for the net of things. Human services are that the segment that is never-endingly made reference to and have some horribly encouraging chances. An outline of this division and trademark of these open doors is that the last piece of the writing audit.

The scholastic writing on plans of action for the net of things is not different, the writing on web of things plans of action in medicinal services is even however that. This makes a propelling investigation chance that might be tended to amid this postulation. Similarly, important is perceiving the troubles, difficulties, and confinements for this thought amid this setting [15].

4 IoT Devices Enabling to Work Together

Guidelines speak to relate degree inalienable test for any surroundings that during which an outsized scope of confounded gadgets got the chance to speak with each other which is unequivocally the situation for the IoT in tending. One expert has depicted the "more noteworthy institutionalization of interchanges convention" as essential to propelling the selection of the IoT [16].

Luckily, models associations are working at present to make pointers for remote interchanges between recognition gadgets and with consideration providers. The Continua Health Alliance (CHA), of that freescale could be a part, could be an alliance of tending and innovation firms that were bolstered in 2006 to find out pointers for down to earth individual well-being arrangements. The association has just printed a gathering of determinations to help ensure capacity. Inside the future, associations that get a Continua-guaranteed gadget can have the genuine feelings of serenity that it will interface with various confirmed gadgets in IoT-driven applications.

Continua's gadget norms are a piece of a greater principles surroundings that highlights data innovation models setup by the arrangement for standardization (ISO) and building gauges set by the Institute of Electrical and physical science Engineers (IEEE®) [17].

In remote innovation, IEEE principles for LANs plot ZigBee and Wi-Fi systems. Models for PANs typify Bluetooth and BLE, which are the IEEE norms identified with the body space organize (BAN). Gauges for cell systems typify CDMA and UMTS/GSM. Restrictive remote systems still play one thing of an assignment in tending conditions regularly and IoT applications particularly, anyway that job gives off an impression of being contracting in light of the fact that the business keeps on moving toward models-based structures.

5 The Time Is Now: IoT in Health Care

The since quite a while ago anticipated Internet of Things upheaval in social insurance is as of now in the air, in light of the fact that the models amid this part give subtle elements. Also, those territory units just the tip of the famous chunk of ice, as new utilize cases still develop to deal with the squeezing need for sensible, open consideration.

Freescale is worked up to be a zone of this upheaval by giving end-to-end process and property answers for IoT-driven medicinal services arrangements, working toward setting up guidelines for these arrangements and quick development for associations expecting to comprehend the upsides of the IoT in human services.

5.1 Focused Applications

At the moment, the new businesses region unit concentrating their valuable suggestion on a chosen downside in medicinal services:

- Up the lifetime of patients with a perpetual condition.
- Simplify/computerize body assignments of the consideration representatives.

Johnson et al. proclaimed accuracy is that the most critical component of a valuable suggestion and along these lines the organization in medicinal services was demonstrated commonly of the three issues, wherever IoT will have the most vital effect inside the writing survey. Regardless these two classes of uses region unit sketchy low-hanging natural products, an administration or item which will be sold-out essentially. The new businesses region unit mindful their present technique has Associate in Nursing termination date and is endeavoring to grow advance PC code administrations to upgrade their contributions. The issue is that these further administrations need self-learning calculations [18].

A possible future situation is that partnerships can at present develop their administrations and grow their specific value recommendation to various issues in human services. This can set aside your opportunity to attempt and do this with progress because of an enough client base will be important to get enough information and extra enhance the administration. Assuming, in any case, partnerships will achieve this progression, the $64,000 possibilities of IoT can surface. What is more, inside the future, the blending of those applications or the persistent expansion of their administration may prompt Associate in Nursing real IoT application that coordinates the definition from the writing survey. As clarified inside the audit, these extra propelled value recommendations would require rectify plans of action in help.

5.2 Health Care and Government Regulations in Belgium

The announcements from (2008) Johnson and van Limburg concerning the social insurance area were to a great extent affirmed for the European nation. It is an elegant market and innovation is implemented gradually. The government laws zone unit is a crucial issue here. The pay frameworks and venture determinations region unit associated with the government and backup plans.

In the writing survey, security was referred to as a noteworthy issue since all the associated articles may as of now be a conceivable danger for unapproved get to. Be that as it may, at the moment this is not the condition for new businesses. The new companies accept comparative measures as various firms in Associate in Nursing IT-condition. The well-being of data in all probability is not a tangle because of there is not huge amounts of self-governing connection between articles inside their application and without a doubt not on a scale stunning in the writing [19].

The equivalent is valid for the client's security. Inside the expressions of nobility Van Pee, one in every one of the interviewees: "Protection might be a test anyway not a genuine issue." In might 2018, the last data Protection Regulation can wind up viable. This can be a fresh out of the box new European direction concerning the use of private data. These laws obviously state what firms will and can not do with the data from their clients. None of the new companies supply data to various associations. The data is not important for various gatherings after they should change in accordance with these protection laws. So now, data evidence is anything but an accommodating asset of income for them [18].

6 The Internet of Things in Health Care (Flanders)

The talked with new companies do not have to relate degree application that coordinates my meaning of IoT. When exploitation the meanings of associated thoughts from the writing survey, the arrangement of Partheas is frequently sorted as close insight. Exploitation RFID labels and screens, the doctor's facility turns into a detecting setting with partner degree algorithmic guideline that figures the normal holding uptime. This technique part computerizes patients stream inside the healing facility with some predefined abilities (shut condition). The machine of Ectosense might be a demonstrative decent gadget associated with a cell phone that sends data to a cloud stage wherever it is broke down. This fulfills a tremendous piece of the IoT definition. Be that as it may, the data originates from one great question, there is no mix {of information of knowledge of information} and in this manner, the information streams primarily one way [18].

In view of the interviewee's remarks, the impression is giving partner degree IoT application that coordinates my definition is not offered at the moment in human services in the European nation. The new businesses that were fixed for meetings had the preeminent likenesses with the anticipated definition from the writing survey.

All the equivalent, it is difficult to demonstrate this supposition since this decision was bolstered the information of the new companies' sites. The investigate for new companies was bolstered well-being school and not the only one on IoT since it is feasible that organizations do not utilize this term to blessing their application, even once it is partner degree IoT application steady with the anticipated definition. Be that as it may, the example of meetings is confined and in this manner, the pursuit was escalated anyway not careful since a few new companies would not out in the open demonstrate their application if it is not completed in any case [17].

Interviewees affirmed that square measure a few times to use IoT applications in medicinal services anyway there square measure still some mechanical confinements. In any case, they trust extra propelled applications can progress toward becoming offered inside the future, anyway, it will take longer than what is commonly expected.

7 Roadmap for IoT

SRI Consulting Business Intelligence sees the occasion of IoT in waves according to Fig. 2. The first wave began with the work of RFID labels to encourage directing, order and misfortune obstacle, as offer chain assistants. Inside the second wave, the vertical-advertise applications e.g. police examination, security, human services, transport, sustenance offer, and archive administration are region unit created. The third wave, which we tend to territory unit by and by encountering, is concerning ubiquitous situating, e.g., finding people and ordinary items. Ensuing wave, that is anticipated to develop in a couple of decade, is the production of a physical-world net, e.g., tele-task and tele-nearness, capacity to watch, and administration inaccessible

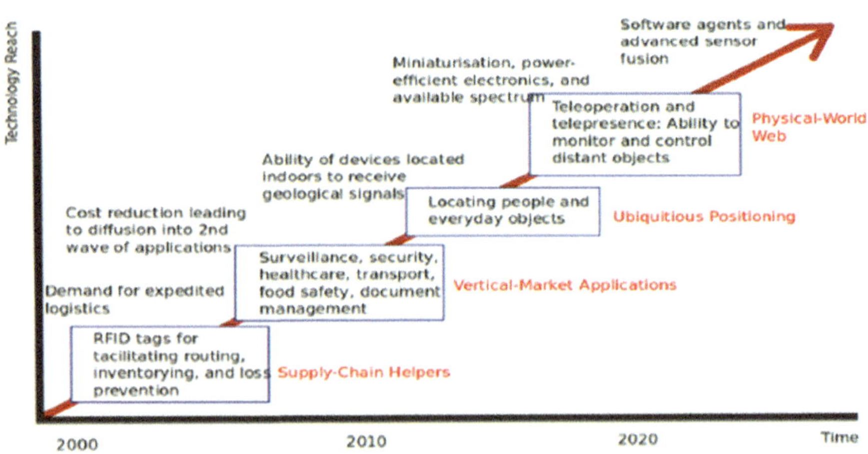

Fig. 2 Roadmap for IoT (SRI Consulting Business Intelligence)

items. Amid this piece of IoT advancement, it is normal that everyone the physical items are consistently incorporated on the whole three dreams [18].

These square measure some of the long run spaces in consideration zone recorded underneath, that is fundamentally determined by the adaptability to watch and administration far off items when.

a. Ingestible identifier

The ingestible identifier is that the finder based for the most part innovation wrapped as a pill. It is made totally of fixings found in nourishment and initiated upon utilization. This is regularly taken on board the recommended drugs, catching the exact time of utilization for interest the consistence to medicine. There are no any battery and no reception apparatus. When the pill with the ingestible identifier is wrapped, a mid-region liquid compound process gives required power supply and enacts the finder. This finder stays inside the patient's mid-region and gives timeframe information with respect to anyway the patient reacts to drugs. The patients do not need to be constrained to persist x-beams, blood tests, or biopsies for his or her specialists to see regardless of whether a medication is working. This might be especially helpful for patients WHO regularly take drugs for incessant gadgets [19].

After initiation, this locator produced and transmits the particular assortment. The fix, expendable and body-worn, catches and transfers your body's physiological reactions and practices. It also gets the particular information from the ingestible locator, identifies fundamental sign, action, and rest, and sends information to your cell phone. At that point, these data sources square measure passed on to the focal entranceway to the anchored data server. This information is any dispersed to required consideration providers as appeared in Fig. 3 [15].

b. Digital drugs

This is relating degree development of the ingestible gadget; here, the medicine itself goes advanced, instead of an additional pill for trailing consistence. Computerized

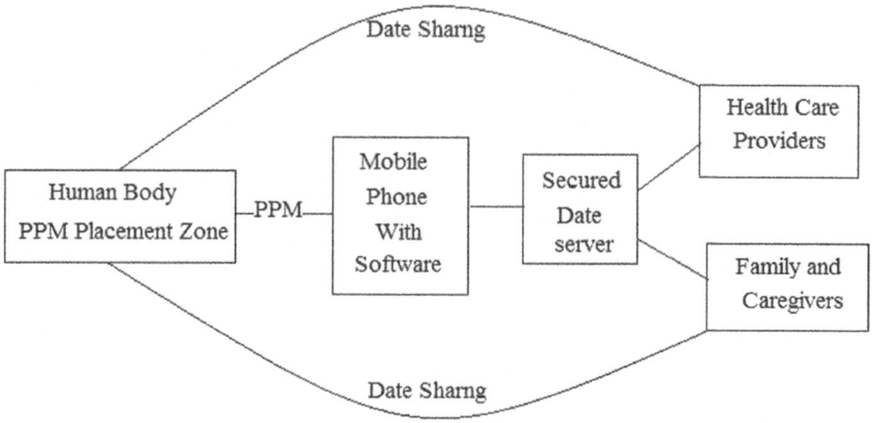

Fig. 3 Ingestible sensor network

medicines are a proportionate physician endorsed drugs devoured nowadays, with one little change: each pill will contain a little gadget that may convey, by means of the advanced well-being criticism framework, imperative information in regards to the prescription taking practices and thus the responsiveness of the body. These ingestible sensors among meds are initiated exclusively on contact with belly corrosive. This aides in giving higher truth and higher harshness of trailing the strength of patients.

This is succeeding phase of constant, wherever divisions can a little while later endorse ingestible sensors, which can remotely report back on the body's critical signs on an every minute of everyday premise. Computerized medicines are underneath improvement and do not appear to be regardless government organization cleared. Directly clinical preliminaries are led in these remedial territories [5]:

- Heart Failure
- Nervous System
- Transplant

c. Customized drugs

As of now, the supply is predicated on populace insights. Patients are isolated into groups sketched out from various perspectives that anyway normally by comparable side effects or by the consequences of fundamental work environment tests (like cholesterin levels).

d. Prophetical investigation

The utilization of prophetical investigation in medicinal services can have the advantage of the converging of different data archives that has the patterns of sustenance propensities. A ton of we as a whole know in regards to an individual or populace, that is, the bigger the picture, a considerable measure of exact the forecasts will be. These models are regularly bespoken bolstered the data focuses, to a specific patient or bunch of patients that at last outcomes in a considerable measure of exact and compelling medicines that are ensured to enhance the general effectuality of the human services framework while at a proportionate time diminishing costs [15].

In the near future, the well-being buddy versatile application is additionally ready to caution you days preceding that you essentially are taking care of show at least a bit of kindness disappointment by detecting beyond any doubt genomic signals current in your circulatory system and making you your heart specialist or to the ER.

8 Conclusion

As made reference to amid this part, all the physical articles can work flawlessly with machine-to-machine and human-to-machine interfaces. This level of interconnection might be an aid for the social insurance. These variables along the edge of the genomic inputs will fabricate it possible to anticipate the well-being patterns and sensitivities of the individual; in this manner, the innovation will give bespoken proposals on fitting physical exercises, eats less carbs, and so forth. These portable specialist buddy applications do not appear to be intended to be the swap for aptitude of the specialists. Proposition of IoT medicinal services gadgets for the creating scene are remote counseling, hand-held symptomatic gadgets for identification pestilence perishes like protozoal contamination and Indian cholera. These gadgets will have the manner in which more extensive achieve contrasted with the standard medicinal guide human services. It is apparent that IoT can encourage new plans of action and new supply models inside the future for each creating and created universes, regardless of the difficulties really young looking right now.

References

1. Lee, C., Yeung, C., Cheng, M.: Research on iot based cyber physical system for industrial big data analytics. In: 2015 EEE International Conference on Industrial Engineering and Engineering Management (IEEM), pp. 1855–1859. IEEE (2015)
2. Rizwan, P., Suresh, K., Babu, M.R.: Real-time smart traffic management system for smart cities by using internet of things and big data. In: International Conference on IEEE Emerging Technological Trends (ICETT), pp. 1–7 (2016)
3. Zhang, Q., Zhang, X., Zhang, Q., Shi, W., Zhong, H.: Firework: Big data sharing and processing in collaborative edge environment. In: 2016 Fourth IEEE Workshop on IEEE Hot Topics in Web Systems and Technologies (HotWeb), pp. 20–25 (2016)
4. Rathore, M.M., Ahmad, A., Paul, A.: Iot-based smart city development using big data analytical approach. In: IEEE International Conference on Automatica (ICA-ACCA), pp. 1–8. IEEE (2016)
5. Minch, R.P.: Location privacy in the era of the internet of things and big data analytics. In: 2015 48th Hawaii International Conference on System Sciences (HICSS), pp. 1521–1530. IEEE (2015)
6. Arora, D., Li, K.F., Loffler, A.: Big data analytics for classification of network enabled devices. In: 2016 30th International Conference on Advanced Information Networking and Applications Workshops (WAINA), pp. 708–713. IEEE (2016)
7. Yen, I.-L., Zhou, G., Zhu, W., Bastani, F., Hwang, S.-Y.: A smart physical world based on service technologies, big data, and game-based crowd sourcing. In: 2015 IEEE International Conference on Web Services (ICWS), pp. 765–772. IEEE (2015)
8. Mishra, N., Lin, C.C., Chang, H.T.: A cognitive adopted framework for IoT big-data management and knowledge discovery prospective. Int. J. Distrib. Sens. Netw. **2015**, 6 (2015)
9. Mukherjee, A., Paul, H.S., Dey, S., Banerjee, A.: Angels for distributed analytics in iot. In: 2014 IEEE World Forum On Internet of Things (WF-IoT), pp. 565–570. IEEE (2014)
10. Bonomi, F., Milito, R., Zhu, J., Addepalli, S.: Fog computing and its role in the internet of things. In: Proceedings of the First Edition of the MCC Workshop on Mobile Cloud Computing, pp. 13–16. ACM (2012)
11. Nandimath, J., Banerjee, E., Patil, A., Kakade, P., Vaidya, S., Chaturvedi, D.: Big data analysis using apache hadoop. In: 2013 IEEE 14th International Conference on Information Reuse and Integration (IRI), pp. 700–703. IEEE (2013)
12. Hashem, I.A.T., Anuar, N.B., Gani, A., Yaqoob, I., Xia, F., Khan, S.U.: Mapreduce: Review and open challenges. In: Scientometrics, pp. 1–34 (2016)
13. Morabito, V.: Managing change for big data driven innovation. In: Big Data and Analytics, pp. 125–153. Springer (2015)
14. Bhardwaj, A., Bhattacherjee, S., Chavan, A., Deshpande, A., Elmore, A.J., Madden, S., Parameswaran, A.G.: Datahub: Collaborative Data Science & Dataset Version Management at Scale. arXiv preprint arXiv:1409.0798 (2014)

15. Hortonworks: Accessed on 3rd June 2017. [Online]. Available: https://hortonworks.com/
16. Sharma, R.: Evolution in smart city infrastructure with IOT potential applications. In: Internet of Things and Big Data Analytics for Smart Generation, Vol. 154. Springer, ISRL (2018)
17. Sharma, R.: Security threats in E-Commerce. In: Handbook of E-Business Security. CRC Press, Taylor & Francis Group (2018). ISBN 9781138571303
18. Sharma, R.: Steps for implementing big data and its security challenges. In: Data Intensive Computing Application for Big Data, Vol. 29. Advances in Parallel Computing series of IOS PRESS (Scopus Indexed) (2018). ISSN 1879-808X
19. Ding, Z., Yang, Q., Wu, H.: Massive heterogeneous sensor data management in the Internet Of Things. In: IEEE International Conferences on Internet of Things, and Cyber, Physical and Social Computing, pp. 100–108 (2011)

Adopting Big Data Analysis in the Agricultural Sector: Financial and Societal Impacts

Rasha Kashef

Abstract Big data analytics (BDA) is constantly formulating decisions in agriculture and transforming the processes by which agriculture operates and controls. The agriculture process can be composed of a sequential flow of stages, starting from planting, spraying, fertilization, collection, and distribution to the end consumer. The variety within agriculture data is distinctly heterogeneous. This big farming data can come in all shapes and sizes from a variety of sources, which composes a hard analytical process for decision-makers. Big data analytical techniques have been applied to each stage in the agricultural operational process. These techniques have been proven to establish both economic gain for farmers and environmental and safety benefits for society at large. This chapter summarizes the state-of-art analytical methods that have been recently used in the agriculture industry, along with their financial and societal impacts.

Keywords Big data · Machine learning · Data analytics · Image processing · Agriculture

1 Introduction

A crucial factor in the profitability of a crop-based agricultural operation is the unpredictable nature of weather conditions, which can render the harvest completely unusable. In these cases, destroyed crops are not only an economic loss for the producer but also a loss for the society in overall sustainability. One solution to the growing demand for food is the use of big data through precision agriculture. Using big data can create easy-to-use and accurate forecasting models to ensure that crops are produced at their full potential, and this is when the idea of predictive analytics comes in. In general, predictive analytics is used to extract trends and predictive insights from data. Traditional weather forecasting models have been built based on deterministic modeling starting with initial conditions, and then entered a

R. Kashef (✉)
Electrical, Computer, and Biomedical Engineering, Ryerson University, Toronto, ON, Canada
e-mail: rkashef@ryerson.ca

P. K. Pattnaik et al. (eds.), *Internet of Things and Analytics for Agriculture, Volume 2*, Studies in Big Data 67,
https://doi.org/10.1007/978-981-15-0663-5_7

supercomputer model to produce a prediction; and later in the 1990s, an ensemble forecasting approach was introduced where multiple systems are combined to process and handle data [1]. The current challenge is to enhance the modeling, the statistical analysis, and visualization technologies, and adapt them to the agricultural industry.

The latest advances in big data analytics around the economic gain of all types of production; yet, societal concerns, such as sustainability and food safety, tend to dictate the trend of big data analytics which is especially pertinent in the agricultural industry. In 2015, venture capital funds invested a total of $64.5 billion across thousands of technology startups, ranging from autonomous car systems to social media companies [2]. During the same period, approximately 7% of this amount was invested in agriculture-related startups. While still a significant amount, it is ironic that an industry critical to sustaining life is not the recipient of more innovation. In part, this is likely due to the traditional nature of the agriculture industry; however, the innovation that is occurring within the industry has the potential to bring about significant environmental, social, and economic social change. The following chapter outlines big data technologies that are impacting the agriculture industry in the aforementioned areas and also analyzes the industry trends that are limiting its innovation. In this chapter, major characteristics of machine learning, big data, and the analytics technologies that are impacting the agriculture industry in the aforementioned areas and also analyzes the industry trends that are limiting its innovation. To pave the way of discussion in this chapter, the list of abbreviation that are used across the entire chapter is summarized in Table 1.

This book chapter is organized as follows: in Sect. 2, machine learning and its categorizations are presented. The notion of big data is illustrated in Sect. 3. An introduction to agriculture data is presented in Sect. 4. Current big data analytics techniques in the agriculture field are presented in Sect. 5. Section 6 provides a discussion of analytics in food safety. Section 7 introduces some analytics software in the agriculture industry. Section 8 discusses the future of analytics in the field. Finally, conclusions are presented in Sect. 9.

2 Machine Learning: Definition, Categorization, and Validation

Machine learning is known as the automated mechanism of discovering information and knowledge from data using intelligent machine that uses either the unsupervised learning, as opposed to supervised learning when it comes to learning from data. Supervised learning is when data has to be manually put in by a user to help categorized and organized clustered data while unsupervised learning allows for the categorization and organization of data without being programmed by a user [3]. Advances in digitization and computing power allowed data scientists to train computers to model predictive events instead of constructing finished models. It includes a collection of algorithms, methods, and tools able to embody intelligence to machines.

Table 1 List of symbols and abbreviations

Abbreviation	Description
ANNs	Artificial neural networks
DNNs	Deep neural networks
CNNs	Convolutional neural networks
NMI	Normalized mutual information
VI	Variation of information
NVI	Normalized variation of information
NLP	Natural language processing
LTI	Labeled parse tree induction
AMI	Adjusted mutual information
NVD	Normalized Van Dongen
CSI	Centroid similarity index
HSV	Hue-saturation-value
RI	Rand index
FGRI	Fuzzy generalized Rand index
DI	Davies-bouldin index
EED	Expected edge density
SI	Silhouette index
DAG	Directed acyclic graph
SOM	Self-organizing maps
GIS	Geographic information systems
RFID	Radio frequency identification
SEO	Search engine optimization
BoW	Bag of words
SVM	Support vector machines

Machine learning tools have been used extensively in several applications across different industries. Machine learning is also used in other industrial applications like pattern recognition, video surveillance, text mining, and even for cancer prognosis and prediction. Given the advancement projected in the next few years, machine learning is expected to be incorporated in more applications. The true potential of machine learning is realized by its modeling tools, which, when trained via a learning procedure with a dataset, uses algorithmic learning to respond to similar data trends. However, it still falls quite a way short of replicating human learning techniques. Where machine learning excels, however, is while crunching large data sets while accounting for several permutations of variables to model an event. Overall, machine learning is characterized as a type of the artificial intelligence field and it increases the overall efficiency of data analysis, as processes that were once done by humans, can now be automated by machines.

Machine learning in big data is preferable to traditional methods, such as Excel. Programs such as excel have a hard time handling big data due to the sheer size of the information, which causes the program to crash. However, machine learning works with big data because it feeds off of it and if you provide machine learning algorithms with more information, they become more accurate [4]. There are a variety of different techniques that are used for big data analysis and machine learning. One method of machine learning is decision tree learning. With decision tree learning, the algorithms use a set of defining characteristics, where they assign data points under these characteristics and produce a set of predictive probabilities [5]. Another method of classification is clustering. Clustering is useful for analysis, as most modern-day analysis only requires a few attributes because having too many attributes will prevent any usable trends to be found. It is useful when the big data points are fairly close together, and you want to discover discernible trends [6].

Finally, another method of machine learning is the artificial neural network (ANN), which is mainly mimicking the human brain and tries to mimic human neural patterns. Information is clustered together in "artificial neurons" and linked together axons, relationships between the neurons. ANN is useful when there are nonlinear relationships for the information and allows multiple relationships between clusters of information [6].

For supervised learning, the commonly used method is k-nearest neighbors. The k-nearest neighbor's method does not split the dataset to the training set and testing set. It simply uses the minimum-distance classification method to classify the result into the nearest user-specified k value. For unsupervised learning, the k-mean method is used to classify the unlabeled data. Compared to the k-nearest neighbor, k-mean method has the same goal which is to separate data into groups. It does not have a set neighbor cluster centers but calculates a center point of the clusters and groups similar data into clusters by minimum-distance classification based on the similarity of the features. The group label will be added after the classification procedure [7].

2.1 Performance Validation

Results validation is the process of approximating how the output fits the structure underlying the input data [8]. Evaluating the clustering results is more challenging than evaluating classifiers, as in supervised learning labels are available, and we can compute performance statistics such as accuracy [9]. Validation metrics are classified into two categories—external validation and internal validation. The external ones evaluate the result based on some supervised information available while internal ones evaluate the result using the information intrinsic to the input data [10]. F-measure, NMI measure, entropy, and purity are popular external measures. Rand index, fuzzy Rand index, expected edge density, and Davies-Bouldin index are internal validation measures. Some of these measures are discussed next.

2.1.1 External Validation Measures

***F*-measure** is mainly defined on the idea of pairing similar classes (or clusters) in two partitions [11]. It depends on a pre-processing stage; in this stage, each cluster is mapped to a class. *F*-measure is widely used in document clustering [12]. The *F*-measure cannot applicable to nested clustering [13]. Moreover, it cannot deal properly with the issue of class size imbalance [14]. **hF-measure** is one of the most promising indices of nested clustering quality based on the experiments on 12 datasets of Polish Wikipedia articles [13].

***L*-measure** is derived from *F*-measure. It is a useful evaluation metric that can be used to measure the quality of a computational lexicon based on clustering concepts. Compared with the *F*-measure, it performs well on a small data sample [15].

Entropy is the weighted average of the sizes of each cluster. It provides a well-designed solution to the matching problem. However, it only measures the homogeneity of a solution [12]. **V** is like other entropy-based criterion that evaluates the level at which both the homogeneity and completeness criteria have been met. It is dependent on how many clusters, the clustering methodology, and the data size. This metric supports solutions having many partitions (clusters) [11]. It has value in the interval [0, 1], and it strongly favors a solution of many small-sized clusters.

VI stands for the variation of information. It is known of two main characteristics; first, it satisfies the metric axioms. Second, it is convexly additive [16]. Nevertheless, the range of scores provided by the VI measure is mainly related to the data size [11]. (The values of VI lie in the interval $[0, 2 \log N]$, where N is the data size).

Normalized Variation of Information (NVI), the value of the NVI good clusterings considered lie in the interval [0, 1]. It maintains the convex-additivity property of the previous VI but not its metric axioms [17]. For the perfect solution, NVI = 0. NVI performs well on large real datasets [16].

Purity only measures the homogeneity of a solution [12]. Clustering solutions with purity values close to 0 are considered bad solutions; a purity of 1 means optimum clustering [18]. The singletons clusterings are considered optimal; although, it has low quality [11]. The matching criteria are the main factor to differentiate between purity, normalized Van Dongen (NVD), the criterion H (CH), and the centroid similarity index (CSI). If the matching solutionis the same, all of the above indexes should obtain a similar clustering result. Various measures known as classification error (or accuracy) are also similar to the purity measure [9].

2.1.2 Internal Validation Measures

The **Normalized Mutual Information** (NMI) measures the shared information that is adopted in both the clustering and the partition. The value of the NMI is between the values of [0, 1]. Values close to 1 show better solutions [19]. NMI equals the adjusted forms of mutual information (AMI) that equals the normalized variation of information (NVI) which also equals the variation of information (VI) under hypergeometric distribution factors, as illustrated by results in [9].

NMI_SM performs well on fuzzy partition generated by fuzzy-c-means, using a collection of 25 artificial and ten real datasets [20]. The artificial datasets contain five ground-truth groups, collected as a mixture of Gaussian probability distributions in the two-dimensional space vector.

Rand Index (RI) measures the similarity between crisp clustering. It corresponds to accuracy even when the class labels are not found. Adjusted Rand index (ARI) is a variation of the Rand index (RI) that is attuned for the chance grouping of elements. The values of the RI ranges from 0 to 1, with 0 showing that the two clusters disagree on any pair of points and the value of 1 denotes that the data clusters have the same points [21]. **Fuzzy generalized Rand index** (FGRI) broadens the scope of the RI for considering the cases of fuzzy partition and fuzzy cluster ensembles [22]. On real-world document clustering data, the **Expected Edge Density** (EED) outperforms the classical cluster validity measures, including Dunn and Davies-Bouldin index [23].

Davies-Bouldin Index (DI) is known as the fraction of the sum of "within-cluster" distance to "between-cluster" separations [24]. It cannot correctly find the optimal clustering among clusterings that can be stemmed from a document cluster, for example. [23]. The experimental results from the comparison of the effectiveness of a multiple classifier intrusion detection system reveal that Davies-Bouldin index is faster than **Silhouette Index** (SI) [25]. **Modified Davies-Bouldin index** is proposed for the rough fuzzy algorithm that adds relative significance to lower versus upper approximations. However, it underperforms on multiclass data because it either increases or decreases monotonically [26].

3 What Is Big Data? And Why?

Big data is defined as large sets of information that exceed current tools to distill this information and provide analysis [27]. Big data is information that is a combination of two main characteristics; size and complexity [28]. The size of big data is not defined by the amount of memory it takes up, rather the number of sources it contains. For example, planes can have up to a hundred thousand different sensors, measuring carbon dioxide levels, speed of the plane, and the outside temperature. For an hour-long plane ride, the total cumulative information from all these sensors only takes up roughly 3 GB. [29]. As well, technology is improving over time, meaning that the amount of memory the information takes up will decrease [27]. The complexity of big (large-scale) data is defined as the relationships and how the data can be arranged with one another. Facebook has 30 billion pieces of unique information shared each month [27]; the sheer size of this data can cause an immeasurable amount of combinations for relational data. Big data analytics includes the methods and devices to synthesize this information into presentable findings.

Analytics allows for the discovery of trends and relationships, which companies and organizations can act on to improve efficiencies, create new opportunities, and remain competitive in their respective industries. There is a multi-step process when it comes to big data analytics. The first step is the data requirements, which are the purpose of acquiring data–for example, trying to find the relationship between age and income of an individual. The second step is the acquisition of data, which can be collected by completing experiments, as well as publicly available information. The third step is data processing, which includes converting the data acquired to a form of presentable finding. The fourth step is extrapolative/interpolative trend analysis, which is the process of using the data acquired to predict trends within and outside of your data range. The final step is modeling, which is the process of turning your information and the data you acquired into a general formula/relationship.

The first big data technique was used in 1890 for the Census Bureau. Herman Hollerith was contacted by the bureau, and he came up with the idea to use punch cards and tabulation machines to process the massive amounts of data [30]. There are also a large variety of big data that can be used. There is text analytics, which is used to analyze social media, emails, news, and documents. The most common application for text analytics is text summarization, producing a simple summary from large sources of text. Audio analytics are currently most commonly used in call centers and health care, mainly in the area of diagnosis, to improve the overall customer experience. Video analytics mainly focus on the analysis of closed-circuit television, to detect security breaches. There is also predictive analytics, which is used to discover trends and extrapolate them. This can be applied to predict the failure of certain products and predict consumer behaviors before an upcoming shopping season [31].

For data analytics, NoSQL, Hadoop, HDFS, and R are just a few examples of programs and tools that allow big data to be analyzed [29]. Hadoop is a popular tool for analyzing big data because it started off as an open-source project, allowing many different insights and perspectives to be implemented in the software. As well, having many people working on the software allowed the program to scale quickly [32].

4 Agriculture-Based Data

The variety within agriculture data is distinctly heterogeneous [32]; this means that all the data in farming comes from various sources. For instance, historical crop patterns typically govern the crops that can be planted in a present year and dictate the quality of the crop in future seasons. Additionally, climate conditions and weather conditions can be strictly predicted based on historical trends and forecasting. These are very important aspects when it comes to pre-planting insights and decision-making. However, beyond this historical data, there is an increasing percentage of data observations being delivered and processed in real-time through sensor data. This is referring to the "Internet of Things", where sensors are continually reporting on the status of a physical object within a process. These are high volume, high velocity, high veracity data that play a major role in agriculture. Sensor data capturing is a relatively new concept that is now being implemented intensively in agriculture. For example, GPS receivers and satellite images can continuously monitor a crop and provide insights into the strength of a field [1]. Moreover, web-based data is merging into the analytics field within the agricultural process as it can indicate food safety once a crop or protein has been delivered to the end consumer.

5 Analytical Models in Agriculture

As the world population continues to grow over 7 billion, feeding people will become more of an issue, and this responsibility falls on the shoulder of the agricultural industry. As the demand for both food quantity and quality rise, improving production efficiency becomes a challenge that must be addressed. A crucial factor in the profitability of a crop-based agricultural operation is the unpredictable nature of weather conditions, which can render the harvest completely unusable. In these cases, destroyed crops are not only an economic loss for the producer but also a loss for the society in overall sustainability. One solution to the growing demand for food is the use of big data through precision agriculture. Adopting big data can create easy-to-use and accurate forecasting models to ensure that crops are produced at their full potential, and this is when the idea of predictive analytics comes in.

5.1 Weather Forecasting Models: Precision Agriculture

The two techniques widely used to build weather forecasting models currently are regression (both linear and nonlinear), and time series (moving average, autoregression, autoregression moving average (ARIMA). On one hand, one can "easily" regress a production decision variable (the dependent variable, such as frequency and level of irrigation required) against weather conditions (the independent variables, such as the amount of natural precipitation, sunlight and temperature) to discover inherent relationships, which can be generalized to reliable practices as experiences accumulate. On the other hand, time series that is primary based on the regression structure, is a very powerful data representation as well in terms of capturing patterns in historical data.

Modeling techniques, such as regression and time series, could be very powerful in terms of making production-related decisions. Challenges still exist in the realm of implementation. One challenge for big data analytics solutions in agriculture is aggregating a large volume of data through multiple sources and then able to properly handle this data. In addition, people (mainly producers) have to be properly trained to use the models and make reasonable business decisions from them, which is a long-term, expensive, and time-consuming task [1].

Taking advantage of precision agriculture, producers can estimate business needs ahead of time and make production plans accordingly. Resource allocation tasks, such as machinery rental or purchase, and the timing and usage of fertilizers, can be projected more effectively, which would contribute to the cost-effectiveness and overall sustainability of the business over time. The effect of weather can have a costly impact on farmers. The failure to predict an upcoming storm can lead to over-irrigation in the days leading up to it, causing water to be wasted and the potential for flooding in some areas. If the storm arrives soon after the farmer recently sprayed crops with pesticide or fertilizer, this can wash away the chemicals, and the farmer will have to repeat the process once the weather clears.

Traditional weather forecasting methods use numeric weather prediction (NWP) mathematical modeling based on Bayesian probabilistic arguments, which have historically led to accurate weather approximations. However, NWP problems are largely constrained by the amount of computing power available, which has made it difficult to provide weather predictions at a high enough resolution for it to be useful for farmers IBM is solving this problem through the use of precision agriculture [33].

5.2 *Neural Network Models*

Artificial neural network (ANN) provides smart alternative prediction methods that are widely applied for agricultural forecasting. The commonly used learning processes in artificial neural networks include the radial basis function (RBF) [34], the perceptron learning [35], the backpropagation algorithm [36], and resilient backpropagation (RBP) [37]. The feedforward backpropagation neural network (FFBP-NN) is the widely known neural nets. The architecture uses a combination of neurons and layers to create a series of connections in a directed acyclic graph (DAG), to provide a classification output. The architecture has three main layers including the input, the hidden (of 1 or more neurons), and the final layers as shown in Fig. 1.

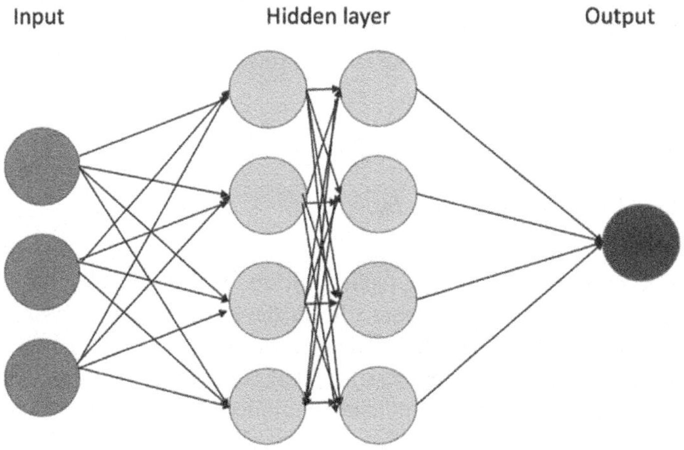

Fig. 1 FF-BP neural network

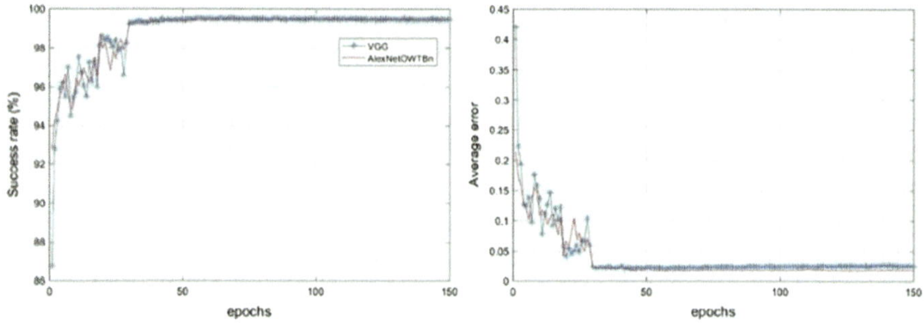

Fig. 2 Success rate using CNN [43]

The ANNs-based learning methods have been used in various fields, such as the counter propagation algorithms [36], the auto-encoder, and the supervised Kohonen networks [37]. The Hopfield Networks [38], the self-organizing maps (SOM) [39], the extreme learning processes [40], the generalized regression-based NN [17], and the adaptive neuro-fuzzy inference systems [41]. Ensemble Neural Networks, and Self-Adaptive Evolutionary Learning Systems [42]. A deep neural network (DNN) is considered as an ANN which is composed of multiple hidden layers and it is classified as supervised-learning NN, semi-supervised NN, or completely unsupervised NN.

DNNs are known for achieving higher accuracy as compared to the traditional ANNs; however, it suffers from high time complexity. The convolutional neural networks (CNNs) are DNNs, where the feature maps can be obtained by applying convolutions in the image-domain. In [43], a CNN model is applied to digital images for recognizing the face of pigs. The CNN model aims at identifying animals without the use of tags of the radio frequency identification (RFID). The model has achieved an accuracy of 96.7%. Another CNN-based model [44] is designed to provide detection and diagnosis of a disease based on images of leaves to categorize healthy and infected leaves in multiple plants. The success rate of the CNN model is presented in Fig. 2.

5.3 Image Processing Models

Besides the effective usage of geographic information systems (GIS) in weather forecasting, which shows a significant role in the agricultural analysis. Common examples of GIS include using satellite images, drone pictures, or thermography to detect field conditions such as temperature, soil condition, and pest damage. The condition helps farmers to create efficient techniques to grow crops which increase food production. On the other hand, the regulator also benefits from the geographic information system, which helps regulators enact effective ordinance by providing specific information in real time. The big data analysis has been in applied in agriculture using both image processing and classification. On the macrolevel, a shared website for agronomists is a good example of using satellite images to monitor the land type and current condition. The result can be used to help fertilizer providers and grain suppliers to have a general idea about the field condition before making decisions.

On the micro-level, farmers can get the real-time field condition information to make planting decisions by reviewing analytical results generated from images taken by drones instead of observing the crops on foot through the whole field.

The common algorithm behind the image processing and classification is that each pixel is converted from the picture into a data array. The data array includes the coordinate of the pixel, the color (in an RGB scale) of the pixel, the reflectance, etc. Each pixel becomes a row in the dataset, and the whole picture will be converted to a data table. After the conversion, all the visionary information becomes numbers which can be processed and classified by a computer [33]. In general, a 1920 * 1080 image contains more than two million pixels, which mean each data table contains a massive amount of data. The next step will be to classify the picture by the data table information. At this step, different machine learning methods will be used based on the study need. If the training data was labeled, the supervised learning will be used; otherwise, the unsupervised learning would be used for classification. The combination of computer vision and machine learning can be applied to identify plant species. These two techniques manifest themselves in a detection–classification two-step process. First, a camera takes a picture of a leaf from the plant under consideration. In a typical application, it must be verified that the picture that there actually is a leaf within the frame before proceeding. This is done by dividing the image into different regions and creating a low dimensional feature vector of each region.

This collection of low dimensional feature vectors is called a “gist”. The gist is subsequently fed into a classification algorithm, for example, a support vector machine (SVM). The classification SVM algorithm is then trained by taking as input labeled images which either contain a leaf or do not, and then creates an optimal hyperplane which maximizes the margin between the two sets of images. If the image contains a leaf, the process continues, and the image is segmented into a binary image (a black–white image), separating the leaf image from the background. First, to improve the segmentation accuracy, the data image is converted to a hue-saturation-value (HSV) space to remove the effects of lighting. In the next step, an expectation-maximization (EM) algorithm assigns probabilities to individual pixels and determines if a pixel corresponds to either the leaf space or non-leaf space, coloring the pixel appropriately. Finally, features from the leaf such as color, shape, and size are extracted from the binary image to construct a vector of features. This vector of features is then used as input to the network to predict the plant species from which the leaf came from.

5.4 Social Network Models

Using social and mobile big data applications could be a way to avoid a situation where a crisis could occur. By using text mining and social network analytics, organizations can have faster responsive actions and therefore limit the consequences associated with food insalubrity. An example of the utilization of those techniques can be seen when a high-risk contaminating disease is being detected by streams of data coming from keywords used in research engines and IP locations [45].

For analytics' purposes, keywords such as indigestion, mold, and insalubrious could be targeted by keyword trackers. Then, a system could flag the entities that sold food to this area depending on the search engine optimization (SEO).

Having access to frequent, low-cost, and online data makes social networks an ideal start to crowd source trends using big data. So, by using the "crowd" as a data source, it is possible for all participants included in the food production process to locate areas with a high probability of food-safety issues at a rapid pace.

The velocity of this data is nearly unfathomable as social network analytics deals with a large volume of data from social media sites every millisecond. The volume also corresponds to this large amount of data, as there is a constant stream of information from sites such as Facebook and Twitter. However, these social media platforms provide valuable insights into food safety and early detection method. In the spread of contaminated food, early detection can quite literally be the difference between life and death. Therefore, having constant surveillance on social media means that the food industry is able to detect when and where certain buzzwords are being used that would relate to an outbreak of food contamination. This way, using big data, the spread of dangerous food can be halted at the source.

In [46], authors have monitored Yelp reviews in the area to try and detect cases of food poisoning. In the way described above, the department used machine learning to mine through thousands of Yelp reviews and attempted to classify if a person or a group was experiencing foodborne illness. On all the reviews sampled, the authors found 468 yelp reviews that indicated a foodborne illness. Of those 468, only 14 were reported to food-safety agencies, which leave room for earlier detection and better response from the industry [47]. To use machine learning, the authors converted each document (review) in Bag-Of-Words (BOW), used three words per BOW (n-gams $= 3$) and then applied logistic regression by classifying if a person was "sick" using "Yes" or "No" [47]. This program was adopted in 2012 by the NYC Department of Health and Mental Hygiene and has been able to correctly identify ten outbreaks of illness due to contaminated food using review sites. The potential for this kind of application is unending as fundamental text mining process can be applied to Twitter and other social databases. In this way, the frequency of keywords can be coupled with the location of the complaints to pinpoint certain areas and food vendors who may be infected.

Also, cell phones have become prevalent across the world, especially in developing countries. From a big data perspective, this gives people access to data on orders of magnitude far greater than we have been exposed to previously. Just as big data has widespread application, so do technologies such as cell phones [33]. Based on the statistics provided by TechAsia, 29 million people use twitter, accounting for 2.4% of the global population [48]. Twitter is a major source of data for text mining that can be used to extract insights about common interests and issues is Twitter. In developing countries, Twitter is being used to derive insights about social well-being in relation to the food price index, which measures the monthly change in commodity prices. Families in countries such as Indonesia are highly exposed to fluctuations in food prices, where a spike in prices could have a devastating impact on the nation's food security.

This means that citizens in these countries are highly aware of events that could lead to food price increases and are constantly sharing their thoughts on food prices through social media channels. In assessing such a scenario, social media acts as an informant to monitor food security.

As a powerful analytical tool, social media analytics can provide insights between conversations, food price, and other external events. Analyzing social media patterns can be a formidable challenge given the large-scale nature of the accessible data. The first step involves creating categories for "tweets" describing food and fuel prices. A supervised algorithm was trained by creating categories such as "positive", "negative", "confused", and "neutral" [49]. A score is then given to each category, and an overall sentiment score is produced. In supervised learning, the algorithm typically has more context around the words and therefore provides to be a more accurate method of interpreting tweets in real time. Typically, there is a lag between the food price index and the actual cost of goods. In analyzing the Twitter data, there was an association determined between the content and the number of Tweets and the final impact on the food price index. The number of tweets increased as food prices increased [2]. This is demonstrated in Fig. 3.

As food prices are also dictated by fuel prices, potential increases in fuel price were related to people's perception of overall food security. Moreover, Twitter conversations are driven up by an increase in food prices as well as other food-related topics such as policy and law [33, 49]. For example, the establishment of a food agency coincided with increased chatter on Twitter.

Furthermore, Twitter can serve as a powerful tool to analyze people's perception of food prices in a real-time environment. It can lead to powerful insights that can be used to extract daily food prices from conversations and essentially create a real-time food price index.

Since there is a lag between the food price index and the actual cost of goods, twitter analytics can be used to nowcast the cost of food [50, 51]. Although twitter analytics serves as a robust tool to establish trends regarding social and economic factors in a country, efforts need to be made to increase the accuracy of these techniques.

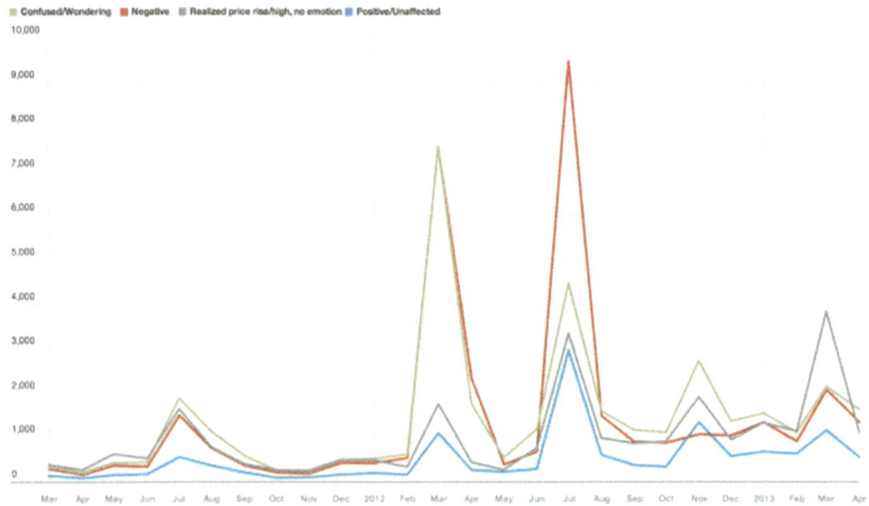

Fig. 3 Public sentiments for food price rise in twitter conversation [49]

6 Food Safety

The typical food chain production starts with a producer and ends with a customer buying the product. However, the journey in between is incredibly varied and depends on the product type, the region, and the legislation around food production. Even if the channels are not standardized, the food must be. Following regulation is what assures the wholesome quality of the food that is consumed. Unfortunately, in some cases, it may show that a single event of contaminated products can heavily impact a well-established business as well as the industry from which it is derived. The future of food safety as the end process in agriculture operations will be in refining the classification of the data. With so much data coming in and being processed in real time as well as the rapid increase in internet circulation, the use of social network analytics will become more pertinent to the food-safety agencies and for those involved in the food industry [52].

7 Business Adoption Software

As stated previously, there are several benchmarks by which one could judge the impact of a big data technique in agriculture. The two main categories include financial and societal impacts. In this case, estimating the environmental impact of farming is becoming an increasingly important process by which to assess the societal sustainability of certain processes. Therefore, as more big data techniques are developed, the natural progression points to enabling agricultural processes to diminish their impact on both water and land. In the case of combining real-time sensor data with historical data and GIS, the applications can be limitless. For example, being able to track wind conditions with fertilizer placement can give an indication of the concentration of fertilizer that is entering nearby water bodies. In doing so, big data applications can merge a whole variety of data to ensure that agricultural processes become more sustainable in the future; some of these software applications are discussed next.

7.1 AgriBigCAT [51]

AgriBigCATis an online software platform which can estimate the environmental impact of agricultural activity by geospatial and big data analysis. This platform not only assists farmers in their decision-making process but also helps government and administration to balance the farmland reclamation and the wetland or forest preserve. The platform allows the acquisition and analysis of the significant parameters of a large-scale in many agricultural systems in real time. The high volume and high variety data are stored in Apache Hive and the analyzed by ArcGIS. This variety of data spans from sensor data to historical data. The architecture of the AgriBigCAT is displayed in Fig. 4.

In terms of velocity, there is a constant stream of sensor data being produced on site and this is being processed and recorded by the program. This leads to many insights during a long time periods. On the data veracity part, the accuracy was guaranteed by cooperating with trustworthy organizations for collecting data. There are several applications of this technology. Furthermore, nitrogen production from farm animal dejection can be tracked and can be coupled with municipal data to ensure that

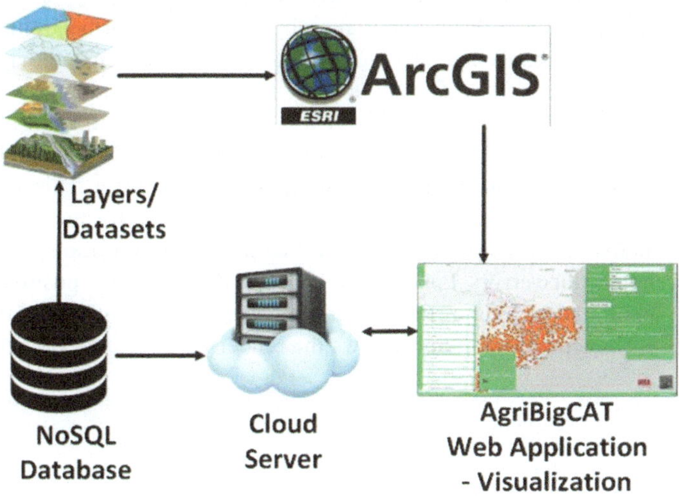

Fig. 4 AgriBigCAT architecture [51]

nitrogen is not being produced over the acceptable limit. On a manure management and planning project, the platform considers all the farms and animals in the system, calculates the estimated emissions under the current weather and temperature condition, and prevents the negative effect created by manure land overdosing.

In general, this system and this brand of big data are useful as it can give guidance to a producer in where they may dispose of manure, among other things. However, there is also inherent usefulness for the policymakers and municipal offices as the data can provide an estimate of the pollution being created, and policies can be set in place to mitigate environmental damage.

7.2 LettuceBot [53, 54]

One commercial application of the image recognition technology is the LettuceBot, which has been designed by Blue River Technology as a smart machine to thin lettuce fields. In order to ensure that a high yield of lettuce is produced, farmers will plant lettuce tightly together, but this makes it difficult to selectively remove weeds and also creates the potential for closely packed plants to choke themselves out. LettuceBot is towed behind a tractor and uses multiple cameras to identify lettuce plants from weeds and the surrounding soil.

An optimization algorithm then determines the optimal lettuce plants to keep based on spacing and size measures, killing any weeds or unwanted lettuce plants by spraying them with herbicide. The LettuceBot significantly reduces the environmental footprint of the farm by reducing the quantity of herbicides that are needed to protect the crops. This prevents over spilling issues that cause chemicals to drain into underground water supplies. The additional benefit is the added cost savings by efficiently managing the use of chemicals. In addition, LettuceBot replaces a process that would have previously been undertaken by hand. With further refinement of this technology, other plants and farming operations can be automated to reduce the impact of overusing herbicides. The application that LettuceBot addresses also leads to some simplifications of the computer vision problem. First, the soil provides a large contrast in color, making it easier to segment a plant from its background in an image. Furthermore, only one type of plant requires identification. This allows both the simplification of algorithm development and the potential for the algorithms to demonstrate better real-time performance. At the same time, variations in plant size and shape can introduce errors. Fortunately, the criticality of a false-positive or false-negative is small, and as long as the rate at which these errors occur is minimized, these errors are acceptable. Figure 5 shows the rendering of LettuceBots thinning a field.

Fig. 5 Rendering of LettuceBots thinning a field [53, 54]

7.3 Deep Thunder [55, 56]

Deep Thunder extends NWP mathematical modeling by taking advantage of the deep neural network (DNN) embedded in the IBM Watson AI [56]. The purpose of this neural network is not only to provide precise weather predictions but to determine the business impact of these predictions for farmers. Localized sensory inputs from the farm are included as inputs to the model to understand the effect of weather on a particular region. As an example, Georgian farmers are typically dependent on irrigation even though they get far more rain than other parts of the country. The porous soil prevents it from retaining the moisture from a heavy rainfall [56]. By using camera inputs on the farmer's property, a machine learning mechanism is trained for estimating the amount of rainfall based on computer vision and combine this information with sensory information from the soil moisture level [57]. The resulting model provides an output for the effect of rainfall on a particular area of the farm. The impact of rainfall is affected by the landscape; a hilly landscape will cause a serious runoff followed by a heavy accumulation, especially in lower-based valleys. By using hydrology and flood modeling, Deep Thunder is able to estimate where the runoff will go and where flooding might occur [56].

The final step for Deep Thunder is to combine the weather forecasts with the sensory data from the farm. The result is a predictive model that is able to pinpoint exactly where the farmer needs to irrigate today and for how long. An output is produced every 12 h to give farmers the tools to make daily business decisions. The predictions provide more efficient use of irrigation and pesticide supplies, ultimately reducing the cost of materials and maximizing crop yield [58]. In addition to reducing costs for the farmer, the cost advantages propagate throughout the entire supply chain. Being able to forecast weather predictions will make agricultural options cheaper to pursue, leading to cost advantages for the end consumer. In Brunei, Deep Thunder is being used to better understand regional conditions on the production of rice. Historically, only 3% of rice was produced internally on the island, with the rest imported. The objective is to use the analytics provided by Deep Thunder to increase this number to 60%, which would reduce the cost of food in the country and improve food security [56]. IBM's Deep Thunder initiative operates on a parallelized, high-performance computing system based on the scalable modeling system architecture. This distributed approach allows the prediction problem to be divided into a hybrid model involving multiple nested layers of prediction, each with a varying resolution level, and the results are combined to provide hyper-local precision forecasting. The initial resolution begins at 16 km and provides weather prediction for a region at a high level. The output is fed into a 4 km resolution layer that creates a narrower prediction grid. The final nest is a prediction grid with 1 km resolution, resulting in a hyper-localized forecast, with 10-min intervals, up to 86 h in advance [59]. Figure 6 shows a sample output of the nested models from Deep Thunder.

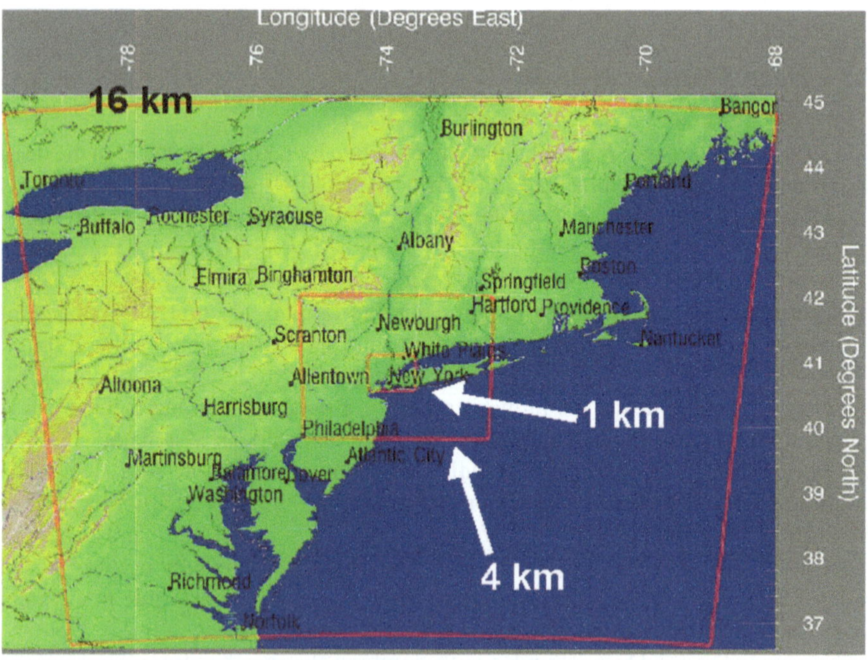

Fig. 6 Output of the deep thunder model (nested layers) [59, 56]

8 Future of Analytics in Farming

The adoption of big data is not only facilitated by the current structure of the food industry. While data collected on one farm might give some insights, it is insufficient to support analytics on their own fully. To gain benefit from analytics using big data, the producers would have to aggregate their data with other organizations in the industry along the "food-to-table" chain because the capital cost of getting such data can be significant.

Some producers are vertically integrated; they either possess the supplier or buyer section of their business; this means that the corporation or cooperative they represent can aggregate the data and make a profit by selling it to other consumers. Having independent producers, suppliers, processors, distributors, and the government collaborating by sharing a variety of data (sensor, streamed, historical) is the biggest challenge the industry has to face right now given its structure [51].

In essence, the problem with implementing some of these big data techniques within the processes of agriculture is that the farmers' choice to aggregate their data. With the scale of singular farm data, there is not enough data available to come up with entirely formative insights. The additional scale from many farmers working in collaboration is when real understandings are made, and real big data applications can be created. Nevertheless, for the farmer, there needs to be an inherent return on investment. It is likely that farmer will only aggregate data and buy remote sensors if they are guaranteed a reward back. However, farming is not sufficiently funded by the government to provide farmers with incentives. Therefore, education must be done so that farmers can see the financial reward in getting behind big data.

However, protecting the environment and ensuring food safety are increasingly important elements of importance for the consumers. The industry ultimately works for the consumer to match demand. The pressure from consumers might help the industry to go over the individual incentives given by the current structure and allow for aggregation of data in a competitive industry. In the future, processors might finance sensors for producers to have access to data related to the quality of their crops, for example. These kinds of partnerships using big data will undoubtedly affect the structure and define which player is going to stay competitive within the industry.

Big data is already having a significant impact on individual farming operations. LettuceBot and Deep Thunder both represent ways in which farmers can harness big data and technology to improve their own activities. The competitive advantage gained by using such techniques is clear within smaller markets and for individual farms, and the drive to adopt such technologies is quickly being spurred by the accompanying economic advantages. Farm managers are able to use the plethora of internet-connected devices and sensors to monitor their operations at increasing levels of precision. Yield measurement monitors, introduced into combine harvesters in the early 2000s, have allowed farmers to track field growth statistics down to the square meter. Soil sensors and moisture monitors can track nutrient levels and moisture levels down to the specific plant. Bringing all this information together has allowed farmers to practice what's known as precision farming, where they break down their fields into tiny sections and treat them all as individual fields.

Companies such as John Deere, Monsanto, and DuPont Pioneer are leading the revolution in precision farming techniques. Tractors can now plant seeds autonomously, and John Deere and Pioneer both offer variable-rate seeding "prescriptions" that take into account the specific production ability of portions of a field and adjust the density of planting accordingly.

Monsanto offers a program called FieldScripts that includes tools like nitrogen advisor; these tools track the available nitrogen at the field, and also it offers projections, recommendations, and alerts [49].

Two big factors stand in the way when it comes to making field data useful, both for improving the individual operations of a farmer and for improving the agricultural industry as a whole. These factors are integration and analysis. With all the different sources of information being made available, integrating these sources in a useful manner, one that facilitates decision-making, is a serious issue for farmers. The ability to analyze all the information in a way that provides useful insights is beyond the practical scope of an average farmer. Additionally, information gathered from a single farm is not nearly as useful as information aggregated from across multiple similar operations. This information can be used to analyze the effect of variations between farming practices and large corporations are better suited when it comes to collecting and aggregating this sort of data from multiple operations, and benefit from the scale when it comes to the cost of analytics work [52]. Large companies like John Deere and Monsanto have all been collecting data and performing the analyses to aid farmers with applications like variable-rate seeding. Due to the competitive nature of corporations, however, information is not being shared—to the detriment of farmers, and the agricultural industry as a whole.

"By 2050 the world population reaches approximately 9.2 billion. Thus, if we continue producing food at the current speed, the world cannot be fed" [49]. The vast amount of agricultural data available leaves opens the possibility for research that can optimize farming practices and revolutionize the industry. Historically, agricultural research has been led by governments and was done in the public sector [51]. Past research was done on small plots to test how different practices affected plant yields and was made public to benefit society as a whole. Over the past couple of decades, however, the majority of agricultural research has shifted to the private sector [51].

Research is being performed by large corporations whose business interests prevent them from sharing what they discover. With all the data available, there is potential for large-scale experiments that benefit the industry as a whole; however, bringing together all of the data in a format that is useful is proving to be a challenge. There are several initiatives being taken to try to bring order to the chaotic world of agricultural data. Multinational corporations are driving the initiative with their projects like Monsanto's Fieldscripts; however, these proprietary solutions are creating issues over data ownership. Farmers are hesitant to provide their data to support the analytics when they are forced to pay a subscription afterward to use the insights gained from it. A number of startups have entered the game in an attempt to bring together a wider range of data sources while giving more power to the farmers as defined next.

- Farm Business Network, a Google Ventures startup, aims prioritize Farmers, this can be established by democratizing information, establishing a non-biased analytics, and producing competitive advantages for farmers [60].
- Farmobile aims to help farmers own their data outright by providing a platform for storing, sharing, and selling agronomic and machine data [59].
- Global Open Data in Agriculture and Nutrition (GODAN) is a joint inventiveness defined by G8 governments to provide a standard for collaboration between partner companies. Their goal is to endorse partnership to employ the increasing size of data produced by innovative technologies to resolve long-standing difficulties and to help farmers and the health of customers [1].

These companies are helping bridge the gaps by integrating data from multiple systems and multiple vendors to fully make use of the technology available to farmers.

9 Conclusion

Big data analytics is continually informing decisions in agriculture and revolutionizing the processes by which agriculture operates. Agriculture can be thought of in several stages, from planting to spraying to the end consumer and supplying food to mass populations. By applying analytical techniques to each step in the agricultural operations process, there can be an economic gain for farmers as well as environmental and safety benefits for society at large. This data can come in all shapes and sizes from a variety of sources. The planting process is defined by forecasting. Big data analytics can inform decisions on prediction and provide insights for how producers should efficiently allocate their time and resources. Continually, big data, in collaboration with geographic information systems, can provide a base for large-scale agricultural mapping. This process can produce several insights on how to treat different crops and how to efficiently supply fertilization techniques based on k-means clustering. This can evolve into the identification of environmental impacts on land and water bodies over time. Overall, the process can be defined by creating value for producers and sustainability for society. This way, policymakers can make informed decisions on limiting policies and municipal bylaws for spraying and planting techniques. Eventually, when the food arrives at a plate of the end consumer, data analytics has been involved in the whole process and can detect any food-safety concerns which assure the customer that they are getting the safest product available. Agriculture is an industry that shoulders the responsibility of sustaining billions of lives every day. Increasing investment in this space will further give rise to innovations that have both an environmental, economic, and social impact. However, it is critical that these innovations work together to help improve farmers' operations rather than make their work more complicated. By breaking down the barriers that prevent the sharing of data across individuals' farms, big data analytics can have a significantly greater impact with the goal of enhancing farm yields at both a local and global scale as discussed in this chapter. This benefits not only farmers but humanity as a whole.

References

1. Bendre, M.R., et al.: Big data in precision agriculture: weather forecasting for future farming. Next Generation Computing Technologies (2015)
2. Jones, S., Rapp, N.: Where $64.5B in VC funding went last year. Fortune, 2 July 2015. Web. 12 Apr. 2017
3. Kotsiantis, S.B.: Supervised machine learning: a review of classification techniques (2007)
4. Dietterich, T.G.: Ensemble methods in machine learning (2000)
5. Kouru, K., Exarchos, T.P., Exarchos, K.P., Karmaouzis, M.V., Fotiadis, D.I.: Machine learning applications in cancer prognosis and prediction (2015)
6. Magidson, J., Vermunt, J.: Latent class models for clustering. Can. J. Market. Res. (2002)
7. Halkidi, M., Batistakis, Y., Vazirgiannis, M.: On clustering validation techniques. J. Intell. Inf. Syst. **17**, 107–145 (2001)
8. Rezaei, M., Fränti, M.P.: Set matching measures for external cluster validity. IEEE Trans. Knowl. Data Eng. **28**(8), 2173–2186 (2016)
9. Rendón, E., Abundez, I., Arizmendi, A., Quiroz, E.M.: Internal versus external cluster validation indexes. Int. J. Comput. Commun. **5**(1), 27–34 (2011)
10. Reichart, R., Rappoport, A.: The NVI clustering evaluation measure. In: Proceedings of the 13 Conference on Computational Natural Language Learning, pp. 165–173) (2009)
11. Rosenberg, A., Hirschberg, J.: V-measure: a conditional entropy-based external cluster evaluation measure. EMNLP-CoNLL **7**, 410–420 (2007)
12. Draszawka, K., Szymanski, J.: External validation measures for nested clustering of text documents. ISMIS Ind. Sess. **369**, 207–225 (2011)
13. De Souto, M.C., Coelho, A., Faceli, K., Sakata, T., Bonadia, V., Costa I. : A comparison of external clustering evaluation indices in the context of imbalanced data sets. Brazilian Symposium on Neural Networks (2012)
14. Dalli A.: Adaptation of the *F*-measure to cluster based lexicon quality evaluation. In: Proceedings of the EACL Workshop on Evaluation Initiatives in Natural Language Processing, pp. 51–56 (2003)
15. Rezaei, M., Fränti, P.: Set matching measures for external cluster validity. IEEE Trans. Knowl. Data Eng. **28**(8), 2173–2186 (2016)
16. Lei, Y., Bezdek, J., Chan, J., Vinh, N., Romano, S., Bailey, J.: Extending information-theoretic validity indices for fuzzy clustering. IEEE Trans. Fuzzy Syst. **25**(4), 1013–1018 (2017)
17. Cao, J., Lin, Z., Huang, G.: Self-adaptive evolutionary extreme learning machine. Neural Process. Lett. **36**, 285–305 (2012)
18. Mary, S., Sivagami, A., Rani, M.: Cluster validity measures dynamic clustering algorithms. ARPN J. Eng. Appl. Sci. **10**(9), 4009–4012 (2015)
19. Jang, J.: ANFIS: adaptive-network-based fuzzy inference system. IEEE Trans. Syst. Man Cybern. **23**, 665–685 (1993)
20. Yeh, C., Yang, M.: Evaluation measures for cluster ensembles based on a fuzzy generalized Rand index. Appl. Soft Comput. **57**, 225–234 (2017)
21. zu Eissen, B., Wißbrock, F.: On cluster validity and the information need of users. ACTA Press, pp. 216–221 (2003)
22. Wu, K., Yang, M.: A cluster validity index for fuzzy clustering. Pattern Recog. Lett. **26**(9), 1275–1291 (2005)
23. Halkidi, M., Batistakis, Y., Vazirgiannis, M.: On clustering validation techniques. J. Intell. Inf. Syst. **17**, 107–145 (2001)

24. Petrovic, S.: A comparison between the silhouette index and the davies-bouldin index in labelling ids clusters. In: Proceedings of the eleventh Nordic Workshop of Secure IT Systems, pp. 53–64 (2006)
25. Brownstein, J., Freifeld, B., Madoff, L.: Digital disease detection: harnessing the web for public health surveillance. New England J. Med. (2009)
26. Manyika, J., Chui, M., Brown, B., Bughin, J., Dobbs, R., Roxburgh, C., Hung Byers, A.: Big data: the next frontier for innovation, competition, and productivity (2011)
27. Rindler, A., McClowry, S., Hillard, R.: Big data definition (2012)
28. Cukier, K., Mayer-Schönberger, V.: Big data: a revolution that will transform how we live, work, and think, p. 22 (2013)
29. Gandomi, A., Haider, M.: Beyond the hype: big data concepts, methods, and analytics (2015)
30. White, T.: Hadoop: the definitive guide (2009)
31. Wolfert, S., et al.: Big data in smart farming. Retrieved from Agricultural Systems (2017)
32. Greengard, Samuel: Weathering a new era of big data. Commun. ACM **57**(9), 12–14 (2014)
33. Mining Indonesian tweets to understand food price crisis. Feb 2014. Web. Apr 2017
34. Rosenblatt, F.: The perceptron: a probabilistic model for information storage and organization in the brain. Psychol. Rev. **65**, 386–408 (1958)
35. Riedmiller, M., Braun, H.: A direct adaptive method for faster back propagation learning: the RPROP algorithm. In: Proceedings of the IEEE International Conference on Neural Networks, pp. 586–591 (1993)
36. Hecht-Nielsen, R.: Counterpropagation Networks. Appl. Opt. **26**, 4979–4983 (1987)
37. Melssen, W., Wehrens, R., Buyden, L.: Supervised Kohonen networks for classification problems. Chemom. Intell. Lab. Syst. **83**, 99–113 (2006)
38. Hopfield, J.J.: Neural networks and physical systems with emergent collective computational abilities. Proc. Natl. Acad. Sci. **79**, 2554–2558 (1982)
39. Kohonen, T.: The self-organizing map. Proc. IEEE. **78**, 1464–1480 (1990)
40. Huang, G., Zhu, Q., Siew, C.: Extreme learning machine: theory and applications. Neurocomputing **70**, 489–501 (2006)
41. Hansen, M., Smith, M., Smith, L., Salter, M., Baxter, E., Farish, M., Grieve, B.: Towards on-farm pig face recognition using convolutional neural networks. Comput. Ind. **98**, 145–152 (2018)
42. Ferentinos, K.: Deep learning models for plant disease detection and diagnosis. Comput. Electron. Agric. **145**, 311–318 (2018)
43. Ju, Y.: Conversion of remote sense picture into data array (2018)
44. Goodfellow, I., Bengio, Y., Courville, A.: Deep learning. MIT Press, pp. 216–261 (2016)
45. Stuart Ward, J., Barker, A.: Undefined by data: a survey of big data definitions (2013)
46. Lukman, E.: Indonesia is social: 2.4% of world's twitter posts come from Jakarta. Tech in Asia—Connecting Asia's Startup Ecosystem, p. 12 (2017)
47. Greenberg, J., Elliott, C.: A cold cut crisis: listeriosis, maple leaf foods, and the politics of apology. Can. J. Commun. **9**, 189–204 (2009)
48. Noyes, K.: Big data poised to change the face of agriculture. Fortune, 02 June 2014 (2017)
49. Kamilaris, A., et al.: Estimating the environmental impact of agriculture by means of geospatial and big data analysis: the case of catalonia. Science to Society (2018)
50. Sonka S.: Big data and the ag sector: more than lots of numbers. Int. Food Agribus. Manag. Rev. (2014)
51. About GODAN. About GODAN|GODAN. GODAN (2017)
52. Blue river technology. http://www.bluerivert.com/
53. Cerqueira, R.F., Mantripragada, K.: Estimating rainfall precipitation amounts by applying computer vision in cameras. U.S. Patent No. 9,436,997 (2016)
54. Kumar, N., et al.: Leafsnap: a computer vision system for automatic plant species identification. Computer Vision ECCV Lecture Notes in Computer, pp. 502–516 (2017)
55. Treinish, L., Praino, A.P.: Applications and implementation of a mesoscale numerical weather prediction and visualization system. In: Proceedings of the 20th Conference on Weather Analysis and Forecasting/16th Conference on Numerical Weather Prediction (2004)

56. Zulkarnain, H.: IBM research: using weather modeling and data analytics to save natural resources (2017)
57. Nowcasting food prices in Indonesia using social media signals (2017)
58. Treinish, A., Anthony, P.: The role of meso-γ-scale numerical weather prediction and visualization for weather-sensitive decision making. Proc. Forum Environ. Risk Impacts Soc. Successes Chall. (2006)
59. Woodie, A.: Farmer's plant for hyper-local forecasts with IBM's deep thunder. Datanami. Tabor Commun. 8 Apr 2017
60. Weston, J.: Support vector machine (and statistical learning theory). Princeton Tutorial. Princeton, USA. Tutorial
61. Zikopoulos, P., Eaton, C.: Understanding big data: analytics for enterprise class Hadoop and streaming data (2011)
62. Meilă, M.: Comparing clusterings: an information based distance. J. Multivar. Anal. **98**(5), 873–895 (2007)
63. Linnainmaa, S.: Taylor expansion of the accumulated rounding error. BIT **16**, 146–160 (1976)
64. Broomhead, D., Lowe, D.: Multivariable functional interpolation and adaptive networks. Comp. Syst. **2**, 321–355 (1988)
65. Specht, D.: A general regression neural network. IEEE Trans. Neural Netw. **2**, 568–576 (1991)
66. Effland, T., et al.: Discovering foodborne illness in online restaurant reviews. J. Am. Med. Inf. Assoc. (2018)
67. Proudly Farmers First. farmers business network (2017)
68. Solano, B.: Our story. Farmobile|Press Page. Farmobile, n.d. Web. 13 Apr 2017 http://press.farmobile.com/
69. Sonka, S.: Big data and the ag sector: more than lots of numbers. Int. Food Agribus. Manag. Rev. **17**(1), 1–20 (2014)

AI-Based Yield Prediction and Smart Irrigation

Deepak Sinwar, Vijaypal Singh Dhaka, Manoj Kumar Sharma and Geeta Rani

Abstract This chapter presents different techniques and applications of Artificial Intelligence for yield prediction and smart irrigation. Timely prediction of irrigation requirements and crop yields is necessary for farmer's welfare and satisfaction. The beforehand prediction significantly contributes to minimizing production cost and maximizing crop yields. The precise prediction of crops' yields is also useful for government, as it is effective in planning various schemes, transport needs, buying mechanisms, storage infrastructure, and liquid position of the economy before actual selling of crop by farmers to market. This chapter acknowledges the past breakthroughs and emerging Artificial Intelligence-based techniques in precision farming specifically for yield prediction and smart irrigation. Artificial Intelligence-based system provides sufficient information about crop yields at an early stage and its associated smart irrigation management system is effective in the judicious use of essential resources such as water and energy for agriculture.

Keywords Yield prediction · Smart farming · Precision agriculture · Smart irrigation · IoT · Temperature monitoring · Soil moisture · Crop · Artificial intelligence

1 Introduction

In recent years, an explosion in applications of Artificial Intelligence (AI) specifically in the domain of health care, weather forecasting, space programmes, automation, agriculture, etc. has been observed. Across the globe, more than half of the population is involved in agricultural practices. In the arena of agriculture, AI is adopted for the planning of farming activities, sales of agricultural products, weather forecasting,

D. Sinwar (✉) · V. S. Dhaka (✉) · M. K. Sharma · G. Rani
Department of Computer and Communication Engineering, School of Computing and Information Technology, Manipal University Jaipur, Dehmi Kalan, Jaipur, Rajasthan, India
e-mail: deepak.sinwar@gmail.com

V. S. Dhaka
e-mail: vijaypal.dhaka@gmail.com

P. K. Pattnaik et al. (eds.), *Internet of Things and Analytics for Agriculture, Volume 2*, Studies in Big Data 67,
https://doi.org/10.1007/978-981-15-0663-5_8

and yield prediction. Farmers are taking advantages of AI and getting higher production than traditional farming techniques. In traditional farming, cultivation was manual process irrigational needs of a crop and its yield was highly dependent on climate factors throughout the globe. The onset of inventions in industrial revolution and improvement in mechanisms involved in agricultural practices gave relief to miserable life of farmers. But still, it is a labor-intensive practice and dependent on mercy of monsoon.

Thus, there is a need of developing a system that automates the agricultural practices such as cultivation and irrigation, predicts yields, and help in better planning for systematizing the demand and supply of agricultural products throughout the globe. A blend of AI and IoT technologies intelligently gathers relevant information about different crops, their growth rate, and irrigation requirements. These technologies are effective in monitoring frequent climate change, weather forecasting, nutrition deficiency in plants, plants' health monitoring, pest control, weed management, etc. Researchers have covered a long way in developing subsidiary technology such as geo-positioning satellites (GPS), satellite imagery and sensors for monitoring of yield affecting parameters such as moisture, soil pH level, temperature, etc. Target experimentation at various stages of cultivation such as water resource management, weed elimination, diseases prediction, pest control, estimation of yield production and effective storage of agricultural products is still starving. These challenges can be addressed by applying the techniques of AI and data science.

A voluminous data about agriculture is collected using various sensors such as remote sensors, proximity sensors, temperature/humidity sensors, etc. Advanced Internet facilities disseminate this data to various channels such as the cloud. The introduction of online data storage at cloud, advancements in techniques of data processing, and increase in computation power become effective in dealing with challenges identified in agriculture. Use of advanced techniques makes agricultural predictions more precise, reliable, and useful.

1.1 Scope of AI in Agriculture

Traditional agricultural practices used by farmers are not adequate to fulfill the increasing demand. To serve this increasing demand, farmers need to adopt the latest advancements in agriculture such as use of state-of-the-art tools/machinery and AI-based techniques. It has been observed that most of the farmers are unaware of uses of pesticides specifically in what amount and when. Due to these practices, fertility of the soil, as well as yield production, degrades in an intensified way. A plethora of researchers has proved the increased gain in yield growth with the help of AI and automation. AI can contribute to agriculture in versatile ways. Authors explain the inclusion of AI in the following seven agricultural practices.

1.1.1 Weather Forecasting

One of the important advantages of AI in agriculture is weather forecasting. With the help of latest AI-based state-of-the-art weather forecasting techniques, it becomes easier for farmers to take appropriate decisions in planning respective crops. Nowadays, many weather forecasting devices are available for predicting the weather. Sehgal et al. [1] presented a visual tool (*ViSeed*) based on long short-term memory (LSTM) [2] which can be used for weather and soil predictions. With the help of such farmer-friendly tools, one can plan his/her crop activities, crop types in an efficient way.

1.1.2 Smart Irrigation

To cope with the shortage of water, there is a critical need for some smart irrigation systems that can irrigate more areas with low consumption of water. A comprehensive review of various techniques of smart irrigation has been presented in Jha et al. [3]. However, there is an availability of various low water consumption-based irrigation techniques, for example, sprinkler systems and drip irrigation systems; but these systems need human intervention up to a great extent. There is a scope to add features to existing systems to develop smart irrigation systems. The system continuously monitors the level of water in a crop, compares the water content available in soil and crop plant with standard need of water. It automatically starts sprinklers or drips as per water requirements of crop. Arif et al. [4] presented ANN-based models for predicting soil moistures.

A Neural Network-based irrigation system (Neuro-Drip) has been presented by Hinnell et al. [5]. On the other hand, an IoT-based smart irrigation management system has been presented by Goap et al. [6], Nawandar and Satpute [7]. AI in this field has contributed a lot in collaboration with the Internet of Things (IoT). For smart irrigation, there is a requirement of gathering of information about level of moisture present in the soil, water content in plants, humidity in atmosphere, temperature, etc. This information can be gathered using soil moisture sensors, temperature sensors, humidity monitoring sensors, etc. These sensors are connected to low price Arduino-based systems for storage of gathered information and executing analysis algorithms for predicting the water requirement of crop at a particular time.

1.1.3 Crop Disease Prediction and Health Monitoring

Traditional methods of monitoring crop health are time-consuming and inefficient especially for larger areas of thousands of acres. Many researchers have developed many AI-based architectures to overcome the challenges identified in the arena of traditional agriculture. Some architectures are surveyed in Kamilaris and Prenafeta-Boldú [8].

AI-based techniques with inclusion of image processing, deep learning, and data analysis provide an easy and effective way for disease prediction and health monitoring. The system captures crop images using high definition(HD) camera-enabled drones, unmanned aerial vehicles (UAVs), or satellite imagery. The captured images are used as dataset for training of Convolution Neural Networks (CNNs) [9], a class of Artificial Neural Network (ANN). CNNs extract useful features from fed images and make predictions about disease(s) in crops. The system is effective in continuous monitoring of health of plants and hence gives better solutions for calculating amount of pesticides to be used and time to use a pesticide.

1.1.4 Crop Readiness Identification

AI-based system captures images of a crop and analyzes them for determining the crop readiness in a particular area for harvesting. The crops can be categorized into different categories on the basis of readiness and other quality parameters before actually sending them to market. For categorization, use of various pattern clustering techniques viz. K-means, fuzzy C-means (FCM), expectation maximization (EM), and hierarchical clustering plays an important role.

1.1.5 Yield Prediction

Yield prediction is an area of interest for researchers for past many decades. Yield prediction requires yield mapping devices, which are still not easily available to farmers [10]. CNNs can solve the problem of yield prediction in an economic and easy way. Many researchers [10–12] have developed models to predict crop yields using AI-based methods which use RGB/normalized difference vegetation index (NDVI) images. Their experimental results on publically available dataset prove the usefulness of their models. A comparative analysis of CNN-based models with AI-based models clearly indicates that CNNs are more advantageous than traditional Machine Learning (ML) and AI-based techniques [13]. Authors will discuss the detailed process of predicting crop yields using CNN in Sect. 2.

1.1.6 Weed/Pest Management

Traditionally, weed management was accomplished with a combination of many techniques viz. mechanical weed control, crop rotation, herbicides, etc. [14]. In addition to the above, various biochemical-based combined weed and pest control solutions are also available in the market, but they all results in reducing the yield productivity. The reduction in crop yield occurs due to the spraying of pesticides frequently and in a constant amount in a field. AI provides an intelligent solution for this challenge.

Recently, one AI-based autonomous robot called "*Agbots*" has been developed by McAllister et al. [14], which can do weed management in the field in an interesting manner. The idea behind AI-based weed management is to identify weeds automatically (by examining camera images obtained from autonomous robots) and perform corrective actions (mechanically weed removal or spraying herbicides) accordingly. Many other AI-based techniques are also available such as See and Spray (by Blue River Technologies [15]), which uses AI to identify and spray individual plants in milliseconds.

1.1.7 Precision Farming/Agriculture

The aim of precision agriculture (PA) is to increase yield production as well as quality by simultaneously reducing the overall cost and environmental pollution [16]. The quantity and quality of a crop depend on many parameters such as soil, weather, irrigation, etc. So, there is a requirement to monitor all these parameters at a regular interval of time. Traditional monitoring techniques are not adequate in accurate and efficient monitoring of these parameters.

So, there is vital demand for an automated system which can perform monitoring of parameters in an effective way. In modern era, proximity sensing and remote sensing dominate the field of agriculture and effectively monitor a plethora of parameters required for better prediction and planning of agricultural practices. Proximity sensing specifically deals with soil using high-resolution data. Remote sensing provides the geographical sensing of fields using various sensors. Thermal remote sensing provides some additional information such as temperature, water status, etc. It also helps us in getting many vegetation indices. Many researchers attempted to present several vegetation indices on the basis of various parameters. Xue and Su [17] presented a systematic reviewed of many vegetation indices. AI in collaboration with these techniques can provide cost-effective solutions for PA.

2 Yield Prediction

Manual record-keeping and monitoring trends in agricultural activities and yields are a tedious and time-consuming task. There are fair chances that farmers are not available all the time to keep track of all agricultural activities. Thus, it results in collection of incomplete and inconsistent data. AI-based solutions are effective to overcome such situations. In the era of technology, this is mandatory to motivate and train farmers for adoption of new tools and techniques. This is highly desirable to reduce the manual efforts, improve quality of crops, and to meet the increasing demand of food for increasing population.

Precision agriculture provides an intelligent solution to the challenges observed in the field of traditional agriculture. Increase in awareness and education make farmers to think about adoption of precision agriculture. Farmers are willing to adopt new strategies of precision agriculture but precise yield prediction is still a challenge due to shape irregularities, illumination conditions, etc.

Continuous monitoring of each activity involved for a crop from the stage of sowing seeds till harvesting contributes toward precision agriculture. The planning of timeline for agricultural activities on the basis of past experience and collected record is demonstrated in Nevavuori et al. [10].

Farmers are preparing timeline and maintaining record of following parameters for better prediction.

- Soil fertility and improving fertility by using manures, compost, organic and inorganic fertilizers, bio-fertilizers, etc.
- Growth rate of a crop.
- Irrigation requirements in terms of amount of water required and time and frequency of irrigation.
- Determining precise amount of pesticide and time of spraying pesticide.
- Identifying type(s) of weed growing with a particular crop, growth rate of weed and determining methods to destroy weed plants without harming the main crop.
- Season wise and crop-wise yield estimation.
- Summarizing timeline for future, based on accumulated knowledge and experience.

2.1 Machine Learning Techniques for Yield Prediction

Review of related literature reveals the existence of a significant research work in the field of Machine Learning (ML) techniques for agriculture. The categories of research work are specifically in yield prediction, disease prediction, crop quality prediction, yield estimation, price estimation and weed detection and eradication. In this section, authors give a brief summary of existing works in the domain of yield prediction using ML-based techniques.

Liakos et al. [18] demonstrate different ML models used for solving real-world problems. Most commonly used ML models are Artificial Neural Networks (ANNs) Deep Learning (DL), Support Vector Machine (SVM), Decision Trees (DT), Bayesian Models (BM), Ensemble Learning (EL), and Dimensionality Reduction (DR). There are numerous ML techniques available based on these ML models. These models are based on one or more ML techniques namely Convolution Neural Networks (CNNs), BackpPropagation Network (BPN), Feedforward Network (FFN), Hopfield Network (HN), Multilayer Perceptron (MLP), Radial Basis Function Network (RBF), Deep Boltzmann Machine (DBM), Deep Neural Networks (DNN), K-Nearest Neighbor (KNN), Linear Discriminant Analysis (LDA), Multiple Linear Regression (MLR), Generative Adversarial Networks (GAN), and Ensemble Neural Networks (ENN).

The process of yield prediction using Machine Learning starts with capturing images of crop. The image capturing process can be carried out by handheld devices

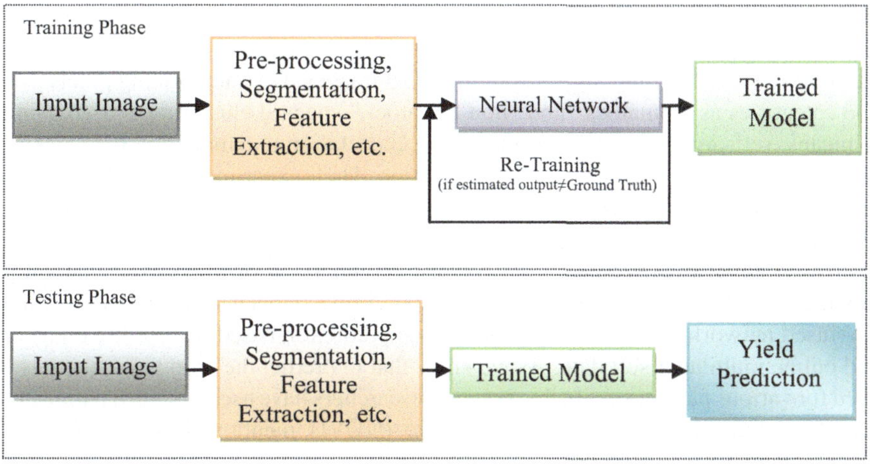

Fig. 1 Architecture of a general yield prediction system

such as mobile phone cameras, digital cameras, and satellite imagery, etc. After capturing, these images are used for providing training to suitable Neural Networks such as ANNs, MLP, and CNNs. The trained networks are used for making predictions in a desired domain. In case there is a mismatch in desired output and ground truth, then there is a need to retrain the model. The retraining continues until the model becomes capable of making precise predictions of yields. Figure 1 demonstrates the architecture of yield prediction using ML.

Let us look at some of the research works carried out by researchers in the field of crop yield prediction using ML techniques.

Wang et al. [19] developed a computer vision-based system which can automatically estimate the number of **apples** in an apple orchard. The system is accurate as well as faster in nature. This system contains a vehicle for moving in the field. The vehicle dismounted with high precision cameras on metal frames. Both vehicle and cameras are calibrated accurately according to coordinate frames. Apple regions are identified with the help of pixel properties. Their model used hue, saturation, and value for identification of specular reflections. Experimental results and analysis show that their model is able to calculate the number of apples accurately and rapidly.

Ramos et al. [20] proposed a method for counting the coffee fruits. Their machine vision system (MVS) is capable of counting and classifying **coffee** fruit. Harvestable and non-harvestable fruits were categorized in this work. They captured images via a mobile-based system and processed the images using linear estimation model to calculate the number of coffee fruits on a branch. MVS has five stages as: (i) acquisition, (ii) segmentation, (iii) fruit boundary/outline detection, (iv) adjustment of ellipses, and (v) detection, classification, and counting of fruits. Promising results were presented on the basis of experimental analysis and validation strategies.

Senthilnath et al. [21] presented a system based on Bayesian Information Criterion (BIC) for detection of **Tomatoes** in the image. They have used unmanned ariel vehicles (UAVs) for obtaining high-resolution RGB images. For categorizing pixels into two categories as "tomatoes" and "non-tomatoes," spectral clustering is used using K-means, self-organizing maps (SOM), and expectation maximization (EM). Morphological operations are used for segmentation. Their experimental results showed that the performance of EM was found better than K-means and SOM in the counting of tomatoes.

Pantazi et al. [12] presented performance comparison of three ML techniques (counter-propagation ANNs (CP-ANNs), supervised Kohonen networks (SKNs) and XY-fused networks (XY-Fs)) for **Wheat** yield prediction based on satellite imagery and online multilayer soil data. As we know that to predict the crop yield, there is a need for various parameters. In this system, researchers have used one such parameter called normalized difference vegetation index (NDVI), mentioned as follows:

$$\mathrm{NDVI} = (\mathrm{NIR} - \mathrm{R})/(\mathrm{NIR} + \mathrm{R}) \tag{1}$$

where NIR is near-infrared and R is red bands, which enables the accurate estimation of yield per area. On the other hand, to measure the soil properties, they have used vis-NIR spectrophotometers. Experimental results consisting of field maps show that the performance of SKNs was found best among three SOM-based models in predicting wheat yield crop.

Amatya et al. [22] developed a machine vision-based system for automatic harvesting of **Cherry** fruits because manual harvesting is very labor exhaustive which causes more than 50% of production cost. A Bayesian classifier is used to classify the detected branches, leaves, cherry, and background. After getting information about branches and cherry, we can harvest them accordingly. The system is capable of detecting branches with an accuracy of nearly 90%.

Sengupta and Lee [23] presented an SVM-based model for detection of immature green **Citrus** fruit in varying lighting conditions. The system is able to detect nearly 80% of citrus fruits in the tree canopy. In addition to the above, lots more work in the area of yield prediction is available. Liakos et al. [18] presented a systematic review of various Machine Learning techniques for agriculture. They have classified these techniques into different categories such as yield prediction, disease detection, weed monitoring, etc.

2.2 Remote Sensing-Based Yield Prediction

Remote sensing brings information about a field, crops, or an object without actually visiting the field. The information is captured with the help of different sensors and is made available in raw, processed, or analyzed form. It works on the principle of properties of objects. The properties may be chemical, physical, structural, energy

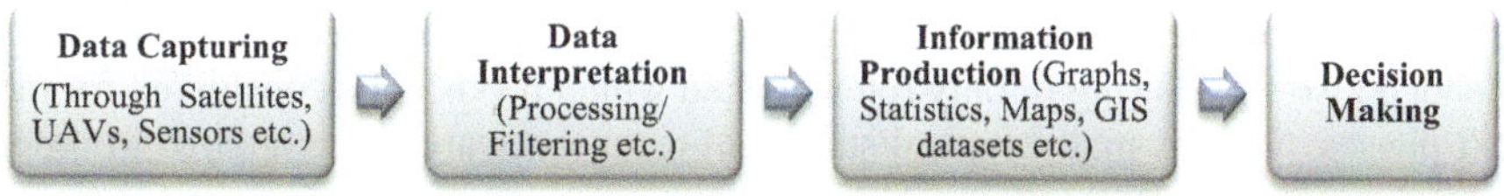

Fig. 2 The Remote Sensing process

Table 1 Various active and passive sensors for remote sensing [24]

Active sensors	Passive sensors
Laser altimeter Lidar Radar Ranging instrument Scatterometer Sounder	Accelerometer Hyperspectral radiometer Imaging radiometer Radiometer Sounder Spectrometer Spectroradiometer

emissions, etc. Sensors sense these properties and send to a storage system. In storage systems, the information is organized in a structured way. This information is analyzed to make decisions for deciding further actions required. Remote sensing techniques have numerous applications in agriculture viz. yield forecasting, damage identification, harvesting time prediction, irrigation estimation, and nutrient requirement prediction, etc. Figure 2 shows the process of remote sensing. It comprises of four basic steps viz. data capturing, data interpretation, information production, and decision making.

2.2.1 Data Capturing

The process of capturing data is done in either of two ways, i.e., active or passive. In the case of active remote sensing, sensors are focusing directly on objects and send the sensed information to the control station which is located on the ground. Active sensors make uses of their own energy sources for object illumination.

Passive remote sensing follows the concept of reflection. Objects are exposed to sunlight and then reflected wavelengths are sensed. The sensed information is further provided to control stations. The electromagnetic spectrum is being used for this purpose. List of some active and passive sensors obtained from [24] is given in Table 1

2.2.2 Data Interpretation

After capturing the required data, the process of processing that data needs to be done in order to obtain the required information. This processing may take a while depending upon the operations/statistical techniques being applied. This process is intended also for filtering of only relevant data which needs to be considered for the next phases. Various mathematical/statistical techniques are being applied in order to get the required information.

2.2.3 Information Production

The next step after interpreting data is the production of the required information. As we know that the processed data can be presented in the form of various visual representations such as graphs, charts, maps, GIS data, etc. We can get various types of information with the help of remote sensing data such as yield modeling/forecasting, nutrient deficiency graphs, area estimation, soil mapping, etc. These graphs/charts are the best way to represent the gathered information, by which we can take effective and timely decisions.

2.2.4 Decision Making

With the help of various visual representations, one can take effective decisions. The objective of these decisions is very clear: maximizing the net profit by increasing the quality as well as quantity of the crop. Also, there is a need to identify risk factors in achieving this objective. Nowadays, the process of decision making becomes easier with the help of various AI-based techniques. Once we have obtained the required information, we can present this information to the information scientists in order to save them in the knowledge base. On the basis of certain parameters AI/Machine Learning (ML) systems are able to predict/provide efficient decisions (such as the necessity of water treatment, moisture prediction, irrigation requirements, pesticide requirements, etc.) which makes farmers away from critical thinking process and obtaining maximum profits out of minimum/required efforts. Traditional ML techniques use feature extraction in the earlier stages, and on the basis of these features, different tasks (crop classification, yield prediction, etc.) were carried out [10]. But traditional ML techniques were not sufficient in producing the optimal decisions. Nowadays, it becomes possible to provide optimal decisions with the help of some advanced paradigms such as Deep Learning paradigm. Among various Deep Learning paradigms, Convolutional Neural Networks (CNNs) prove to be efficient in image analysis/classification. By combing Remote Sensing with Deep Learning techniques, we can have effective agricultural decisions that can maximize the farmer's profit.

2.3 Big Data-Based Yield Prediction

With the advancement of Information and Communication Technology (ICT), massive volumes of data are being generated. Big data techniques play important roles in analyzing these massive volumes for smart farming and yield prediction. The process of big data starts by implementing "data chain," which is the organized sequence of activities from capturing to decision making/marketing [25] mentioned as follows:

1. Capturing of data (Identifying sources of data, types, and purposes)
2. Storage requirements (Identifying storage location, permissions/authenticity, etc.)
3. Transfer from one location to another
4. The transformation from one form to another
5. Analysis (Identifying methods, purposes, and output)
6. Decision Making (Identifying strategies, risk assessment, etc.)

Big data can play important roles in predicting the amount of yield prediction. Here, the emphasis is on analysis rather than big or small. An agricultural data without proper analysis is of no use until it is being analyzed properly. However, big data in agriculture is still in an early stage, and there is a need to do more work on big data with respect to agricultural applications. As mentioned above in the first step, we may require the capturing of data from multiple sources. If data is gathered from multiple sources, definitely it will not be in a uniform shape. It is the task of data analysts to prepare and arrange data in a uniform shape which in future can be transferred/transformed.

There are numerous data analysis techniques [26] available, but for analyzing agricultural data, we have to focus on some specific techniques such as scalable vector machines, K-means, NDVI, Wavelet/Fourier transformations, etc. In addition, to predict the crop yield, big data can do more with agricultural data such as weather forecasting, soil/land analysis, food availability, etc. In this regard, Kamilaris et al.[26] presented a systematic yet remarkable review of various big data techniques for agriculture.

3 Smart Irrigation Systems

Efficient utilization of water is a challenging task especially for places where availability of water is a major concern. For gaining high yields, one has to irrigate the field when needed or we can say that the exact amount and time of irrigation must be known. Nowadays with the help of new innovations in technology, we are living in the world of advanced irrigation systems called "smart irrigation." The word smart means, the sensors are able to sense the water requirements in plants. This ability is achieved by combining multiple technologies viz. automation, sensors, and knowledge (AI). Even though these systems are not costly in nature, due to lack of awareness, most of the farmers are adopting a traditional way of irrigation which results in inefficient utilization of water and low crop yield. Thanks to the researchers who have developed such low-cost smart irrigation systems. One such system is shown in Fig. 3.

Fig. 3 AI-based smart irrigation system [35]

The automation behind these smart irrigation systems is achieved using smart embedded systems. These smart embedded systems can be formed with the help of various technologies/devices such as Raspberry Pi, Arduino, power unit, temperature sensors, moisture sensors, Machine Learning, and Internet of Things (IoT) as shown in Fig. 4.

Jha et al. [3] presented a review of such irrigation systems in a systematic way. Smart irrigation systems are able to save a massive amount of water, which can be utilized for other important purposes of mankind.

Smart irrigation systems also ensure the reachability of water to each and every plant in an exact amount which results in maintaining the good health (by preventing them from dehydration and excessive/irregular irrigation) of the plant and makes away from diseases. It has been observed that some of the diseases in plants are due to improper irrigation which is mostly due to the adaptation of traditional irrigation systems. Continuous observations of sprinklers/drip modules are also necessary at regular intervals to prevent them from failures. Sometimes these systems need recalibrations due to degradation in performances or damages caused by animals/environment. In this situation, some part of the field is highly irrigated while some part remains unirrigated. The main challenge in designing a smart irrigation system is the nonuniformity of land and crop types. Some of the smart irrigation systems are discussed in the upcoming section.

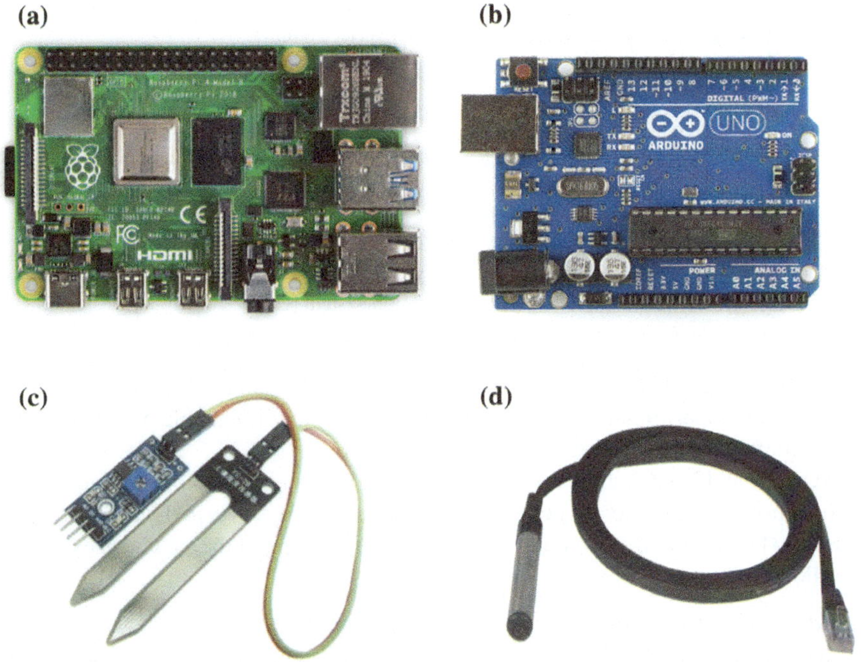

Fig. 4 Various devices for smart implementing smart irrigation systems: **a** Raspberry Pi, **b** Arduino, **c** soil moisture sensor, **d** temperature sensor

3.1 Existing Smart Irrigation Systems

Many researchers work in the area of smart irrigation systems so far. Let us discuss some of them as follows:

Suresh and Umasankar [27] developed a smart irrigation system which works on the solar power supply. The system is able to minimize electricity costs by utilizing solar power. It is composed of an automated water flow system which works by sensing the moisture level in the soil.

The intention behind the development of this system is to save both electricity as well as water in the Indian scenario. The block diagram of solar-powered smart irrigation is given in Fig. 5, which consists of solar panels, battery, converter module, and water pump. The system works on solar-powered supply, which is used to turn on the stepper motor to feed the water from well to storage tank on the basis of sensed moisture. The storage tank is mounted with a water outlet valve which is electronically controlled by moisture sensing circuitry. The moisture sensors are mounted inside the field of where cultivation of the crop is going on. Sensors are able to convert the moisture content to equivalent voltage which is compared with the

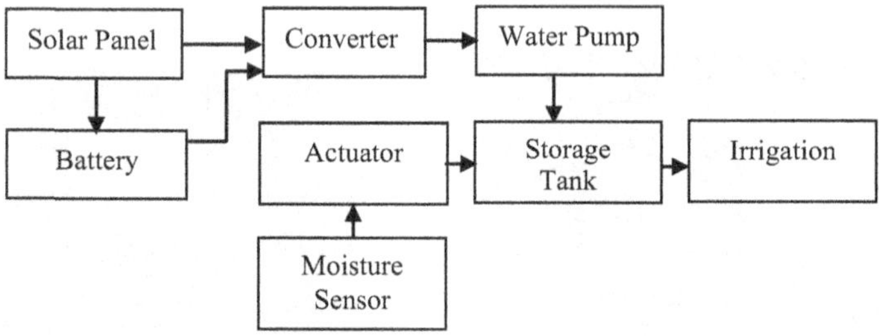

Fig. 5 Architecture of a solar-powered smart irrigation system [27]

reference voltage. The voltage difference level is used to find out the water requirements. Their experimental results showed that the system is capable of irrigating the field with the minimum usage of water and energy.

Kehui et al. [28] designed a system using wireless sensor network for monitoring of moisture and water height in the rice field. The system was tested on real time and proved to be feasible for irrigating rice fields. The working of the system is explained as follows:

1. Moisture data is calibrated to the data center.
2. The existing expert data available with the data center is being compared with real-time sensed data.
3. Irrigation instructions are being sent to the base station on the basis of irrigation requirements.
4. The base station sends these requirements to the irrigation control system (ICS).
5. ICS opens the electronic valve and closes after irrigation.
6. The process works in a cycle as per the irrigation requirements.

Vellidis et al. [29] developed a prototype for scheduling irrigation to cotton fields on the basis of soil moisture and soil temperature as shown in Fig. 6. This smart irrigation system works on the principle of smart nodes consisting of a combination of a sensor array, RFID tag for sending information to the central computer, and electronic circuitry. The tags used in these nodes have the capacity to transmit signals up to the range of 0.8 km. The smart nodes will send the sensed information at user-specified intervals. The information in each transmission is the unique node identification and 12 bytes of sensor values (two temperature values and three sensor values). The nodes are provided with a power backup of 9 V lithium batteries. The microcontrollers are programmed for efficient usage of power during information transmission and sensing moisture/temperature levels. The system was tested in a real environment which offered real-time monitoring of water in the soil. According to these values, users are able to schedule the irrigation requirements (which were not automated in this system).

Nasiakou et al. [30] presented a software system for reducing the irrigation cost by combining the smartness of renewable energy sources and irrigation as shown in Fig. 7. The system contains common information and common bus which is

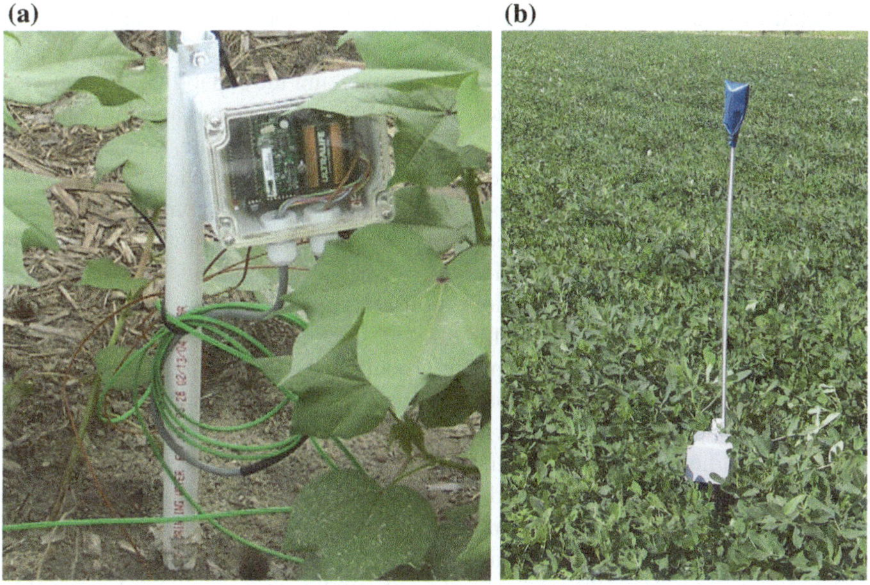

Fig. 6 Smart irrigation system [29], **a** smart sensor array for cotton field, **b** modified sensor array for peanut

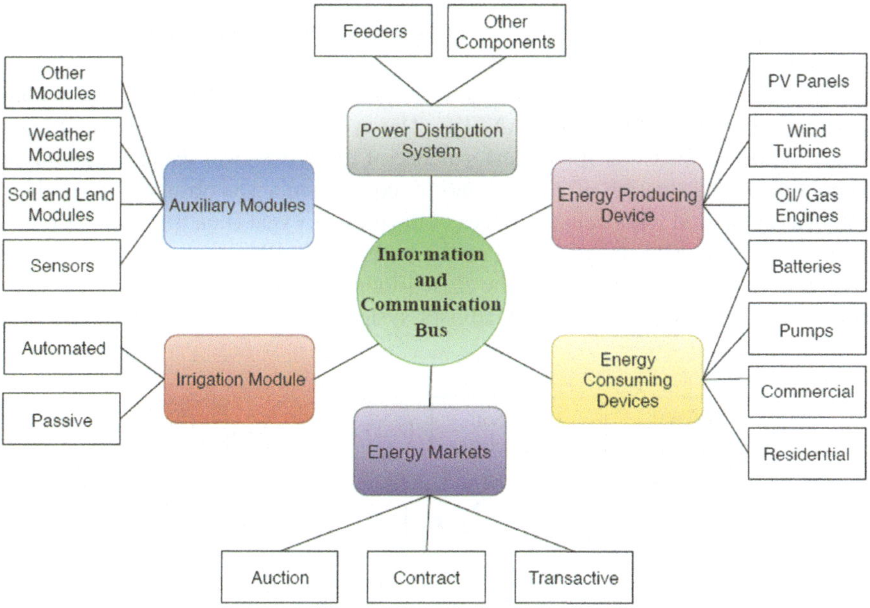

Fig. 7 Component view of energy-efficient irrigation system [30]

connected with six components as irrigation modules, energy-consuming/producing devices, power distribution system, energy markets, and auxiliary modules.

The main goal of designing this system is analyzing the capabilities of batteries and renewable energy sources to handle both residential as well as agricultural load. As shown in Fig. 7, the six basic components (irrigation module, power distribution module, energy consumption devices, energy-producing devices, energy markets, and auxiliary module) of their systems are connected to a common bus called information and communication bus. Every module has its designated purpose in contributing to the irrigation system. Power distribution module is responsible for distributing electricity flow to the whole system. It calculates the overall energy flow in terms of the voltage vector of different nodes. Electricity generating module is responsible for implanting various energy-producing devices, i.e., engines, turbines, PV panels, and batteries. Energy consumption module on the other hand provides possible energy-consuming devices which can be categorized into two categories, viz. residential/household devices and commercial devices. The energy market is responsible for organizing energy bids between producers and consumers. The main module of the system, i.e., irrigation module is capable of irrigating plants in two different ways, viz. automated and passive. All other components of the system such as temperature sensors, soil moisture sensors, humidity sensors are implemented in the auxiliary module of the system. Simulation of the system was achieved on R4 25:00 1 and IEEE-13 bus feeders equipped with 68 and 80 houses, respectively. The simulation gives promising results in terms of satisfying the irrigation requirements of both agricultural and residential design with low energy consumptions.

Navarro-Hellín et al. [31] presented a smart irrigation decision support system (SIDSS) for estimating the weekly irrigation requirements using sensors and Machine Learning techniques (PLSR and ANFIS). The system was tested on citrus plants located in the southeast of Spain. Performance of the system was tested with human expert values. A reduction of 22% irrigation requirements was observed using their system as compared with weather values of the previous year (i.e., irrigation requirements of the year 2015 are predicted on the basis of 2014).

Figure 8 shows the prototype of the closed-loop irrigation system which is able to send feedback to a typical irrigation advisor system (shown in dashed box). Earlier,

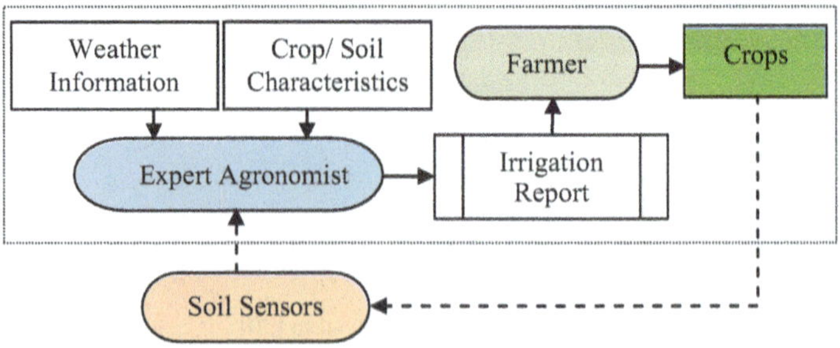

Fig. 8 Closed-loop irrigation advisor system [31]

the decisions were taken by expert agronomist on the basis of weather information and characteristics of soil/crop. The expert's knowledge is being used in their proposed system, which incorporates two Machine Learning algorithms along with expert knowledge to take irrigation decisions as shown in Fig. 9. They have developed their own wireless module (as shown in Fig. 10) for collecting information about soil/crop from a lemon field. The device is equipped with GPRS/GSM modem for accessing it from anywhere. The advantage of these modules is that they are powered with a solar power system and are capable of providing sufficient information for decision support systems. The collected information is being stored in the databases for making irrigation decisions along with previous knowledge of experts using Machine Learning algorithms.

Weather stations, on the other hand, are sending the following information:

Relative humidity (RH), temperature (T), rainfall (RF), wind speed (WS), vapor pressure deficit (VPD), dew point (DP), global radiation (GR). This information

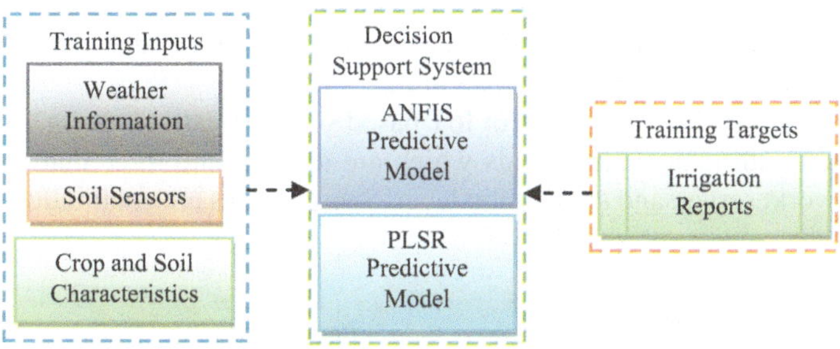

Fig. 9 SIDSS with training inputs and targets [31]

Fig. 10 Wireless information collection device with GPRS/GSM modem installed in the southeast of Spain for the lemon crop by Navarro-hellín et al. [31]

is made available to the public on the SIAM website [32]. The goal of SIDSS is to take final decisions on the basis of information collected from different sensors and expert's knowledge regarding water requirements or time requirements for constant water irrigation. They have incorporated two Machine Learning techniques, viz. partial least square regression (PLSR) and adaptive neuro-fuzzy inference system (ANFIS). PLSR is a statistical model which is needed for pursuing knowledge between predictor and response variables, whereas ANFIS is capable of generating fuzzy inference rules on the basis of given input/output dataset. The system was tested to three different locations and found to be suitable for scheduling irrigation requirements on the basis of continuous monitoring of soil/crop characteristics along with expert's knowledge.

Recently, Goap et al. [6] developed an IoT-based solution for smart irrigation using open-source technologies. The system is fully functional which takes advantages of its integration with the cloud using web-based services. The system works on the principle of sensed values (i.e., soil moisture, ultraviolet radiations (UV), soil/air temperature, and humidity) of from sensor nodes. The proposed system is intelligent enough in providing real-time irrigation decisions and web-based visualizations. The block diagram of their system is shown in Fig. 11.

Figure 11 shows the architecture of IoT-based smart irrigation system. The system is embedded with many components which are divided into three main units, viz. sensor network, server, and application layer. Sensor nodes are continuously sensing

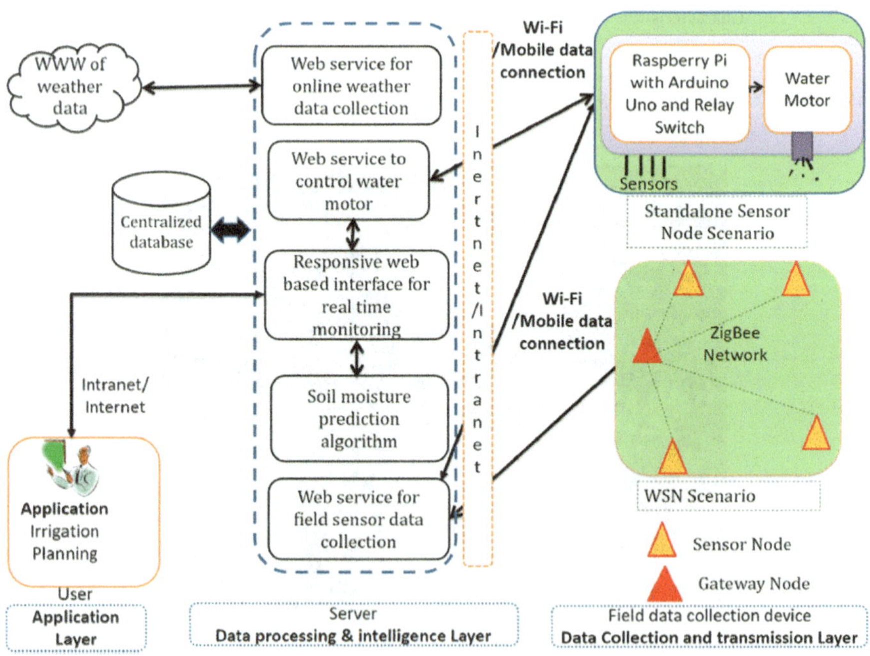

Fig. 11 The architecture of IoT-based smart irrigation system using open-source technologies [6]

the required values from the fields and sending them to the data processing server for storing them to a centralized database system. The centralized database along and other web-based data are collectively used for making irrigation decisions on the basis of an intelligent algorithm. The algorithm makes use of both supervised and unsupervised techniques which gives better accuracies while minimizing the mean squared error for predicting the soil moisture level of upcoming days. Values obtained from sensor units are being supplied to support vector regression (SVR) (modified SVM) model for providing training. After getting training, SVR is used to make predictions on soil moisture differences (SMD). For improving the accuracy of predictions, K-means clustering is used. The final predicted values are used for making irrigation decisions. The system gave promising results while tested on real environment.

Gu et al. [33] presented a software-based solution called Root Zone Water Quality Model (RZWQM) for scheduling irrigation requirements. The system is able to identify irrigation requirements on the basis of crop computed on the basis of agricultural models because it seems harder to calculate this stress in real time. Irrigation timings are suggested on the basis of water stress (WS) calculations by model, whereas irrigation depth is calculated on the basis of soil moisture threshold. The model captures weather inputs from databases of weather websites (historical and four days ahead of data) and on-site sensors. Water requirements for irrigations are calculated on the basis of Eq. (2).

$$\mathrm{IR}_{(t0)} = K\left\{\sum_{(i=1)}^{N}\left(\theta_{(fci)} - \theta_{(t(0)i)}\right)\cdot D_{(i)}\right\} - P_{(t0+4d)} \quad (2)$$

where

$IR_{(t0)}$ is the required water supply for irrigation

i represents layers of soil

N represents the deepest rooting depth derived from crop biomass

$\theta_{(fci)}$ represents volumetric θ at the field capacity of ith layer of root

$(t_{(0)i})$ represents volumetric θ on the irrigation day at ith layer of root

K represents the proportion of irrigation losses which needs to be done again

$P_{(t0+4d)}$ represents expected accumulative rain and data of the next four days.

All existing irrigation calibrations are being removed and RZWQM is then recalibrated on the basis of calculated WS. GUI of RZWQM irrigation scheduling software is shown in Fig. 12. The system is able to save the water requirement with a trivial reduction (0.03–3.81%) in crop yield. Experimental results showed that the system is able to save water and maintain the crop yields.

Nawandar and Satpute [7] proposed an IoT and Neural Network-based low-cost smart irrigation system. IoT is used for connecting various devices, whereas Neural Networks are used for irrigation decision-making purposes.

The system is user-friendly in the nature that it provides various crop-related details like estimated irrigation requirements, soil statistics, zone-wise irrigation alert, and remote data monitoring through message queue telemetry transport

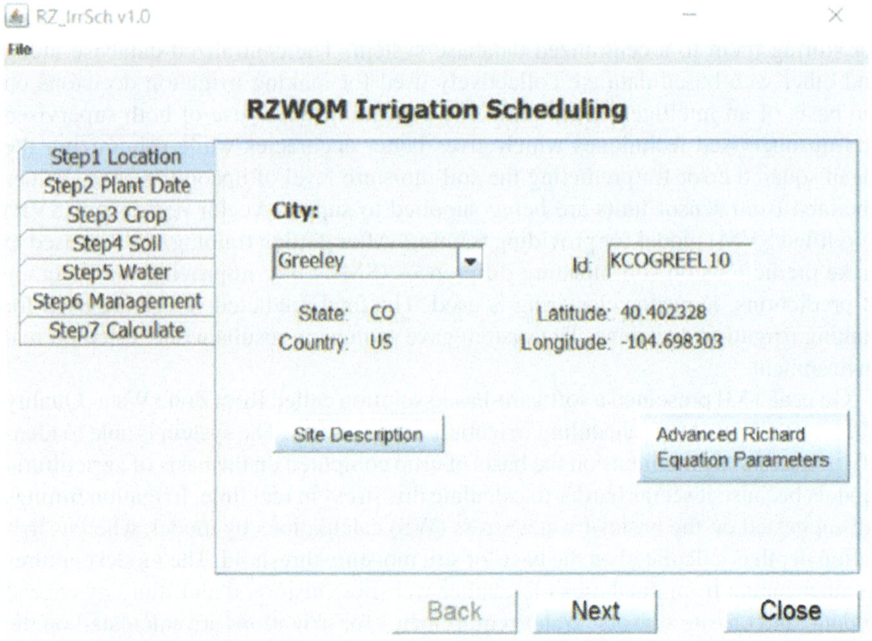

Fig. 12 The interface of RZWQM scheduling software [33]

(MQTT) protocol. The architecture of the proposed low-cost smart irrigation system is shown in Fig. 13.

The system consists of a unified sensor pole (USP) which is called as brain of this system. The purpose of including USP is to take intelligent decisions about irrigation process. USP senses real-time data (i.e., temperature, humidity, and soil moisture) from smart nodes and sends them to both irrigation unit (IU) and sensor information unit (SIU). SIU is capable of logging sensor data to website through 802.11x. A

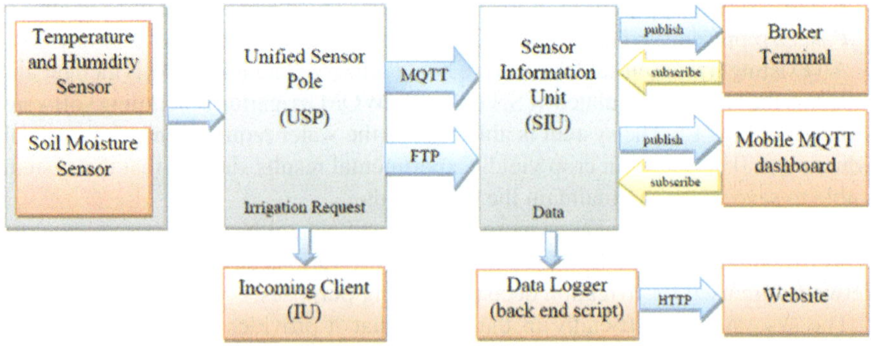

Fig. 13 Architecture of IoT and Neural Network-based low-cost smart irrigation system [7]

Fig. 14 Installation of IoT-based smart irrigation system [7]: **a** plant, **b** moisture sensor, **c** humidity and temperature sensor, **d** USP

HTTP server hosts these data to a webpage for remote monitoring. Installed module is shown in Fig. 14, which costs ≈$13. USP is responsible for providing sensed values at various regular intervals. The position of smart node is decided on the basis of water availability. Low water availability spaces cause massive logging of same data at regular intervals. So, the area which is having average availability of water is chosen for the installation of these sensors, because it is able to sense the water requirements of nearby areas. IU on the other hand is responsible for making intelligent decisions about irrigation requirements. USP has the capability to work in either admin or user mode. It works in admin mode until it times out. USP is connected with irrigation unit which instructs stepper motor via relay to start or stop.

MQTT broker sends sensor values through messages in an energy-efficient way. Examples of some famous MQTT brokers are Mosca, RabbitMQ, HiveMQ, Cloud-MQTT, etc. The principle of sending message is publish–subscribing mechanism, in which sensor nodes are publishing data to MQTT on topics. Subscribes on the other hand are end users, which use the published topics. For taking intelligent irrigation decisions, a Neural Network is provided training of real-time data as well as crop database inputs. The system showed promising results after installation. Figure 15 shows the crop growth after installation.

Fig. 15 Crop growth after installation of IoT-based low-cost smart irrigation system [7]

3.2 Architecture of Proposed Smart Irrigation Systems/Automated Irrigation Infrastructure

As discussed in the previous section, various smart irrigation systems have been proposed by many researchers which are overcoming the drawbacks of traditional irrigation systems (drip irrigation, direct pump/canal irrigation, sprinkler irrigation, tube well irrigation) that need continuous monitoring of field. Most of these smart irrigation systems are capable of automated irrigation of crop field on the basis of sensor values. Sensor nodes are the most essential part of these systems. Other important components include but not limited to: controller unit, wireless medium, GSM module, Arduino, display unit, relay, Raspberry Pi, water pump, sprinklers/drip modules, etc. The difference in these systems arises due to two things, viz. variety/capability of sensor nodes and their organization with other components. Temperature sensors are responsible for observing the amount of temperature adequate for a particular crop, whereas moisture sensors are capable of sensing level of moisture in soil. It is the responsibility of system designer to obtain the values from these sensors continuously or at some specified intervals. Some systems were found to be energy efficient in the sense that wireless nodes must not sense and send the information continuously; rather, they were calibrated to do these tasks at user-specified interval thresholds. A controlled unit is continuously observing these sensed values and enables water pump operations (On and Off). The positioning of sprinklers/drip modules should be mounted specifically according to the plant/crop and land structure. There is no general irrigation system available which can satisfy the irrigation requirement of all types of crops/fields, because the structure and irrigation needs are not same for all crops. Considering these versatilities in the field/crop, intelligent irrigation systems should be capable of making decisions on real-time sensed values as well as

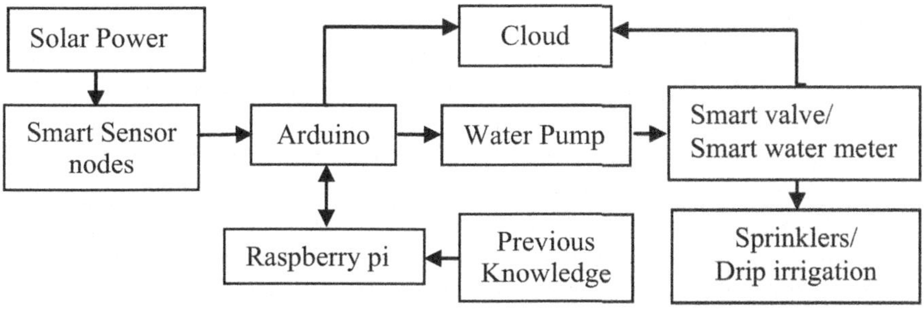

Fig. 16 The architecture of the proposed smart irrigation system CS-HYSIS

analyzing the previous knowledge. After observing various irrigation systems in the previous section, an architecture of a smart irrigation system named cloud and solar power enabled hybrid smart irrigation system (CS-HYSIS) has been proposed as shown in Fig. 16.

3.2.1 Working Principle

As shown in Fig. 16, the system contains smart sensor nodes which are provided power supply from solar panels. The choice of solar panels may vary according to the weather conditions (if the system is being installed in a cloudy/rainy environment, definitely advanced controllers for solar panels need to be installed). The sensor nodes contain three sensor units mentioned as follows:

- Temperature sensor
- Soil moisture sensor
- Humidity sensor

These smart sensor nodes are directly connected to the Arduino controller. The controller enables water pump to turn on or off on the basis of real-time sensor values and existing knowledge (trained inside a Neural Network) from Raspberry Pi. The concept of existing knowledge has been incorporated in this system to irrigate the plants during unavailability/failure of sensor nodes. Because sensor nodes may stop working due to various reasons, i.e., power failure, fault in a part of sensor, physically damaged by any animal(s), etc. This feature prevents plants from dehydration in case of failure of smart sensor nodes. Existing knowledge can be obtained from previous data or from agricultural expert. Pretrained network plays an important role in providing overall predictions about irrigation requirements. The controller is able to make decisions either on the basis of real-time values (obtained from smart sensor units)/pretrained network or from both. It analyzes both existing pieces of knowledge and real-time values in order to make appropriate decisions (turning on or off the water pump) on the basis of user-specified thresholds (i.e., moisture level, humidity level, and temperature).

The most important part of this system is that it is able to send all irrigation-related data to the cloud, for making analysis and future purposes. User can see the current and past irrigation usages anytime on a smart web-based or app-based portal. The interface may provide several additional features, i.e., start/stop irrigation, total water consumptions in a particular amount of time/area, real-time sensor values, etc. Smart valve/smart meter will count the units of water consumptions and send these details to the cloud. With the help of this methodology, we can make sense about total water consumption/requirements for a particular crop and monthly/weekly consumption of water. These values can be used for making future plans/decisions about water requirements.

The position of sprinklers and drip modules is calibrated manually according to the situations of different crops/fields. Nowadays, the sprinklers are so smart in the nature that they can itself irrigate the field smartly. Traditional sprinklers are not able to irrigate the fields properly. Wi-Fi enabled sprinklers provide the facility for controlling the sprinkler nozzles using smartphones. The direction of the nozzles and speed of water spraying can be controlled through smartphones. Even some smart sprinklers are able to make their own irrigation schedules as well. For more information on smart sprinklers, please refer to the review of latest sprinklers which is provided in Top 8 Best Smart Sprinkler Review [34]. The choice of sprinkler systems is also challenging for implementing smart irrigation systems in order to do optimized irrigation.

The proposed system is able to achieve precision agriculture by doing optimized irrigation and can provide several advantages as follows:

- Cost minimization:
  - Optimum utilization of energy (obtained through solar power systems
  - Optimum utilization of water (sprayed through smart sprinklers as per requirements).
- Increase in crop growth:
  - Achieving good health of plants by providing them optimized irrigation in order to prevent them from dehydration and excessive/irregular irrigation.
- Reduction in manpower requirements for irrigation.
- Providing complete irrigation data through cloud computing.
- Efficient utilization of existing knowledge through Artificial Intelligence.

4 Conclusion

Yield prediction and smart irrigation systems using AI-based techniques are demanding in precision agriculture. This chapter examines the critical parameters for predicting both crop yields and irrigation requirements in a systematic way. After carefully examining many existing systems, we have provided effective prototypes and solutions to meet the expectations of the current scenario. It has been observed that AI-based techniques play a major role along with other hardware components such as Raspberry Pi, soil moisture sensors, temperature sensors, humidity sensors, etc. in predicting the crop yields and irrigation requirements. If both yield prediction and smart irrigation systems are well equipped with AI-based technologies, then these can prove the effectiveness in minimizing the overall agricultural cost, increasing growth in the economy and minimizing wastage of essential resources such as water, energy. These techniques are effective in reducing human efforts and fasten the planning of agricultural practices. But minimizing the cost of system deployment and training of uneducated and non-techno-savvy farmers is still challenging.

References

1. Sehgal, S., et al.: Crop planning using stochastic visual optimization, 2017. IEEE Visual. Data Sci. VDS **2017**, 47–51 (2018)
2. Hochreiter, S., Schmidhuber, J.: Long short-term memory. Neural Comput. **9**(8), 1–32 (1997)
3. Jha, K., Doshi, A., Patel, P., Shah, M.: A comprehensive review on automation in agriculture using artificial intelligence. Artif. Intell. Agric. **2**, 1–12 (2019). https://linkinghub.elsevier.com/retrieve/pii/S2589721719300182
4. Arif, C., Mizoguchi, M., Mizoguchi, M., Doi, R.: Estimation of soil moisture in paddy field using artificial neural networks. Int. J. Adv. Res. Artif. Intell. **1**(1), 17–21 (2013)
5. Hinnell, A.C., et al.: Neuro-drip: estimation of subsurface wetting patterns for drip irrigation using neural networks. Irrig. Sci. **28**(6), 535–544 (2010)
6. Goap, A., Deepak Sharma, A.K., Shukla, Rama Krishna, C.: An IoT based smart irrigation management system using machine learning and open source technologies. Comput. Electron. Agric. **155**(May), 41–49 (2018). https://doi.org/10.1016/j.compag.2018.09.040
7. Nawandar, N.K., Satpute, V.R.: IoT based low cost and intelligent module for smart irrigation system. Comput. Electron. Agric. **162**, 979–90 (2019). https://doi.org/10.1016/j.compag.2019.05.027
8. Kamilaris, A., Prenafeta-Boldú, F.X.: Deep learning in agriculture: a survey. Comput. Electron. Agric. **147**, 70–90 (2018)
9. Lecun, Y., Bottou, L., Bengio, Y., Haffner, P.: Gradient-based learning applied to document recognition. Proc. IEEE **86**(11), 2278–2324 (1998)
10. Nevavuori, P., Narra, N., Lipping, T.: Crop yield prediction with deep convolutional neural networks. Comput. Electron. Agric. **163**(April), 104859 (2019). https://linkinghub.elsevier.com/retrieve/pii/S0168169919306842
11. Crane-Droesch, A. Machine learning methods for crop yield prediction and climate change impact assessment in agriculture. Environ. Res. Lett. **13**(11) (2018)
12. Pantazi, X.E., et al.: Wheat yield prediction using machine learning and advanced sensing techniques. Comput. Electron. Agric. **121**, 57–65 (2016). https://doi.org/10.1016/j.compag.2015.11.018
13. You, J., et al.: Deep gaussian process for crop yield prediction based on remote sensing data. 31st AAAI Conference on Artificial Intelligence, AAAI 2017: pp. 4559–4565 (2017)
14. McAllister, W., Osipychev, D., Davis, A., Chowdhary, G.: Agbots: weeding a field with a team of autonomous robots. Comput. Electron. Agric. **163**(September 2018), 104827 (2019). https://doi.org/10.1016/j.compag.2019.05.036
15. Chostner, B.: Smart Machines for Weed Control and Beyond (2017)

16. Chlingaryan, A., Sukkarieh, S., Whelan, B.: Machine learning approaches for crop yield prediction and nitrogen status estimation in precision agriculture: a review. Comput. Electron. Agric. **151**(May), 61–69 (2018). https://doi.org/10.1016/j.compag.2018.05.012
17. Xue, J., Su, B.: Significant remote sensing vegetation indices: a review of developments and applications. J. Sens (2017)
18. Liakos Konstantinos, G., et al.: Machine learning in agriculture: a review. Sensors (Switzerland) **18**(8), 1–29 (2018)
19. Wang, Q., Nuske, S., Bergerman, M., Singh, S.: Automated crop yield estimation for apple orchards. Exper. Robot. Springer Tracts Adv. Robot. **88**, 745–758 (2013). http://link.springer.com/10.1007/978-3-540-77457-0
20. Ramos, P.J., Prieto, F.A., Montoya, E.C., Oliveros, C.E.: Automatic fruit count on coffee branches using computer vision. Comput. Electron. Agric. **137**, 9–22 (2017). https://doi.org/10.1016/j.compag.2017.03.010
21. Senthilnath, J., et al.: Detection of tomatoes using spectral-spatial methods in remotely sensed rgb images captured by UAV. Biosys. Eng. **146**, 16–32 (2016). https://doi.org/10.1016/j.biosystemseng.2015.12.003
22. Amatya, S., et al.: Detection of cherry tree branches with full foliage in planar architecture for automated sweet-cherry harvesting. Biosys. Eng. **146**, 3–15 (2015). https://doi.org/10.1016/j.biosystemseng.2015.10.003
23. Sengupta, S., Lee, W.S.: Identification and determination of the number of immature green citrus fruit in a canopy under different ambient light conditions. Biosyst. Eng. **117**(C), 51–61 (2014). http://dx.doi.org/10.1016/j.biosystemseng.2013.07.007
24. NASA. Remote Sensors. EARTHDATA powered by EOSDIS. https://earthdata.nasa.gov/learn/remote-sensors, 9 Aug 2019
25. Wolfert, S., Ge, L., Verdouw, C., Bogaardt, M.: Big data in smart farming—a review. Agric. Syst. **153**, 69–80 (2017). https://doi.org/10.1016/j.agsy.2017.01.023
26. Kamilaris, A., Kartakoullis, A., Prenafeta-Boldú, F.X.: A review on the practice of big data analysis in agriculture. Comput. Electron. Agric. **143**(September), 23–37 (2017). https://doi.org/10.1016/j.compag.2017.09.037
27. Harishankar, S., Umasankar, V.: Solar Powered Smart Irrigation System (September 2015)
28. Kehui, X., Xiao D., Luo X.: Smart water-saving irrigation system in precision agriculture based on wireless sensor network, pp. 170–75 (2010)
29. Vellidis, G., et al.: A real-time wireless smart sensor array for scheduling irrigation. Comput. Electron. Agric. **61**, 44–50 (2008)
30. Nasiakou, A., Vavalis, M., Zimeris, D.: Smart energy for smart irrigation. Comput. Electron. Agric. **129**, 74–83 (2016). https://doi.org/10.1016/j.compag.2016.09.008
31. Navarro-hellín, H., et al.: A decision support system for managing irrigation in agriculture. Comput. Electron. Agric. **124**, 121–131 (2016)
32. SIAM: agricultural information system of Murcia (2019). http://siam.imida.es/apex/f?p=101:1:138900487927405
33. Gu, Z., et al.: Development of an irrigation scheduling software based on model predicted crop water stress. Comput. Electron. Agric. **143**(January), 208–221 (2017). https://doi.org/10.1016/j.compag.2017.10.023
34. Top 8 Best Smart Sprinkler Review (2019). https://www.helpadvisors.org/best-smart-sprinkler-systems/, 22 Aug 2019
35. Microsoft Cloud Perspectives (2018). Feeding the World with AI-driven Agriculture Innovation

IoT-Based Smart Tree Management Solution for Green Cities

Bilal Shabandri, Sahith Reddy Madara and Piyush Maheshwari

Abstract This paper discusses key challenges faced by transformation of smart cities and integration of green technologies, initiatives, and existing systems that are in place to solve for these challenges. With overpopulation, fast urbanization, and environmental changes, powerful administration of trees and plants is required like never before. We present an Internet of things (IoT)-based smart tree management architecture for cities which is able to track various characteristics of a given tree such as the air quality, sunlight level, sound pollution level, and other important characteristics for efficient city planning, afforestation plans, and higher quality produce. We divide the solution into three sections: The Sensor Node, The Cloud Integration, and User Experience. We have developed a working prototype using a Raspberry Pi and a mobile application consisting of an intuitive user interface that lets users and farm owners get information about a certain tree and view its information and alerts. We utilize low-power wide-area network modules for the connectivity of the various sensors to the cloud. A map view is also developed which consists of various filters that let city authorities understand the distribution of healthy and damaged trees across the city accompanied by an extensive customized dashboard. A prototype of a green smart tree management system based on the Internet of things is discussed and implemented. The paper describes the functions and actual characteristics of each system component in detail. The experimental results suggest the feasibility of the application to enable the development of smart cities across the world which will play a key role in decreasing pollution levels across the city and increasing afforestation

B. Shabandri
Department of Electronics and Telecommunication,
Amity University Dubai, Dubai, United Arab Emirates
e-mail: bilal.fas@gmail.com

S. R. Madara (✉)
Sharjah Academy for Astronomy, Space Sciences and Technology,
University of Sharjah, Sharjah, United Arab Emirates
e-mail: smadara@sharjah.ac.ae

P. Maheshwari
Department of Computer Science and Engineering,
Amity University Dubai, Dubai, United Arab Emirates
e-mail: pmaheshwari@amityuniversity.ae

P. K. Pattnaik et al. (eds.), *Internet of Things and Analytics for Agriculture, Volume 2*, Studies in Big Data 67,
https://doi.org/10.1007/978-981-15-0663-5_9

in areas that require it the most. This solution can be useful in spreading awareness among citizens on the importance of afforestation and proper management of trees in the city and can help farmers in rural areas to avoid hours of manual inspection and labor spent in inspecting the health of their farms.

Keywords IoT · Tree management · Cloud computing · Green city · Smart tree system

1 Introduction

A city can be classified as smart and green only through vital arrangement of information and communication technology foundations and administration to accomplish maintainability strategy goals. However, this requires that trees must be included. Not only do trees just constitute green space valuable to differentiate urban contamination impacts or give ecosystemic advantages to inhabitants, but also they can likewise be utilized as bioindicators and their association in correspondence systems can symbolize to a huge commitment to construct/build a keen, green city. Smart tree management utilizes sensors to screen the strength of trees over the city to monitor the air contamination and advance planting of more trees with the assistance of web and mobile application. With a quickly evolving atmosphere, scene, and city framework because of human maneuvers, successful administration of plants is required like never before [1]. Advancements, for example, radio-frequency identification and remote wireless systems can interchange objects to residents as well as trees. The smart city idea can be viewed as an expansion of smart gadgets that discourse singular needs and smart frameworks that address the necessities of a household/gathering of building clients. Greening metropolitan spaces is a key movement that practically each management seeks after [2].

The protection of environment is an originator to a sustainable, adjusted, and comprehensive development. The responsibility to ensure a high caliber and prosperous life for the future age cannot be met unless the present pressing environmental issues are tended to in an appropriate and auspicious way. We believe that a connected smart management solution for trees over the city is a standout among the most naturally cordial approaches to enhance the air quality and biology in urban zones, in this way empowering urban areas to have much lower carbon impressions. This smart tree can be utilized for the tourists and residents. An augmented reality encounter lets users simply point to a specific tree and view its health and the air quality details. There is additionally an option to alert the authorities if a specific tree is obstructing the street or causing inconvenience. With a connected tree framework, a campaign for boosting green propensities can be remunerated. Moreover, users can assert different sorts of saplings to upgrade greenery and cultivate an adoration for nature in the psyches of new ages. Consequently, users can examine specific trees with their mobile applications and win tree saplings when they answer a couple of trivia questions concerning the area around a tree. The need for this solution stems from the continuous temperature rise caused by a dangerous atmospheric deviation. This adds to the water deficiency of the United Arab Emirates, due to dry season, rising sea level, and aridity. Ecological issues in the UAE are caused by the misuse of characteristic assets, quick populace development, and high energy necessities. The farmland of the United Arab Emirates, portrayed with its enormous dried land, accidental precipitation, and high temperatures are as of now getting to be bone-dry [3, 4].

In the recent years, the number of deaths due to respiratory diseases has become significant. A major contributor to this statistic is due to abundant unmonitored pollution across urban cities. Figure 1 obtained from the UAE Ministry of Health shows the main causes of deaths in the United Arab Emirates. Respiratory diseases rank extremely high and are responsible for majority of deaths in the country [5]. Globally, epidemics caused by contamination were responsible in 2015 for an expected 9 M premature deaths—16 percent of all casualties around the world, which is three times greater number of deaths than AIDS, tuberculosis, and intestinal sickness combined, and fifteen times more than all wars and different types of violence. It kills a greater

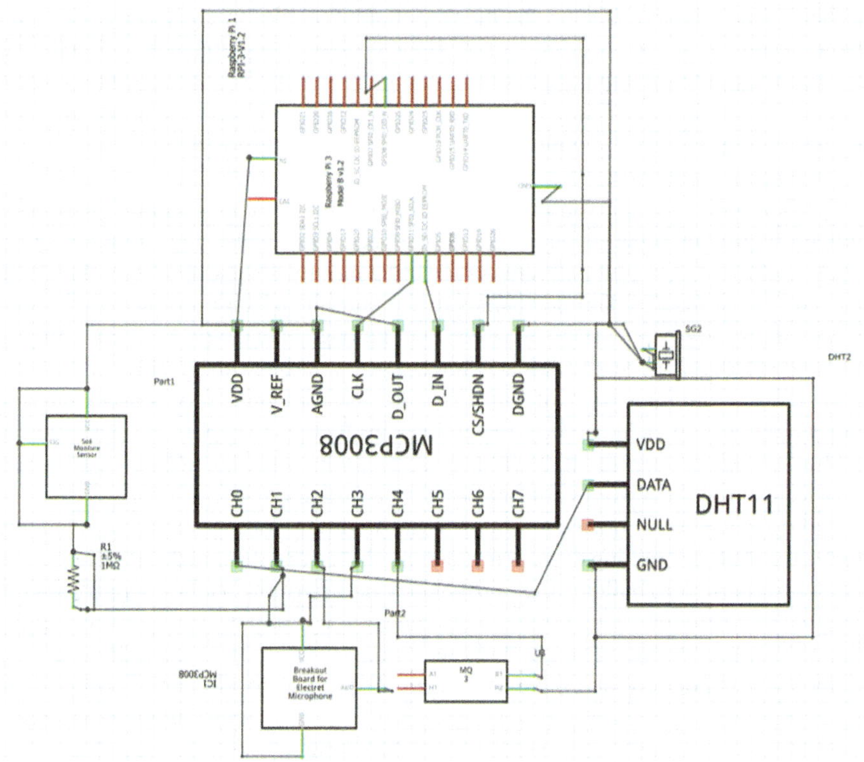

Fig. 1 Circuit diagram and components of prototype

number of individuals than smoking, poverty, and natural disasters. In a few nations, it accounts for one out of four demises [6].

The rest of the paper is organized as follows. Section 2 describes the related work carried out in the field of IoT and green cities. Section 3 discusses the prototype implementation and is subdivided further into the hardware components and mobile application. Section 4 explains the system architecture and details the implementation of such a system on an industrial scale along with test results and screens from the user interface of the mobile application. Finally, Section 5 presents our conclusion and future prospects with the project.

2 Related Work

Before delving into the points of interest of our Internet of things-based smart tree management framework, we present an audit of a portion of the current framework in vogue relating to smart tree system. Enormous research has been conducted in the improvement of models in the extensive rural area. Applying models on farms has been generally sporadic despite the potential favorable circumstances that could result. In spite of the fact that Internet of things will be an essential innovation for the smart cultivation itself, their utilization in conjunction with future Internet (FI) offers a reason for another age of Farm Management Information Systems empowering keen ranches end up plainly dynamic hubs in business to business (B2B) arrangements and farming worth chains [7].

Conventional garbage administration utilizing human is not extremely powerful with any legitimate supervision. Research has been done in utilizing IoT-based innovation in checking the status of canister toward gathering the garbage once threshold reached. There were certain downsides in the current IoT-based framework which prompted the improvement of IoT-based smart garbage management system utilizing ultrasonic sensors. The venture created toward savvy garbage management got parcel of scope for future improvement. One of the key parts of smart city is smart garbage management which includes checking garbage in city toward expulsion appropriately and furthermore arranging off. There is no appropriate supervision from corporation toward checking the movement of garbage accumulation [8].

Radio-frequency identification innovation represents one of the most standouts among the most fascinating devices in machine-driven and designed parts. It signifies another framework to build process proficiency in supply chains. While picking a tag for a tree radio-frequency identification framework, it is important to consider different perspectives, including size and shape, term, protection from outer physical and compound components, and separation from the per-user/reader, breakdowns close metals and fluids, satisfaction of nearby directions, and memory limit. When managing tiny trees, for example, those commonly found in nurseries, radio-frequency identification label imbedding inside organs of little size requires particular strategies, an echo in regard to labels imbedding in Citrus spp. at nursery arrange is a case of a plant labeling system that was dependable, rock-solid, and secure. Regardless of whether not straightforwardly related to plants, radio-frequency identification labels could be executed in biomonitoring frameworks with a specific end goal to ensure a constant information correspondence [9].

Soils are vital to the capacity of every single imaginable biological system and to nourishment and fiber generation. An ignored part of soils is their capability to moderate ozone-depleting substance discharges. Albeit demonstrated practices exist, the usage of soil-based ozone harming substance alleviation exercises is at a beginning time and precisely evaluating discharges. This will encourage the execution of atmosphere smart soil administration arrangements, through top and trade frameworks, item inventory network activities for low-carbon customer items, and national and worldwide greenhouse gas moderation approaches; it will likewise advance more economical and atmosphere strong agrarian frameworks [10].

Previous work in the field has been done such as the work of Chua [11], which provides a tree MIS used to improve the efficiency of specialist consultants such as trained company personnel. It is geared more toward areas with high pedestrian and traffic flows, such as crowded streets and it focuses on risk mitigation due to a collapse of trees in urban areas. Our work is geared toward easing the work of farmers in rural areas and we achieve this by employing additional sensors that give a more holistic picture of a particular tree in the farm. As the worldwide population rises, progressive decrease of energy sources and their resulting higher cost, environmental change and air contamination are a portion of the primary issues that the urban areas without limits should adapt to survive, changing into smart cities and concentrating on green information technology and smart mobility.

Trees and individuals can be present together in urbanized spaces and can be beneficial to each other. Trees can control microclimates, for example, air temperature and humidity diminish the power of bright light through screening and deliver an abnormal state of oxygen while engrossing carbon dioxide during the day. The development of trees is compelled by limited planting and confined headroom or is influenced by development activities. In this manner, individuals have a commitment to enable trees to develop in urbanized spaces by, for instance, checking their status and actualizing well-being therapeutic activities [12].

3 Prototype Implementation

3.1 *Hardware Components*

The proposed method consists of various sensors which are connected to the Raspberry Pi 3. For this prototype, we show a Pi connected to the Internet via an Arduino Shield consisting of narrowband IoT (NB-IoT) [13] and LTE Cat-M1, category M1 (LTE-M) [14] modules for connectivity. The NB-IoT is a development of the LTE framework and works with a transporter data transfer capacity of 180 kHz [13–15].

Figure 1 consists of various modules and sensors used for the prototyping device. It includes the DHT11 temperature and humidity sensor, MQ135 gas sensor, YL-69 moisture sensor, LDR, SEN-09198 piezo vibration sensor, and an MCP3008 for 8 channel analog input as the Raspberry Pi does not directly support analog inputs.

(a) **Raspberry Pi 3 Model B**—The Raspberry PI3 Model B+ retains the same mechanical footprint as its predecessors while improving in performance. Its 1.4 Ghz quad core 64 bits processor coupled with 1 GB of RAM ensures unprecedented speed. It has Wi-Fi 2.4 and 5 Ghz wireless connectivity as well as Bluetooth 4.2. It comes with a faster RJ45 plug, four USB ports and a Micro SD card slot in a compact format.
(b) **Sound Sensor**—To detect sound pollution in the vicinity, a module is used to identify the point in time in which the sound has crossed a selected threshold. Sound is detected via a microphone and fed into an LM393 op-amp. This threshold is set using the in-built variable resistor. To display the crossing of the limit, an LED on the module illuminates and the resulting output is set to low.
(c) **Piezo Vibration Sensor**—To detect if tree has sudden movements indicating impacts, a piezoelectric sensor (7BB-20-6L0) is used. It works as a transducer by measuring changes in weight, speeding up, temperature, strain, or power and changing over them to an electrical charge.
(d) **LDR**—To detect if the tree is receiving enough sunlight, a light-dependent resistor (LDR) is used. An LDR is sometimes called a photoresistor or a photodetector as it changes resistance based on the amount of light that falls on its surface. It is produced using semiconductor, cadmium sulfide (Cds), or cadmium selenide (CdSe). These two substances are semiconductors covered in a clay sheet as a base.
(e) **Temperature Sensor**—To ensure that adequate water is reaching the tree, DHT22 is used as a sensor for measuring temperature and humidity. The estimation unit will be exhibited in a computerized flag/signal form.
(f) **Air Quality Sensor**—To detect air quality and pollution levels around the tree, the module fills in as an air quality detection gas sensor, this is delicate to gas unsafe to human, connected to measure NO_x, alchohol, NH_3, benzene, CO, CO_2, etc. The module is additionally utilized for controlling climate conditions and air cleaners in building structures. The estimation/measurement unit is introduced as an analog flag/signal. In the present work, three sensor shapes were utilized, comprising of MQ-2, MQ-135, and MQ-136 due to its gas measurement contrasts.
(g) **Moisture Sensor**—To detect the soil moisture level, the SEN13322 is used. It has two large/expansive uncovered cushions which work as probes for the sensor, together acting as a variable resistor. The more water that is in the dirt/soil means the better conductivity between the cushions will be and will bring about a lower opposition/resistance, and a higher SIG out.
(h) **Carbon Sensor**—To ensure that the tree is intaking carbon dioxide ensuring tree's health.

Combining all the sensors, connectivity. Figure 2 depicts a test setup of the smart

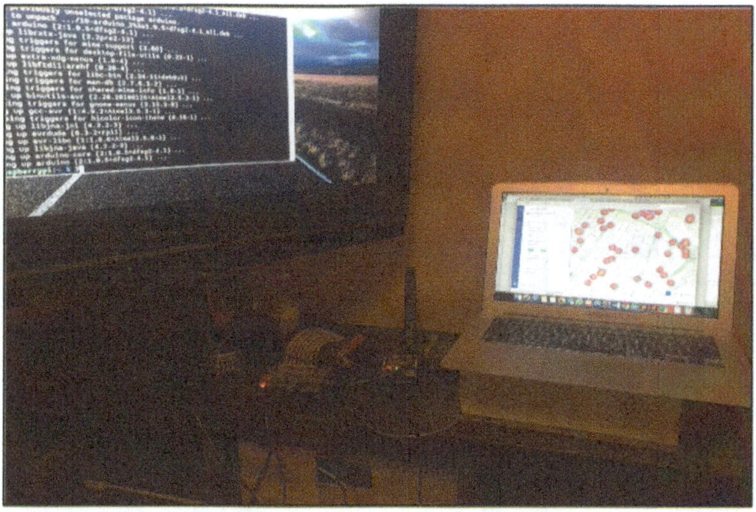

Fig. 2 Test set up of the IoT monitoring device

tree monitoring device connected with the various sensors sending data to the cloud platform through the NB-IoT shield to create a map view of the pollution points across the city.

3.2 *Mobile Application*

A mobile application has been developed for the user which consists of three main features or tabs, namely—'Stats,' 'Schedule,' and 'Support.' In the Stats tab, Fig. 3a shows the dashboard of the mobile application with a custom user interface giving options to monitor the water, air, sunlight, or overall health of the farm. Figure 3b shows intelligent insights and predictive analytics based on various parameters collected from the data from the individual trees. Figure 2c shows an in-depth view of a selected tree in the farm and displays useful information such as alerts, monthly reports, location, and infections.

The Schedule tab allows the farmer or the user to set reminders on his phone to visit a specific tree so that he can water it, harvest it, or apply medication to it. The Support tab allows the user to get in touch with the equipment providers through a live chat in case a particular device outputs incorrect/inaccurate value.

Smart trees can likewise be used for city authorities. An outline of all the associated trees is given on an adaptable guide that gives clients a chance to channel considering air quality, populace density, water level, light level, and different parameters. This information can enable the specialists to draft afforestation intends to quicken greenery over the city. There is additionally a chatbot for the city specialists to screen

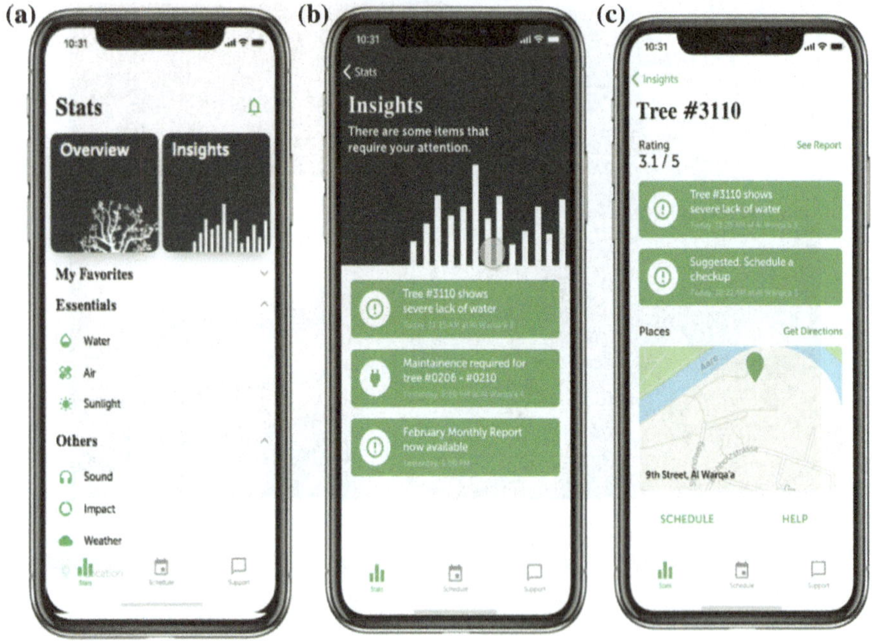

Fig. 3 Mobile application: **a** dashboard screen, **b** intelligent insights screen, **c** detailed tree-specific insights

and team up more effectively. Moreover, there is an Amazon Alexa application to discover the health of trees at a specific area with the use of voice. This information can enable the authorities to draft afforestation plans to quicken greenery over the city. We are using trees to reduce air pollution as well as sound pollution which protects the overall health of the population; on top of everything, we are monitoring the trees when they are about to fall through movement sensor, to avoid hazard accidents where it might hit a car or injure a person.

4 System Architecture

Now, we present a smart tree management (STM) solution based on Raspberry Pi 3, LPWAN modules and sensors. Figure 3 shows the system architecture of STM which consists of a Raspberry Pi 3, carbon sensor, temperature sensor, humidity sensor, light-dependent resistor (LDR), air quality sensor, piezo vibration sensor, and sound sensor.

Figure 4 represents the distributed implementation system for the smart tree management solution. Each tree is equipped with NB-IoT and LTE-M sensors that communicate directly with the IoT cloud platform. The issue with traditionally creating

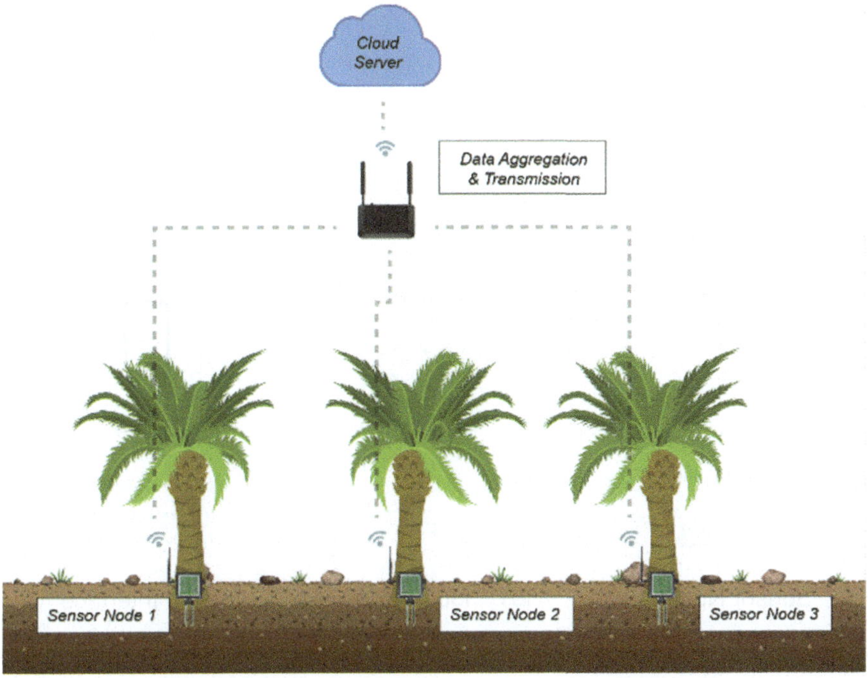

Fig. 4 Distributed implementation of monitoring devices

multiple gateways for assets was that devices that can connect to 4G are quite expensive in an industrialized IoT use case with thousands of sensors constantly sending and receiving data. Moreover, the process of connecting via Wi-Fi or a gateway is quite complicated. With the help of LTE-M and NB-IoT solutions [16], such low-density sensors and devices can communicate directly with the cloud and provide for a plug-and-play solution that allows for greater scalability of the system.

Figure 5 shows the block diagram and working of the entire solution. Data flows from thin devices directly to the IoT cloud server where information is analyzed and segregated. It is then transmitted to a mobile application through the IBM Cloud [17] API's with an intuitive and clean UI for users to take action. The working of the system is divided into three parts: (A) Sensor Node, (B) IoT Cloud Server, and (C) User Interface.

4.1 Sensor Node

This consists of the industrial device that measures various essential parameters about the health of a tree. This sensing is done with the help of an onboard MCU that

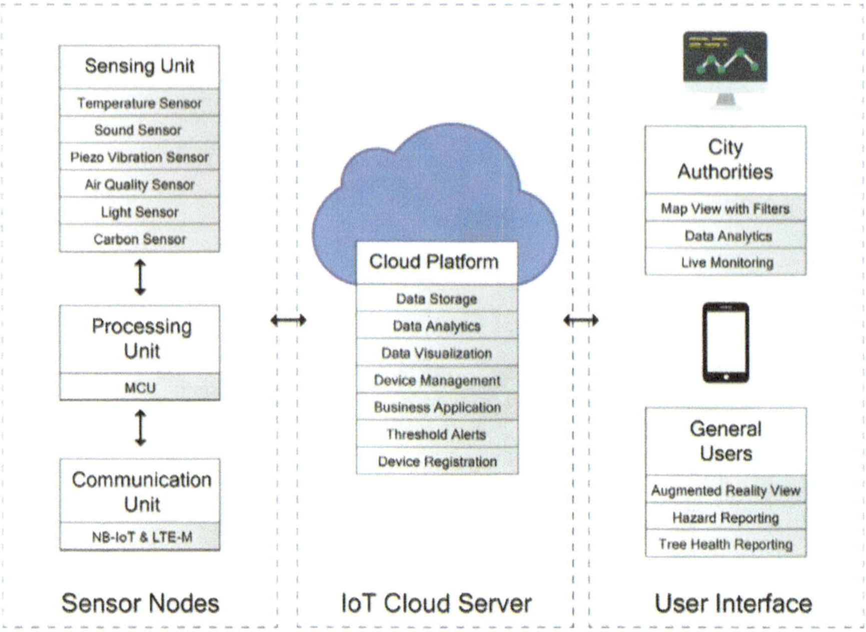

Fig. 5 Block diagram of an industrialized system

performs necessary calculations, regularizations, and data cleaning. This data is then sent to the IoT cloud platform directly through the LPWAN communication unit.

- **Sensing Unit**: The custom industrialized monitoring device consists of environmental sensors and NB-IoT/LTE-M modules. The lower portion of the device consists of a moisture sensor that is pushed into the ground to help monitor the amount of water received by a specific tree. The sensor is also connected to LED indicators that help to adjust the sensitivity of detection so that each device can be manually calibrated to the specific amount of water required for being detected as 'watered.' The devices are designed to be modular and 'plug-and-play' compatible with a variety of types of trees. The device is powered by batteries and can have a battery life of up to 10 years.
- **Processing Unit**: LPWAN applications are best suited for devices with non-critical data with low processing power. Thus, a device with minimal processing power is sufficient for this particular use case. The unit receives the values from all the sensors and stores them in the appropriate variables and prepares for Message Queuing Telemetry Transport (MQTT) which is used for transfer of the data to the cloud [18].
- **Communication Unit**: The device communicates via the MQTT protocol with the help of subscribe and publish keys.

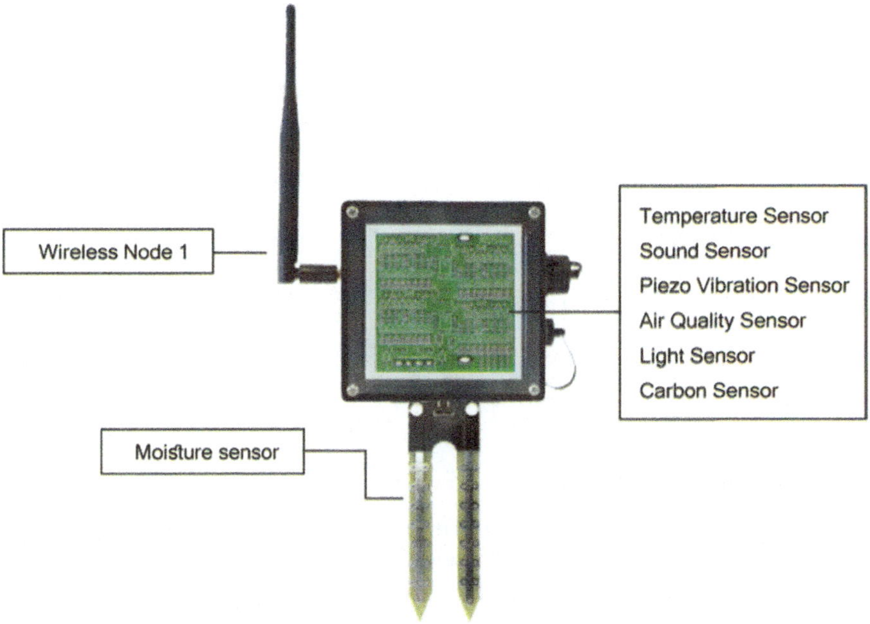

Fig. 6 Prototype unit consisting of various sensors

- The interval of packet transmission can be adjusted in the device management platform so that battery life can be optimized. The GNSS capability in the NB-IoT/LTE-M module provides autonomous geospatial positioning with global coverage. During installation, the devices require a one-time setup for initializing the name of the tree, setting the exact coordinates, and adjusting the sensitivity of various parameters which is calibrated by a trained professional (Fig. 6).

4.2 IoT Cloud Server

The process of cloud enablement includes device authentication and identity, setting up device templates, device dashboards, alert monitoring, and managing the database. This can be achieved by setting up three components:

- Node-RED application
- Internet of things platform service
- Cloudant NoSQL database

Figure 7 describes a high-level overview of the system and lists the various sensors required for connectivity along with the various cloud services and frameworks used in the implementation of the system.

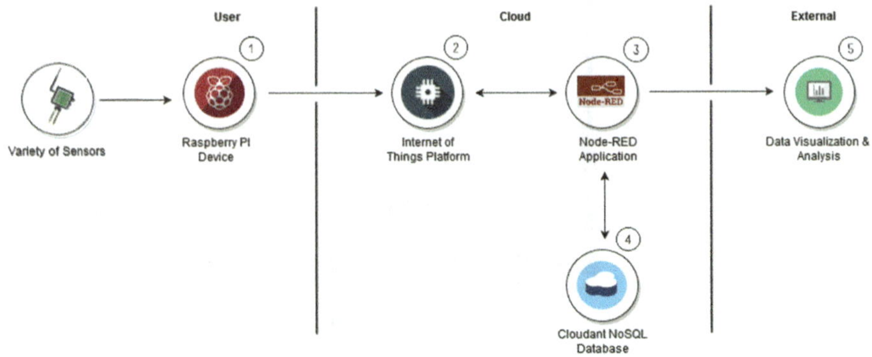

Fig. 7 Overview of prototype implementation

(1) Setting up Device Identity

For the prototype, we used Node-RED on the IBM Watson IoT platform which allows for the management of multiple IoT devices. Custom function nodes are used for parsing the incoming data streams and to ensure that messages are not lost if there is a huge influx of data. In the device information section (Fig. 8), we configure detailed information regarding the device identity such as serial number, hardware version, manufacturer, firmware version, etc. This metadata can be useful for device updates and debugging purposes. After this, we define an authentication token that is either system generated or manually defined by us.

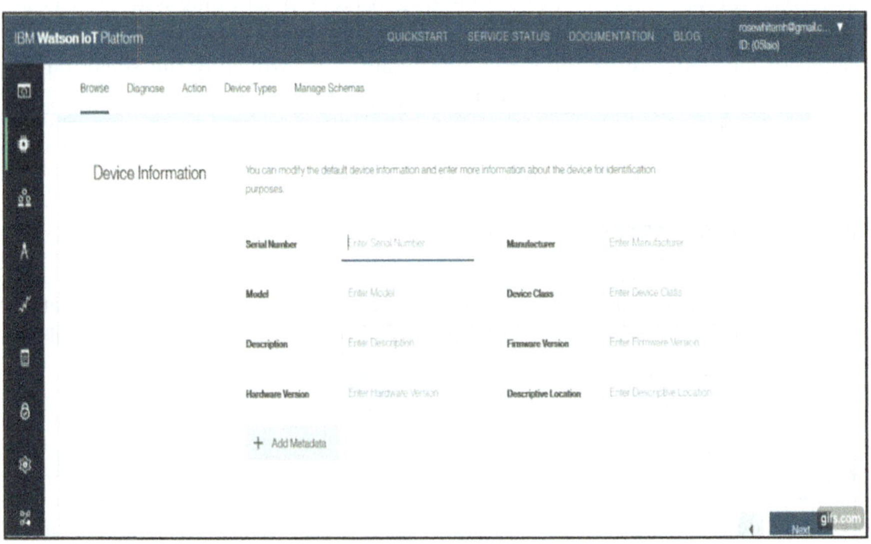

Fig. 8 Updating device information and metadata

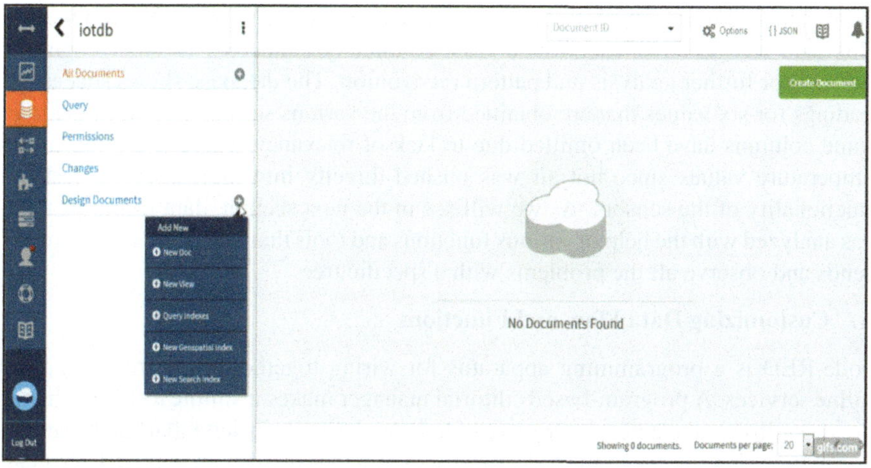

Fig. 9 Cloudant NoSQL DB for data storage and analysis

(2) **Connecting to the Database**

As shown in Fig. 9, IBM Cloudant is a NoSQL JSON record store that is upgraded/optimized for handling of overwhelming workloads of simultaneous reads and writes in the cloud; a workload that is regular of expansive, quickly developing web and mobile applications. You can utilize Cloudant as a completely managed DBaaS running on IBM Cloud.

For the database, we use a Cloudant NoSQL DB for all our data storage and communication with the Watson IoT platform. One way to ensure that we define the categories of data we are using from the database is by defining functions, which can be considered as a form of a special document called 'design document.' We created a design document with a simple view, which is a blueprint for how our functions are defined and how the data will be categorized. This view grabs all documents with the key corresponding to the timestamp defined for each tree/device and the value corresponding to the entire document object. Instead of the timestamp, a unique ID can also be generated for each data transfer request if the application requires.

4.3 Experimental Results

We used Python to establish an MQTT connection between the Raspberry Pi and the Watson IoT platform. This can be achieved by downloading the ibmiotf library and entering various credentials details such as deviceID, orgID, authentication token, etc. This is followed by initializing the device client and connecting and sending a data point with a value to the cloud as an event. Appropriate delays can be introduced between message publish requests as required. Finally, this data is sent to a real-time

IoT database (Fig. 5) where each message is stamped with a unique ID or timestamp. This data can be downloaded as a CSV or directly connected to other cognitive services for further analysis and pattern recognition. The database shows successful readings for six values that are obtained from the various sensors placed in the tree. Some columns have been omitted due to lack of relevancy. There is a jump in the temperature values since hot air was pushed directly into the sensors to test the functionality of the sensors. As we will see in the next section, data from this table was analyzed with the help of various functions and tools that make it easier to predict trends and observe all the problems with a specific tree.

(a) **Customizing Data Flow and Functions**

Node-RED is a programming apparatus for wiring together hardware, APIs, and online services. A program-based editorial manager makes it simple to wire together streams utilizing the extensive variety of hubs/nodes in the palette that can be sent to its run time in a single click. Node-RED also provides us with an app URL that can be used to display the front end in an active URL that can be accessed remotely on any device. The IBM IoT node allows us to consume any data received by the IoT platform service. The connection to the Cloudant database is also done at this node. The data is then sent over to individual functions that contain single parameters regarding the health of the tree, which is passed on to individual nodes that are responsible for generating visual displays of the data in various forms such as line charts, area graphs, pie charts, etc. (Fig. 10).

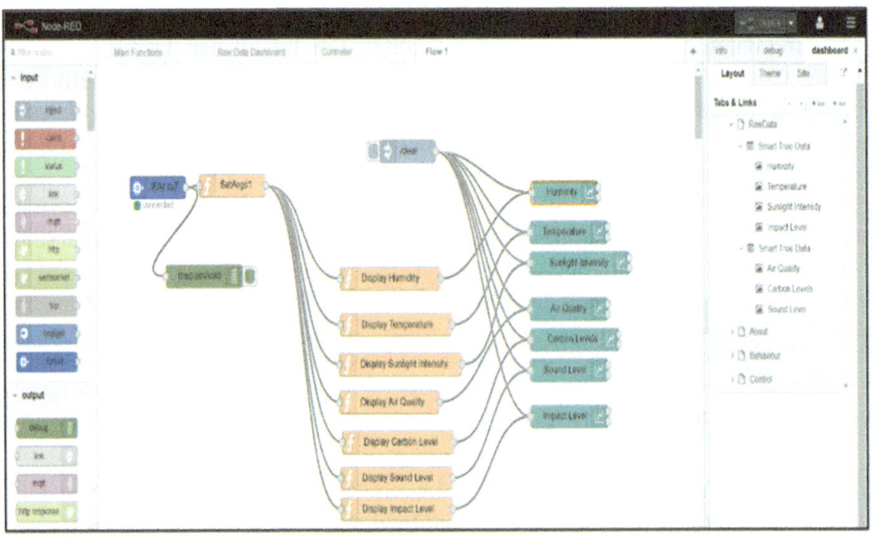

Fig. 10 Node-RED flow for dashboard view

4.4 User Interface

(a) **Mobile Application**

The mobile app interfaces with the IBM Bluemix platform by invoking REST API's with IBM API Connect used to discover APIs, test them, and get code snippets to make it easier to call them from the mobile apps.

- **Augmented Reality View**: Our mobile application provides an interface for the user where they can access an augmented reality view of smart trees in the public. They must simply point their smartphone cameras to a tree, and they can get information about the health of the tree for instance the amount of light it is receiving, the moisture content, the air quality around it, and many other characteristics of the tree (Table 1).
- **Hazard Reporting**: In case of a scenario where a certain tree is blocking the path of vehicles or posing a threat to the pedestrians, the users have an option of simply selecting the particular tree and reporting it as hazardous. This data is then verified by the relevant local authorities and teams are quickly dispatched to solve the problem.
- **Tree Health Alerts**: In addition to hazard reporting, users can also report trees that look unhealthy—i.e., they are not watered regularly, they are affected by bugs, or if there are polluting sources nearby. These reports are sent to relevant authorities so that they can take the required action to solve the problem. The heat map of air pollution levels across the city is shown in Fig. 11.
- **Community Integration**: Along with reporting, various campaigns can be launched to encourage people to plant more trees. For instance, using the augmented reality interface, every week a contest can be held where community members have to take pictures of particular trees across the city to win saplings that they can plant. Similarly, other interactive activities can create a closely knit community that contributes to afforestation in the city.

(b) **Web Application**

The web application, as shown in Fig. 12, is meant as a control panel or dashboard for city authorities or officials for monitoring tree hotspots in the city. This consists of an overall map view of the city as well as a data analytics and monitoring section. Unlike the mobile app, access to this application is restricted to certain users who are provided with historical and real-time datasets of smart trees across the city.

- **Map View with Filters**: The map view consists of the culmination of all connected trees across the city. This is displayed in the form of a heat map based on all the measured parameters. With a simple toggle, city authorities in the city can detect areas with maximum pollution, areas with the least trees, and regions with the most complaints. Thus, individual trees can be monitored, and a variety of antipollution campaigns can be launched to suggest the most appropriate area for tree plantation cultivation.

Table 1 Real-time sensor values

S. No.	Humidity	Temperature	Intensity	Air quality	Carbon level	Sound level	Impact level
1	39	29	761	221.8994142	32.11447715	1.841570345	25.87355163
2	39	29	761	221.8994142	32.11447715	1.841570345	25.87355163
3	39	29	761	221.8994142	32.11447715	1.841570345	25.87355163
4	39	29	761	221.8994142	32.11447715	1.841570345	25.87355163
5	39	29	761	221.8994142	32.11447715	1.841570345	25.87355163
6	64	40	750	232.7778636	53.60739733	1.542790012	8.215529875
7	64	40	750	232.7778636	53.60739733	1.542790012	8.215529875
8	64	40	750	232.7778636	53.60739733	1.542790012	8.215529875
9	64	40	750	232.7778636	53.60739733	1.542790012	8.215529875
10	64	40	750	232.7778636	53.60739733	1.542790012	8.215529875
11	64	40	750	232.7778636	53.60739733	1.542790012	8.215529875
12	64	40	750	232.7778636	53.60739733	1.542790012	8.215529875
13	64	40	750	232.7778636	53.60739733	1.542790012	8.215529875
14	30	30	750	259.3853959	111.6718109	1.530899469	9.285961972
15	30	30	750	259.3853959	111.6718109	1.530899469	9.285961972

Fig. 11 Heat map of air pollution levels across the city

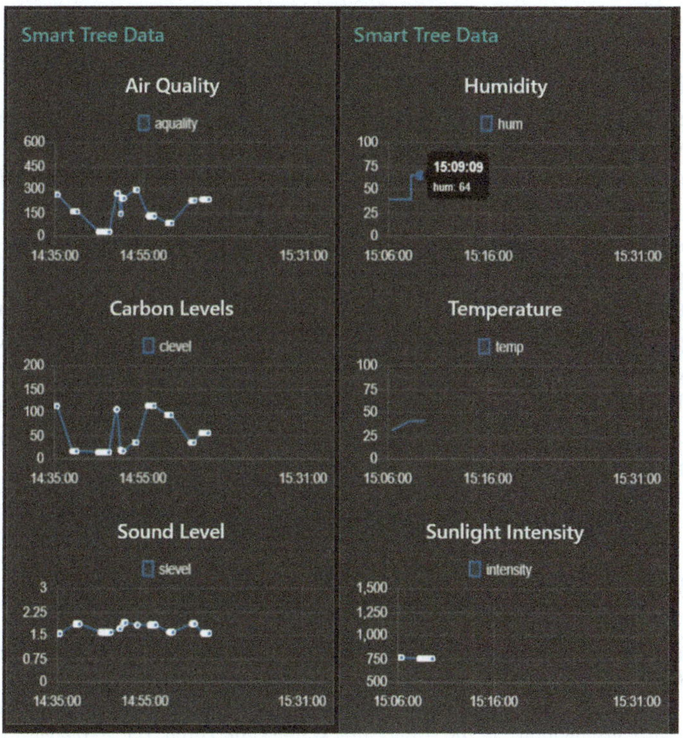

Fig. 12 Dashboard with real-time tree health data

- **Data Analytics and Live Monitoring**: The web app allows real-time information viewing of tree health data across the city. Individual trees can be assessed at specific time frames and patterns can be obtained. This information can be extremely useful for predicting the best environment for tree growth and good behavior from citizens can be incentivized and measured with active goals such as reducing carbon levels in a specific timeframe or decreasing level of sound pollution in an area. This sound data generated across the city can also be used to draft out a sound density map of the city. This can be used in innovative ways of energy production using sound.

5 Conclusions and Future Prospects

In this paper, a prototype of a green smart tree management system based on the IoT is discussed and implemented. The paper describes the functions and actual characteristics of each system component in detail. The experimental results suggest the feasibility of the application to enable the development of smart cities across the world which will play a key role in decreasing pollution levels across the city and increasing afforestation in areas that require it the most. In the future, we can expand the prototype to simulate an actual city infrastructure by extending the network of connected trees and improvements can be made in the design and casing of the model to avoid damage due to rain, harsh weather conditions, or human interference. Finally, we conclude that this solution can be useful in spreading awareness among citizens on the importance of afforestation and proper management of trees in the city and can help farmers in rural areas to avoid hours of manual inspection and labor spent in inspecting the health of the farm. Extended applications of the research could include employing CCTV cameras to remotely monitor multiple sites in the city. Computer vision and machine learning could be utilized to provide predictive analytics about damage that could occur to the tree and preventive action can be taken before the actual damage takes place.

References

1. Anguluri, R.: Role of green space in urban planning: outlook towards smart cities. Urban For. Urban Greening **9**(2) (2017)
2. Brown, P.R.: The role of citizen activists in urban infrastructure development, growing greener cities: urban sustainability in the twenty-first century. In: Birch, E.L., Wachter, S.M. (eds.), pp. 152–169. University of Pennsylvania Press (2011)
3. Kirk, S.: Energy performance resilience of UAE buildings to climate change. Int. J. Environ. Sustain. **7**(1), 90–97 (2018)
4. Hamdan, N.M., Alawadhi, H., Jisrawi, N.: Particulate matter pollution in the United Arab Emirates: elemental analysis and phase identification of fine particulate pollutants. In: Proceedings of the 2nd World Congress on New Technologies, Budapest, Hungary, 18–19 Aug 2016

5. Barakat-Haddad, C., Zhang, S., Siddiqua, A., Dghaim, R.: Air quality and respiratory health among adolescents from the United Arab Emirates. J. Environ. Public Health **1**, 1–13 (2015)
6. Al-Maskari, F., Bener, A., Al-Kaabi, A., Al-Suwaidi, N., Norman, N., Brebner, J.: Asthma and respiratory symptoms among school children in United Arab Emirates. Allergie Immunol. **32**(4), 159–163 (2000)
7. O'Grady, M.J.: Modelling the smart farm. Inf. Process. Agric. **4**(3), 179–187 (2017)
8. Dwivedi, P.: IoT based smart garbage management system. Int. J. Adv. Trends. Comput. Sci. Eng. **6**(4), 71–76 (2017)
9. Luvisi, A., Lorenzini, G.: RFID-plants in the smart city: applications and outlook for urban green management. Urban For. Urban Green. **9**(2), 1–18 (2014)
10. Paustian, K.: Climate-smart soils, nature. Int. J. Sci. **532**, 49–57 (2017)
11. To, W.M.: Tree MIS caring for ecological assets in smart cities. IEEE Comput. Soc. **18**, 50–55 (2016)
12. Chua, M.: Tree management system. Proc. De La Salle Univ. Research Congress, no. SEE-III-021 (2014)
13. Wang, Y.P.E.: A primer on 3GPP narrowband internet of things. IEEE Commun. Mag. **55**(3), 117–123 (2017)
14. Chiti, F., et al.: Optimized narrow-band M2M systems for massive cellular IoT communications. IEEE Global Commun. Conf. (GLOBECOM) (2016)
15. Nokia Networks.: Future Works LTE-M—optimizing LTE for the internet of things. novotech.com/docs/default-source/default-document-library/lte-m-optimizing-lte-for-the-internet-of-things.pdf
16. Rapeepat, R., et al.: NB-IoT system for M2M communication. IEEE Wirel. Commun. Netw. Conf. Workshops (WCNCW) (2016)
17. Derhamy, H., et al.: A survey of commercial frameworks for the internet of things. IEEE 20th Conference on Emerging Technologies and Factory Automation (ETFA) (2015)
18. Hunkeler, U., et al.: MQTT-S—A publish/subscribe protocol for wireless sensor networks. 3rd International Conference on Communication Systems Software and Middleware and Workshops (COMSWARE 08) (2008)

Data Analysis and Prediction Using Big Data Analytics in Agriculture

K. Aditya Shastry and H. A. Sanjay

Abstract Big data analytics analyzes large sized data to discover useful and hidden patterns along with correlations/associations and other beneficial insights. Even though rapid developments in technology have taken place, and traditional agriculture practices are being followed in several countries (especially developing countries). In this regard, big data analytics possesses the capability to positively transform the agricultural domain. The agriculturists are heavily dependent on traditional practices to select suitable crop to grow and how to grow the crop in efficient and productive way. These traditional methods usually comprise of walking in fields, choosing soil samples for soil moisture analysis, and detecting plant diseases by observing the leaves manually. It is in fact significant for the agriculturists (especially farmers) to understand the importance of technology for gathering crop information in large quantities for transforming the collected data into meaningful knowledge. Integrating big data with the agricultural domain can bring about radical changes in how agriculture is being done currently. The technique of growing crops effectively can be well understood by the appropriate utilization of big data analytics. The three extensive areas which are transforming agriculture are accurate crop forecasts, precision farming, and generating high-quality seeds. This chapter is composed of four sections. The first section provides an overview of analytics on big data in agriculture. This is followed by the second section which will highlight the different big data techniques in agriculture. The third section describes two case studies that apply techniques for big data analysis in the agricultural domain. The fourth section analyzes the various research areas related to agriculture that can be explored by the academicians and researchers. The summary of the chapter is provided at the end.

Keywords Big data analytics · Agriculture · Prediction · Analysis

---

K. Aditya Shastry (✉) · H. A. Sanjay
Department of Information Science and Engineering, Nitte Meenakshi Institute of Technology, Yelahanka, Bangalore, Karnataka 560065, India
e-mail: adityashastry.k@nmit.ac.in

H. A. Sanjay
e-mail: sanjay.ha@nmit.ac.in

P. K. Pattnaik et al. (eds.), *Internet of Things and Analytics for Agriculture, Volume 2*, Studies in Big Data 67,
https://doi.org/10.1007/978-981-15-0663-5_10

1 Introduction

Agriculture represents the sector which is responsible for providing food to every human being in the world. However, the people who do agriculture are not given much importance. After failing institutions, time has indeed come for technology to take over the change. Emerging technologies need to be incorporated in agriculture as new issues or challenges are faced everyday by the different agriculture stakeholders especially farmers.

Agriculture in many countries lacks institutional attention, supports from banks in terms of loans and farmer welfare schemes, and suffers from a myriad of disasters like depleting groundwater levels in rural areas, climate change, unpredictable monsoon or lack of it, droughts, floods, unfair price-fixing policies of produce, migration of farmers toward the cities for pursuing better paying jobs, and more.

Big data analytics analyzes large sized data to discover useful and hidden patterns along with correlations/associations and other beneficial insights. Even though rapid developments in technology have taken place, traditional agriculture practices are being followed in several countries (especially developing countries). In this regard, big data analytics possesses the capability to positively transform the agricultural domain. The agriculturists are heavily dependent on traditional practices to select suitable crop to grow and how to grow the crop in efficient and productive way. These traditional methods usually comprise of walking in fields, choosing soil samples for soil moisture analysis, and detecting plant diseases by observing the leaves manually [1].

However, it is critical for the agriculturists (especially farmers) to understand the importance of technology for gathering crop information in large quantities for transforming the collected data into meaningful knowledge. Integrating big data with the agricultural domain can bring about radical changes in how agriculture is being done currently. The technique of growing crops effectively can be well understood by the suitable utilization of big data analytics. The three extensive areas which are transforming agriculture are highlighted below [2]:

- Accurate crop forecasts: The major apprehension of farmers is how the crops that they have sown will grow. In previous years, crop forecasts have been inaccurate. The modern farmers can predict the crop yields more accurately using technologies that utilize high-end algorithms. These algorithms analyze huge amount of crop and weather information to arrive at the predictions. Farmers can nowadays predict the yield even before a seed is sown in soil. The big data analytics delivers valuable insights that empower the farmers to harvest at appropriate times and reap maximum yields. The farmer's stress about worrying about the crop can be considerably reduced by accurate crop forecasting.
- Precision farming: It represents the modern method to grow crops. Several of the agricultural methods may be automated using the Internet of things (IoT). Through precision farming, many farmers particularly in developed countries are profiting when compared to the farmers who are not using the technology. The quality of crops has also improved by the use of precision farming.

- Generating seeds of high quality and minimizing hunger: Latest studies demonstrate that starvation in the world is increasing due to the exponential growth of population and global warming as seen in Africa where around 20 million humans are facing the real threat of starvation. Though social activists across the globe are putting their effort toward avoiding the food crisis, the actual solution may lie in the proper application of big data analytics in the agricultural domain. The plant data is being analyzed from past decades by several researchers and agricultural scientists in the hope of cultivating effective and healthy crops. Recent innovations in the analytical techniques for big data have led to the faster, larger, and improved growth of crops. Global hunger can be eliminated by the use of chemically engineered seeds assisted by big data analytics.

2 Big Data and Related Techniques in Agriculture

This section discusses the role of big data in agriculture. Some of the popular methods for analyzing big data are also discussed. Before explaining the big data techniques, we briefly look into what is big data.

2.1 Role of Big Data in Agriculture

According to [3], the big data is characterized using the following five dimensions:

- V1 denotes the volume of data. V1 indicates the data size gathered for analysis.
- V2 signifies the window of time between which the data is relevant and useful. For instance, certain data has to be analyzed within specific time durations else the data loses its significance. Specific examples include identifying pests and diseases in animals within their lifetime.
- V3 represents variety of data. The variety in data may occur due to multiple data sources (e.g., videos, remote sensing data, images), multi-temporal (e.g., data collected during different times or dates), and multi-resolution (e.g., spatially diverse image resolution data). Data coming from different application domains also possesses variety.
- V4 denotes veracity of data. V4 comprises of the potential, quality, reliability, accuracy, and overall confidence of data.
- V5 indicates valorization of data which is its ability to spread appreciation, innovation, and knowledge.

Even though the five "V"s are able to describe big data, the big data analysis needs not to satisfy these five dimensions [4]. Generally, big data can have less accuracy and lack stability making the fourth V weaker. According to [5, 6], another "V" for visualization can be added for meaningfully demonstrating or illustrating the data using complex data structures and information-rich scenarios.

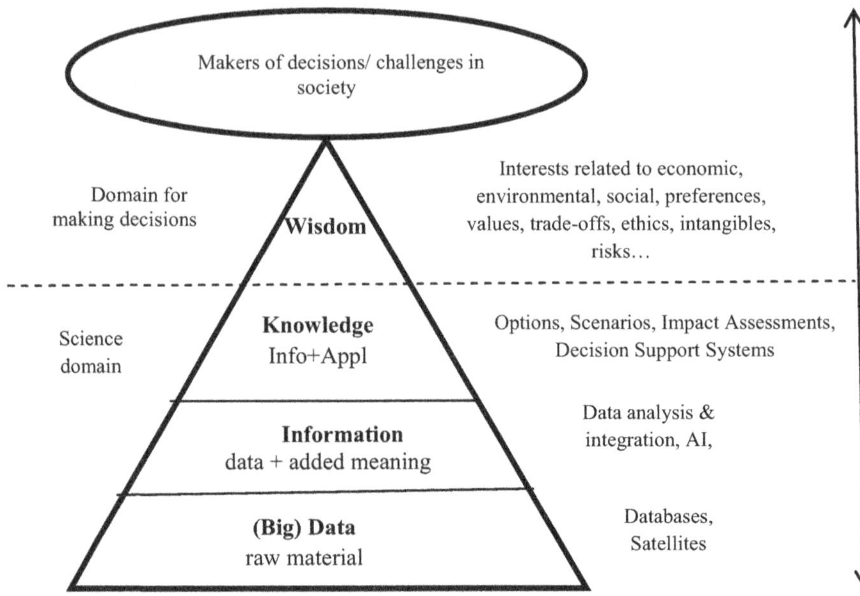

Fig. 1 DIKW hierarchy [1]

As per [7], big data deals more with the capacity to cross-reference, search, visualize, and aggregate large dataset faster. It largely emphasizes the ability to mine information and insights which were expensive and not technically feasible [1, 8].

The relationships between information, wisdom, knowledge, and data are depicted in the DIKW hierarchy where D stands for data, I for information, K for knowledge, and W for wisdom. It basically conveys the concept that information is described in data terms, knowledge with respect to information, and wisdom in terms of knowledge. Figure 1 depicts DIKW hierarchy.

Figure 1 depicts how a lower-level entity (data) is transformed into a higher-level entity (information). The processes involved in the transformation is also identified and described. The concept is to take decisions based on evidence which requires "wisdom." The wisdom is combined with knowledge available having fewer assets such as values, ethics, interests, and preferences that are tangible. From data, the knowledge base is derived. In this respect, data is considered as the raw entity to generate information along with meaning. The enrichment of information is done by creating knowledge utilizing and combining policy and decision applications such as instance integrated models, assessments of impact or support systems for decision making which are contextual in nature.

The systems related to agriculture and environment research are dynamic in nature and comprise of complex communications among living species and products which are perishable (like humans, plants, agricultural products, animals) and their environment. The elaborate information related to entity status and behavior and their environmental conditions is needed for describing such systems. These systems can hold actual status, as well as conditions related to history or future. The data related to spatial and temporal entity behavior is critical since the living systems possess variability in spatial and temporal conditions. Furthermore, detailed knowledge about the different subsystems is required for understanding these interactions. The subsystems may include climate, water, biological, and soil.

With regard to big data, the DIKW hierarchy demonstrates the process of transforming raw data which is less useful to knowledge which is more useful to the end users. The knowledge represents the pieces of data which are useful for providing meaningful decisions. This reliable and timely information is utilized by end users who assume that the resources have been combined using the most suitable sources from pool of big data, inferred and processed according to their decision context. Obviously, such assumptions face a vast challenge to the community composed of ICT experts, experts in domain, and data scientists who are involved in this work. More cooperative approach and expertise with hybrid bottom-up and top-down approaches are required for making decisions regarding the present context in big data analytics [2].

Table 1 depicts the association among the number of works done in different agriculture areas and the three V's of big data. The third column gives the number of papers that are giving solutions in the respective agricultural areas. Majority of the studies have involved topics related to availability of food and security, insurance and finance of farmers, changes in climate and weather, and research based on management of lands and animals. The average ratings for first three V's related to volume, velocity, and variety with respect to each agricultural area are provided by the authors [1] from columns 4 to 6 in Table 1.

The authors [1] used low (L), medium (M), and high (H) indicators for estimating the first three "V"s of big data based on survey of 34 papers. These papers were compared among them, and relative rankings in the form of L, M, or H were given. The papers that were surveyed had data volumes ranging from medium to high and

Table 1 General areas of agriculture in which big data is used [1]

Agriculture area	V1	V2	V3	No. of papers
Climate and weather change	M	M	H	4
Land	H	L	M	5
Research in animals	M	H	L	4
Crops	M	M	L	3
Soil	M	L	L	2
Weeds	L	H	L	1
Availability of food and security	M	L	M	4
Research in plant and animal varieties	M	L	H	1
Decision making by farmers	H	M	H	2
Insurance and finance for farmers	H	M	M	5
Remote sensing	H	M	M	3

variety ranging from medium to low. The projects related to weeds and animals were the exceptions that involved high velocity since urgent actions are required when proof of diseases and weeds are found. High information variety was required by projects related to climate change and weather along with the apps related to decision making by farmers since the regular forecasting of change in climate, weather, etc. aids the farmers in doing his routine tasks. The applications linked with remote sensing consumed highest volume of data since images have large sizes. The projects related to land management along with availability and security of food, analysis of soil, and biodiversity required the lowest velocity.

Lastly, soil, research related to animals, weeds, crops, and remote sensing possess the least variety. From this, we can infer that these agricultural domains do not need or do not have sufficient access to variety of data for resolving the domain-specific issues. Table 2 depicts the associations of agricultural applications with the "V" features of big data for the surveyed papers.

Data in large volumes is required for applications related to forecasting of weather, security of food, estimation of yields of crops, and mapping of land. High velocity is required for applications related to identification of diseases in animals, poor nutrition in plants and decisions on productivity of farmers along with the quality, and safety of food which have to be performed in real time (near). These applications require decisions which have to be taken tactically faster instead of strategic planning for longer terms (lower velocity). The fourth dimension of big data which is the variety of data is required by applications related to the estimation of insurance and productivity of farmers. The fifth V related to high data reliability is required by applications dealing with diseases of animals/plants, culling of herd, improvements in crop productivity, and forecasting of yields. The number of publications and commercial applications related to analytics of big data in agriculture is less, and hence, it can be observed that it is still in early development stage.

2.2 Sources of Big Data and Techniques

It is worthwhile to note the sources from which big data originates for analysis to be performed. Some of the common big data sources are as follows:

- Agricultural fields, i.e., from sensors on ground related to devices for chemical detection, sensors that are bio in nature, stations associated with weather, etc.
- Data collected from third-party companies and governmental organizations such as yearbooks related to statistics, reports from governments, guidelines, and regulations from public bodies, alerts, etc.
- Distributed sources such as online repositories and services of web
- Data from sensors which are airborne like satellites, planes, aerial vehicles which are unmanned
- Web data from companies through web services which are online, social media feeds, etc.

Table 2 Usage of big data in diverse agricultural applications [1]

Big data dimension	Description of "V"	Agricultural applications
V1	Large data volume	Forecasting weather [9], culling of daily herd [10], productivity improvement of farmers [11], protection and insurance of small farmers [12], financing of farmers [13], estimations of food security using remote sensing [14], classification of land use and land cover [15], sharing of data for earth observations [16]
V2	High velocity for time critical applications	Prediction of weather, fermentation of wine [9], quality and safety of animal food [17], discrimination of weed [18], recognition of diseases in animals [19], improvement in productivity [11], financial transactions of farmers in secluded regions [20], data sharing of earth observations [16]
V3	High variety for data originating from heterogeneous sources	Identification of management zones [21], estimation of food in developing countries [22], evaluation of wildlife population [23], insurance and protection of farmers [12], improvement of farmers productivity [10], understanding of farmers sustainability and operational efficiency [24], tolerance of crop's drought [25, 26], science of climate [27]
V4	High veracity for applications where data reliability and accuracy is critical	Dairy herd culling [10], quality and safety of animal food [17], weed discrimination [18], animals' disease recognition [19], food availability estimation in developing countries [22], small farmers' insurance and protection [12], farmers' productivity improvement [11], data sharing of earth observations [16], evaluate wildlife population viability [23]

The data from above-mentioned sources differs in velocity and volume and are mostly heterogeneous in nature. The data can be accessed differently like in web services, repositories that are static, live feeds, archives, files, etc. The data is also characterized by different formats and types.

The big data sources along with the related technologies are shown in Table 3 for each agricultural domain.

Every application of agriculture requires diverse big data sources to resolve the associated problems. The commonly used big data sources include static repositories, geospatial, and remote sensing using satellites for most of the agricultural domains. The ground sensors are utilized by research related to soil, animals, and crops. Similarly, weather stations are used for applications related to climate change and weather, mapping of lands, decision making for farmers, finance, and insurance. Most of the works have combined data from variety of sources to address specific problems.

2.3 Tools and Techniques for Analyzing Big Data in Agriculture

Column 4 of Table 2 provides particular approaches that are utilized in different agricultural areas. Among the 34 papers surveyed by author [1], ML is utilized in 13 papers, cloud-based platforms is used in nine papers, image processing in eight papers, simulation and modeling in seven papers, analysis based on statistics in six papers, and vegetation indices (NDVI) in six papers. Apart from this, five papers used online services and four papers in geographical information systems (GIS). The land applications related to remote sensing among many others use hybrid techniques except certain applications related to soil.

The following tools are utilized in the respective applications as per the survey done by authors [1]

- For agricultural prediction, the machine learning tools are used by Ma et al. [28] and Mucherino et al. [29].
- The authors [9] and [30] used clustering in agriculture.
- Armstrong et al. [31] and Meyer et al. [32] utilized classification techniques in agriculture.
- The work by Chi et al. [3], Ishimwe et al. [33] used image processing techniques like harmonic and Fourier analysis, decomposition of wavelets, and fitting of curves in agriculture.
- Remote sensing techniques along with image processing were utilized in Karmas et al. [6] in which the input to the ML model was the image processing output.
- Large-scale storing is possible in agriculture using cloud platforms along with MapReduce as demonstrated in the work by Hashem et al. [5]. The Geographical Information Systems (GIS) were utilized by Lucas and Chhajed [34] for geospatial problems.

Table 3 Big data sources with techniques for analyzing the big data in agriculture [1]

Agricultural domain	Big data sources	Big data analysis techniques
Climate and weather change	Weather stations, surveys of historical static information (climate and weather data), remote sensing through satellites, data related to geospatial	Techniques of machine learning related to scalable vector machines (SVM), analysis based on statistics, cloud platforms, GIS analysis, MapReduce analytics
Land	Remote sensing using airplanes, radar data, data in geospatial zone, historical datasets linked to characterization of land, phenology of crops, temperature and rainfall, cover maps of global tree, elevation, sensors in cameras related to imaging of multispectral data, weather stations	Techniques of machine learning related to SVM, K-means clustering, random forests, randomized trees, vegetation indices related to NDVI
Research in animals	Historical information of animals and soils related to their physiological characteristics, ground sensors related to grazing action, intake of food, heat, production of milk, weight), camera sensors	Methods of machine learning related to decision trees, SVM, and neural networks
Crops	Sensors placed in ground (metabolites), remote sensing (satellite), historical datasets (utilization of land, information of land country wise, and data of statistics on yields).	Machine learning (SVM, clustering based on K-means), Wavelet based filtering, Fourier transform, NDVI vegetation indices
Soil	Sensors placed in ground (salinity, conductivity of electricity, humidity), cameras (photosensitive), historical databases	Machine learning (K-means clustering)
Weeds	Remote sensing (plane, drones), information based on historical (digital library related to plant and weed images, data specific to plants)	NDVI vegetation indices, ML techniques like ANN and logistic regression) and image processing techniques

(continued)

Table 3 (continued)

Agricultural domain	Big data sources	Big data analysis techniques
Availability of food and security	Surveys, historical information and databases (e.g., CIALCA, ENAR, rice crop growth datasets), GIS geospatial data, statistical data, remote sensing (synthetic aperture radar)	Machine learning (neural networks), analysis linked to statistics, simulation and modeling, analysis related to networks, analysis of GIS geospatial data, image processing
Research in plant and animal varieties	GIS geospatial data, historical information and databases (SER database of wildlife species	Statistics (Bayesian belief networks)
Decision making by farmers	Datasets containing historical information (e.g., survey data of US government), remote sensing related to satellites and drones, weather stations, human based sensors, data based on web, data linked to geospatial domain such as GIS, social media based feeds	Platforms which are cloud based, services based on web, applications of mobile, statistical analysis, modeling, simulation, benchmarking, big data storage, message-oriented middleware
Insurance and finance for farmers	Web-based data, historical information, weather stations, humans as sensors (crops, yields, financial transactions data)	Cloud platforms, web services, mobile applications
Remote sensing	Remote sensing related to planes, drones and satellites, datasets containing historical information such as datasets of MODIS surface reflectance, datasets of surface land containing images, datasets of WMO weather, heights of reservoirs originating from the altimetry of the reservoirs, data based on web, data related to geospatial regions such as maps and imaging	Platforms linked to cloud, analysis of data based on statistics, analysis of geospatial data using GIS maps, processing of images, indices of vegetation related to NDVI, systems capable of making decisions, storage of big data, portals based on community and web, analytics using MapReduce, applications of mobile, computer vision, and AI

Database management systems (DBMS) that implement array data model are suitable for storing big data which are heterogeneous in nature. NoSQL platforms can be utilized for managing and storing huge volumes of unstructured data. The big raster datasets are stored using the array DBMS that are specifically created for this purpose.

Furthermore, vegetation indices (VI) have been utilized frequently in remote sensing applications for soils or crops mapping. The VI is represented as combinations of reflectance surface at 2/more wavelengths used to focus on a particular vegetation property. The NDVI is the most used VI which denotes a graphical indicator utilized for analyzing the measurements of remote sensing and to estimate whether the observed target possesses green vegetation or not. The message-oriented middleware assists the event-based systems for notifying alerts faster such as prediction of natural hazard, etc.

Table 4 illustrates the common and specific software's utilized in big data analysis based on the analysis type to be done.

A new practice is to apply image analysis for approximating agricultural problems on the images derived from remote sensing. There was a rapid increase in the usage of Landsat satellite data since 2008 due to free availability of the data. Remote sensing is widely used in agriculture since it is not destructive and information may be collected methodically over large geographical boundaries. Recent works by De Leeuw et al. [35] have been on the product delivery of insurance materials due to damage of crops, estimation of risks due to fire and floods or from surplus rain and drought [12].

Table 4 Popular software tools utilized for big data analysis in agriculture [1]

Category	Software tools
Tools related to image processing	Toolkits related to OpenCV library, IM, and VTK
Tools linked to ML	TensorFlow of Google, Weka tool, Oryx, Flavia, SHOGUN, R, mlPy, Apache Mahout, Mlpack, and Mlib
Platforms based on cloud for storing big information along with computation and analysis	Cloudera, EMC Corporation, IBM InfoSphere BigInsights, IBM PureData system for analytics, Aster SQL MapReduce, and Apache Pig
Databases associated with GIS systems and big data	MiraMon, Autodesk, MapInfo, ArcGIS
Middleware linked with message-oriented systems	PostGIS, Apache HAWQ, Oracle GeoRaster, MongoDB, HadoopDB, Google BigTable, Cassandra, MonetDB/SciQL, Hive, Oracle GeoRaster, SciDB
Simulation and Modeling	RabbitMQ and MQTT
Tools related to statistics	R, NorsysNetica, and Weka
Analysis related to time series data	MATLAB, Stata, BFAST, RATS

The coarse resolution satellites like the AVHRR, MODIS, MERIS, and SPOT-VEGETATION have become the major data sources for covering wide areas. The ancillary data, viz. GIS data, field sensors, and historical data, is combined with remote sensing to considerably improve the analysis performed particularly for predictive applications like identification of crops, or differentiating between grasslands. As per the survey made by Kamilaris et al. [1] relatively fewer works exist in the domain of analytics of big data for agriculture, indicating that it is still in early stage. However, immense potential exists in this domain due to the diverse stakeholders involved and the rapid dynamics involved in the system. New start-ups like [20–38] have come up for delivering service to the agriculture stakeholders via the utilization of big data analytics. The authors [5] and [30] observe that new business models may be developed in agriculture because of rapid progress in cloud technologies which enable large storage and computation facilities.

3 Case Studies on Big Data Analytical Techniques Used in Agriculture

This section describes the different case studies used for big data analytics in agriculture. Recent and relevant works are considered.

3.1 Case Study-1: Big Data Querying

The domain related to agriculture and environment consists of datasets which possess diverse and large formats for content, storage, and structure with respect to research purpose. Recently, many datasets are made open source by making them a part of cloud related to Linked Open Data (LOD) which has formed data sources that are heterogeneous and distributed in nature. However, accessing the LOD is not easy and is not able to meet with the ever-evolving changes in data science. It requires extensive big data querying facilities. In this case study, we study a research project called SemaGrow that was initiated by ER FP7 for providing efficient ways of accessing distributed data. Its main objective was the development of infrastructure and algorithms for querying the distributed data sources such and the nodes of LOD in an effective and efficient manner. It utilized the concepts of Semantic Web such as SPARQL query language and RDF that facilitated the reuse and data sharing among applications and scientific communities. It permitted the data interconnection in the cloud related to LOD.

The particular challenges of agricultural data were handled by the SemaGrow system by utilizing different use cases through sequence of data pilots related to the agricultural domain. The merging of big data related to spatiotemporal datasets which were very dissimilar along with its discovery was some of the challenges handled by SemaGrow. The modeling of climatic conditions related to agriculture was one of the use cases handled by SemaGrow which inherently dealt with the adaptation of the system to climatic changes. Large files having multiple dimensions were utilized for storing the parameters of climate that were needed for modeling the regional parameters. For this case, the spatial coverage related to different regions and coverage of temporal conditions for long term was utilized. In many research experiments, scientists try to produce redundant data for experimentation purpose. This leads to inefficiency of resource usage and poses serious problems in regions with limited storage, computing, and networking facilities. The SemaGrow was able to generate lightweight datasets utilizing efficient querying methods related to merging, thematic, temporal, and spatial conditions. For instance, the SemaGrow system was able to generate a combined dataset of soil, temperature, and precipitation parameters for an agricultural scientist in a specific region of Ghana with limited storage and networking facilities.

SemaGrow mainly aimed at providing seamless access to big heterogeneous data sources to the agricultural scientists allowing them to combine and download heterogeneous datasets. The problems related to data integration were reasonably resolved, thus reducing the time and labor involved in the manual integration of big data. The system was able to provide valuable information on time to the stakeholders which aided their decision making. It focused on the layers of data and information of the DIKW model to reduce the gap among these levels by coordinating and processing the big data sources at the information level. Subsequently, it was able to reduce the efforts and resources needed for generating information from raw data and solving the integration challenges among heterogeneous data sources.

3.1.1 Methodology

The system of Trees4Future Clearinghouse was modified to work utilizing the technologies of SemaGrow for better big data querying capabilities for real-world agricultural applications. For this reason, the back-end of Trees4Future was substituted with SemaGrow stack and RDF databases comprising of data and metadata. Hence, the application was able to perform semantic queries on metadata and actual data. A set of heterogeneous data nodes which are distributed can be efficiently queried by the SemaGrow Stack which can be found at the URL https://github.com/semagrow/semagrow. The Stack is a query processor related to the federated SPARQL. A query planner is included that optimizes the execution of query by utilizing the metadata about the nodes. The Sevod vocabulary located at the Web site http://www.w3.org/2015/03/sevod was established in the project in which the VoID vocabulary is extended by including statistical information similar to histograms of database. The data nodes that are slow are handled by utilizing the reactive software model in the Stack. Thus, without the knowledge about the heterogeneous schemas the Stack provides querying of the triple stores present underneath in a transparent fashion.

By utilizing the initiatives for data coordination taken by ISIMIP and AgMIP, the triple stores were established. To demonstrate the influence of global warming at several levels, the Inter-Sectorial Impact Model Inter-comparison Community Project was established. The format of NetCDF files was utilized as per the conventions of climate and forecast for making the I/O data from ISIMIP available. Better economic and crop models are produced by AgMIP for assessing the impact of climate for subsequent generations by connecting the communities related to economic, climate, and crop. Current information technology is being utilized for this purpose. The JSON format is utilized for representing data in AgMIP by utilizing the ICASA Variable List with regard to its metadata. The triplification of limited datasets into triple stores is performed for the data from dissimilar sources. This enables the SPARQL to query the data.

Finally, to resolve spatial queries like point-in-polygon a spatiotemporal triple store called Strabon present at URL http://strabon.di.uoa.gr is added to the combined nodes. Middleware software is utilized for the connection between the front-end of the demonstrator and the Stack instance of SemaGrow. It functions as a pre-processor of results of raw results of query into a response that can be handled by the demonstrator. It also functions as a translator for translating the requests of URL into the appropriate queries of SPARQL for the Stack. To efficiently serve the requirements of the users using the system, the Strabon creates usable NetCDF files.

3.1.2 Results

Less number of users used the demonstrator application for the purposes of testing. The users gave positive feedback regarding its functionality. However, better performance was expected by the users in case of searches related to metadata which took approximately 5e30 s. Similarly, the download provided by the system was very slow (taking days) and the users expected the download to complete within 30 min. This requires up scaling the infrastructure to a massive level. However, research experts are of the opinion that although the system takes several days, it is faster compared to the manual approach which involves repeated consultation, cooperation with remote and local experts which are still more time-consuming. Hence, the hours or days taken by the fusion queries are faster than the manual process.

3.2 Case Study-2: Data-Driven Discovery

The Trees4Future project available at the URL www.trees4future.eu represents European research project whose objectives are to combine, develop, and enhance the infrastructures related to key forest genetics and research in forestry. The forestry data is made available to a broader audience of agriculturists and decision makers through this project. To do their work, the researchers of forestry collect data using their peer and scientific networks. However, recently the distribution of forestry data is being performed particularly in the domain of genetics. Nonetheless, data storage for relevant data is done in silos including private or local databases. Also, documentation of metadata is not done.

In several instances, data is not documented by researchers or data is made private since the associated results of research are not published or due to the chance of data misuse. Also, data sharing through networks which are trusted mostly offers no benefits. Hence, valuable data for agricultural research purposes is not available easily. Forestry modelers form a research community in the Trees4Future project. Integrated modeling is required to meet the present day challenges related to agriculture such as change in climate. Model combinations which explain several subdomains and provide coverage to temporal and spatial resolutions are required for estimating the climatic changes along with determining the adaptation methods for climatic conditions. To establish these integrated models, the Trees4Future has linked ForGEM [39], EFISCEN [40], and Tosia models. The estimation of responses with respect to genetic adaptation on population and individual tree levels are done by the ForGEM model. The development of resources with respect to forests is projected by the EFISCEN model on European and regional scale. The impacts on the production chains of forestry wood with respect to social impacts, environmental, and economic changes are analyzed by the Tosia model.

The problems with integrated modeling are mainly related to heterogeneous data that come to various sources such as genetics, soil, statistics, market, and climate. It is hard to access this kind of data for forestry purposes. This problem is resolved by the Trees4Future project which is capable of providing technical solution that aids the publication, discoverability, and documentation of forestry data by establishing an infrastructure for forestry data. Besides, this infrastructure is able to provide benefits for data sharing and documentation related activities.

The issues related to quality and trust also needs to be handled in this project. From the perspective of the data owner, the data should not be misused after sharing. From the consumer point of view, correct and complete data should be provided. The data variety issues are to be handled since data can originate from various heterogeneous resources and subdomains. The metadata has to be properly linked with the actual data semantically.

3.2.1 Methodology and Implementation

The development of data discovery and search was done for improving the data access. The chief aim was to provide wide access to already documented datasets as well as to the datasets that were inaccessible. Improved facilities were provided to the end users to search the datasets easily and discover the published forestry data. This search mechanism was user-friendly for both the end users of the system and the data publishers/data owners.

The workflow established in Trees4Future project is illustrated in Fig. 2.

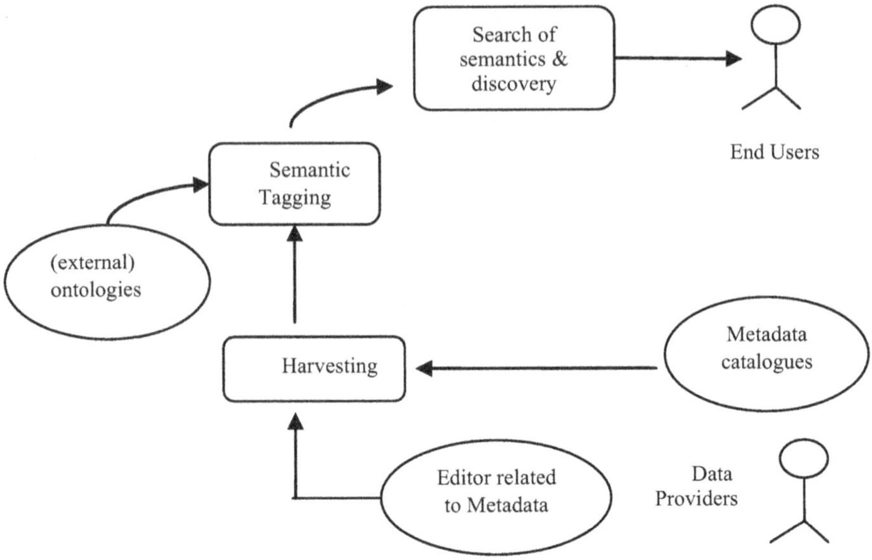

Fig. 2 Publication process workflow developed in Trees4Future [2]

- To aid metadata specific to forestry, a brief schema of metadata is required which is centered on Dublin Core standard.
- The metadata records are published via the OAI-PMH protocol, which consists of a metadata editor (online) and embedded repository for delivering an endpoint that is standardized and harvestable.
- The datasets are conceptualized and linked with external ontologies like AGROVOC, genetic traits by the forestry ontology.
- A mechanism for annotation, harvesting, and triplification for assisting the gathering of metadata using the metadata forestry schema and metadata (standardized) is provided.
- INSPIRE and ISO schemas are provided that are utilized to decompose the metadata into concepts of ontology using the NLP strategies. These are then stored in RDF database and linkage between external ontologies and derived dataset concepts are done.
- A mechanism and user interface for efficient searching of datasets that are registered by utilizing the semantics stored in the RDF store.

3.2.2 Results

The infrastructure that was implemented provided efficient mechanism for discovering research data related to forestry and provided access to a wider audience. The infrastructure is capable of handling the heterogeneous data of different variety that is required for integrated modeling in forestry such as the use case on climate adaptation that was discussed before. The major advantage is that the system is able to integrate the scattered forestry data under one roof. More than 300 datasets from data repositories of Europe are published by the system. An access point is also made available for small companies and individuals for publishing their metadata. Furthermore, opportunities to publish reference datasets are provided to fresh modelers for experimentation on integrated modeling.

The linkage of concepts of metadata with concepts of external ontology is complicated owing to the shortage of amount and quality of the metadata. The automatic selection of datasets that suit the needs of researchers like in technical domains related to simulation and modeling is also performed by the system. The complex nature of scientific datasets is not captured properly due to the absence of depth and structure in metadata. Frequently utilized and important fields such as lineage do not offer the required structure for addressing the data production processes which are complex in nature. The structured addressing of data is prevented by the lack of depth in the metadata structure. In the presented system, this issue was resolved by linking the fragments of AGROVOC vocabulary that were isolated with comprehensive subdomain with semantics that is specific. Apparently, this presents an elaborate and customized mechanism and is not a generic solution.

4 Research Areas Associated with Analytics of Big Data for Agriculture

This section describes the different open problems with solutions that exist for the analysis of big data for agriculture.

4.1 Open Problems

Certain problems are created due to the application of techniques related to analysis of big data in the agricultural domain. Some of the key problems are listed below

- Monopoly of agriculture data by only big agriculture organizations can take place due to the expense involved in data storage and its distribution. This leads to a high farmer dependency on large agricultural organizations as observed by the author [41]. The dissemination of technology to smaller organizations may be restricted due to the monopoly of big data by the larger organizations [42].
- The ownership issues due to the ambiguity created as to who will be the data owner along with the related monetization issues may arise as suggested by Nandyala and Kim [43]. The misuse of farming information by seed organizations and farms may arise as observed by Shin and Choi [44]. Unnecessary speculation in markets

related to commodity may be created by hedge funds due to the availability of real-time agricultural data [21].

- Several issues related to the security, access, and accuracy of big data during its analytics and collection may arise as elaborated by Nandyala and Kim [43] and Sykuta [41].
- Furthermore, digital divide amid developing and developed countries is created because of unbalanced technology access like software, computing power, bandwidth of Internet, and lack of experts [12]. Also, the crucial benefits of big data may be obtained only if the farmers are well educated with sufficient digital exposure which is lacking in developing countries suggest [30–45].
- The developers of products have access to limited ground truth information [46]. But, the information on ground truth is essential for the product evaluation and services operating under several settings related to weather or physical conditions. Likewise, the large data visualization is quite complex.

4.2 Obstacles for Adopting Analysis Techniques for Big Data in Agriculture

Some of the obstacles that come in the way of using techniques related to the big data analysis are listed below:
Lack of human expertise and resources [45]

- Lack of reliable infrastructures for gathering and analyzing big data [20]. Infrastructures are required for the efficient storage of big data and management of high-dimensional datasets.
- Lack of technical expertise and skills in the analysts for the accurate analysis of big data [12, 22].
- The authors [43] and [16] pointed out that there is an absence of proper governance and structure associated with big data.
- To provide fair share among the agriculture stakeholders attractive business models are required as suggested in the works [35–48].
- Owing to the absence of definite semantics, the reuse and knowledge of big data have become difficult for organizations and researchers.

The authors [1] observed in their study (Table 1) that focus is provided to large volumes in case of big data such as the applications related to changes in climate and weather, identification of land, decision making of farmers, finance and insurance, and remote sensing. However, the other aspects of big data such as velocity, veracity, valorization, and variety are just as significant as the volume of data. Ignoring these features has led to the narrow and skewed perception regarding big data in agriculture.

Furthermore, the systems related to remote sensing for management of farms still suffer from various practical difficulties like the gathering and distribution of images within the specified time duration, errors related to sampling, and the absence of resolution data related to spatial. The challenges related to interpretation of images, extraction of data, and the impact of conditions related to weather also prevail. Lastly, the common obstacles observed by many agricultural researchers are the lack of the data (or some part of it) along with its unreliable nature. The relevance of the data with regard to time and variety also has come across as major challenges faced by analytics of big data in agriculture.

4.3 Addressing Open Problems and Overcoming Barriers

From the perspective of society and political views, several agriculturists (especially farmers) around the globe began to establish cooperative societies and online groups in order to share their experiences about farming along with the knowledge of big data. Proper policies regarding ownership of data need to be in place in order to safeguard the copyrights of data owners and govern the user access to data. Furthermore, policies to manage the security and management of data are required in order to secure the big data and increase its impact by utilizing appropriate tools. Large monetary investments are needed for building agriculture cloud infrastructures for storing, visualizing, and analyzing the data related to agriculture. It is also required for the analytics to be performed at a faster speed and larger scale. Agricultural data should be available at reasonable rates or better be made open source for aiding valuable research. User-friendly platforms may be created by utilizing aggregation, reduction, and appropriate data analysis.

For data interoperability, technologies related to data semantics such as RDF, linked data along with ontologies may be utilized. Certain open-source tools like maps for identifying crop types and calendars, indices of vegetation, models to predict yields, estimation of crop area, and prediction of seasonal weather may be useful in the agriculture domain. To assist large-scale operations along with analysis of varied data, it should be possible to merge these tools. As discussed before, big data in agriculture needs to be made easily available for extensive research to happen. With regard to this, several institutions and organizations along with agricultural universities throughout the world have started providing free access to large-scale agricultural data covering various domains of agriculture.

4.4 Possible Application Areas Related to the Big Data Analysis in Agriculture

The probable areas of the application of analytical techniques in big data for resolving several problems related to agriculture are discussed in this subsection. Some of the potential application areas are listed below:

1. To meet global standards for agriculture, effective platforms facilitating supply chain stakeholders to have access to products of high-quality and processes need to be developed [14, 45].
2. Better yield prediction tools along with forecasting accurate demands need to be implemented in the future [22].
3. Accurate information on the usage of herbicides, fertilizers, and pesticides needs to be provided to farmers for better crop growth [45].
4. To improve food tracking in supply chain equipments related to scanning in plants, tracking of shipment, and monitoring the purchases of customers need to be developed.
5. Foodborne diseases need to be prevented via the utilization of big data analysis strategies in agriculture.
6. Efficient techniques for optimization are required since the agriculture products tend to deteriorate after harvesting [48]. This is very much needed to reduce the losses and enhance the quality of food products [17]. The popular optimization methods like genetic algorithms [17], neural networks [49], and meta-heuristics are being applied to processing of food.
7. Efficient remote sensing techniques are very much essential for mapping large-scale lands to crops. This is required for the effective monitoring of cultivated lands around the globe which can lead to better productivity as suggested by Schuster [21], Becker-Reshef et al. [47], Barrett et al. [50], Waldhoff et al. [51].
8. Advanced scientific and simulation models for imitating the environmental process need to be developed. These models are beneficial for policy makers in making agriculture-related decisions to achieve sustainability in physical ecosystems [16, 27, 52].
9. To effectively capture the interaction between environment and plants, screening techniques possessing high throughput are required. These techniques need to provide higher accuracy and precision with the capability of analyzing the data in a quantitative fashion [6, 53].
10. Automatic identification and removal of weeds [38], identification and prevention of pests [54], and crop harvesting [51] may be done by developing agricultural robots. These robots have the capability to progressively change agriculture from its existing uncertainty to a more certain future.
11. Specific genetic engineering known as "genome editing" may be utilized to modify the genome of crop or animal up to the level of each genetic letter as observed by Hartung and Schiemann [37]. It is able to replicate the mutation process on which the breeding of crops is reliant. Also, it does not cause the generation of cross-genetic animals or plants [55]. The research on epigenetics can be supplemented by this technology.

The above-mentioned applications are able to produce massive amounts of big data that could be utilized by creators of policies in the future to balance the production and consumption of agricultural products. Furthermore, making the agricultural data open source would generate valuable research and development opportunities for smart farming in the agriculture sector.

5 Conclusions

A detailed discussion about the different techniques of big data analysis and its applications in agriculture is provided in this chapter. Detailed results regarding the techniques employed in various applications are also elaborated. The reader of this chapter can be benefitted by the knowledge about the current projects going on in the domain of agriculture by utilizing the techniques of big data analytics. The application of appropriate characteristics of big data for diverse scenarios along with the sources of big data and analytical techniques for big data are elaborated. The open research problems along with the challenges faced for applying big data techniques in agriculture are also discussed. As discussed, the openness and availability of software and hardware, tools and techniques for analysis of big data, and increased accessibility of open-source agricultural datasets can bring about radical changes in how farming is done. Many start-up companies are coming up for performing analysis of big data in agriculture. The development and research opportunities in agriculture can be boosted by adopting open standards for datasets. The food shortage can be effectively resolved by balancing the demand and supply of food through big data analytical techniques. This in turn can protect the physical ecosystems and assist in safeguarding our natural resources.

References

1. Kamilaris, A., Kartakoullis, A., Prenafeta-Boldú, F.X.: A review on the practice of big data analysis in agriculture. Comput. Electron. Agric. **143**, 23–37, ISSN 0168-1699 (2017). https://doi.org/10.1016/j.compag.2017.09.037
2. Lokers, R., Knapen, R., Janssen, S., van Randen, Y., Jansen, J.: Analysis of big data technologies for use in agro-environmental science. Environ. Modell. Softw. **84**, 494–504, ISSN 1364-8152 (2016). https://doi.org/10.1016/j.envsoft.2016.07.017
3. Chi, M., Plaza, A., Benediktsson, J.A., Sun, Z., Shen, J., Zhu, Y.: Big data for remote sensing: challenges and opportunities. Proc. IEEE **104**(11), 2207–2219 (2016). https://doi.org/10.1109/JPROC.2016.2598228
4. Rodriguez, D., de Voil, P., Rufino, M.C., Odendo M., van Wijk, M.T.: To mulch or to munch? Big modelling of big data. Agric. Syst. **153**, 32–42, ISSN 0308-521X (2017). https://doi.org/10.1016/j.agsy.2017.01.010
5. Hashem, I.A.T., Yaqoob, I., Anuar, N.B., Mokhtar, S., Gani, A., Khan, S.U.: The rise of "big data" on cloud computing: review and open research issues. Inf. Syst. **47**, 98–115, ISSN 0306-4379 (2015). https://doi.org/10.1016/j.is.2014.07.006
6. Karmas, A., Karantzalos, K.: Athanasiou, S.: Online analysis of remote sensing data for agricultural applications. s.l., OSGeo's European conference on free and open source software for geospatial (2014)
7. Wolfert, S., Ge, L., Verdouw, C., Bogaardt, M.J.: Big data in smart farming—a review. Agric. Syst. **153**, 69–80, ISSN 0308-521X (2017). https://doi.org/10.1016/j.agsy.2017.01.023
8. Sonka, S.: Big data: fueling the next evolution of agricultural innovation. J. Innov. Manag. **4**, 114–136 (2016). https://doi.org/10.24840/2183-0606_004.001_0008
9. Tripathi, S., Srinivas, V.V., Nanjundiah, R.S.: Downscaling of precipitation for climate change scenarios: A support vector machine approach. J. Hydrol. **330**(3–4), 621–640, ISSN 0022-1694 (2006). https://doi.org/10.1016/j.jhydrol.2006.04.030

10. McQueen, R., Garner, S.R., Nevill-Manning, C.G., Witten, I.H.: Applying machine learning to agricultural data. Compt. Electron. Agric **12**(1) (1995)
11. GSMA.: mAgriProgramme. [Online] Available at http://www.gsma.com/mobilefordevelopment/programmes/magri/programme-overview Accessed 2019 (2014)
12. Syngenta Foundation for Sustainable Agriculture.: FarmForce [Online]. Available at http://www.farmforce.com/. Accessed 2018 (2016)
13. Global Envision.: Unleashing Ugandan farmers' potential through mobile phones. [Online] Available at https://www.mercycorps.org/research-resources/unleashing-ugandan-farmers-potential-through-mobile-phones. Accessed 2017 (2006)
14. RIICE Partnership.: Remote sensing-based Information and insurance for crops in emerging economies. [Online] Available at: http://www.riice.org/. Accessed 2019 (2014)
15. Wardlow, B.D., Egbert, S.L., Kastens, J.H.: Analysis of time-series MODIS 250 m vegetation index data for crop classification in the US central great plains. Remote Sens. Environ. **108**(3), 290–310 (2007)
16. Nativi, S., Mazzetti, P., Santoro, M., Papeschi, F., Craglia, M., Ochiai, O.: Big data challenges in building the global earth observation system of systems. Environ. Modell. Softw. **68**, 1–26, ISSN 1364-8152 https://doi.org/10.1016/j.envsoft.2015.01.017 (2015)
17. FernándezPierna, J.A., Baeten, V., MichotteRenier, A., Cogdill, R.P., Dardenne, P.: Combination of support vector machines (SVM) and near-infrared (NIR) imaging spectroscopy for the detection of meat and bone meal (MBM) in compound feeds. J. Chemom. **18**(7–8), 341–349 (2004)
18. Gutiérrez, P.A., López-Granados, F., Peña-Barragán, J.M., Jurado-Expósito, M., Hervás-Martínez, C.: Logistic regression product-unit neural networks for mapping Ridolfiasegetum infestations in sunflower crop using multitemporal remote sensed data. Comput. Electr. Agric. **64**(2), 293–306, ISSN 0168-1699, (2008). https://doi.org/10.1016/j.compag.2008.06.001
19. Chedad, A., Moshou, D., Aerts, J.M., Van Hirtum, A., Ramon, H., Berckmans, D.: AP—animal production technology: recognition system for pig cough based on probabilistic neural networks. J. Agric. Eng. Res. **79**, 449–457 (2001). https://doi.org/10.1006/jaer.2001.0719
20. Akinboro, B.: Bringing mobile wallets to nigerian farmers. [Online] Available at http://www.cgap.org/blog/bringing-mobile-wallets-nigerian-farmers. Accessed 2018 (2016)
21. Schuster, J.: Big data ethics and the digital age of agriculture. Am. Soc. Agric. Biol. Eng. **24**(1), 20–21 (2017)
22. Frelat, R., Lopez-Ridaura, S., Giller, K.E., Herrero, M., Douxchamps, S., Djurfeldt, A.A., Erenstein, O., Henderson, B., Kassie, M., Paul, B.K., Rigolot, C., Ritzema, R.S., Rodriguez, D., van Asten, P.J.A., van Wijk, M.T.: Proc. Nat. Acad. Sci. Jan 2016, **113**(2), 458–463 https://doi.org/10.1073/pnas.1518384112
23. Marcot, B.G., Holthausen, R.S., Raphael, M.G., Rowland, M.M., Wisdom, M.J.: Using Bayesian belief networks to evaluate fish and wildlife population viability under land management alternatives from an environmental impact statement. Forest Ecol. Manag. **153**(1–3), 29–42, ISSN 0378-1127 https://doi.org/10.1016/S0378-1127(01)00452-2 (2001)
24. Field to Market.: Fieldprint calculator. [Online] Available at https://www.fieldtomarket.org/fieldprint-calculator/. Accessed 2018 (2015)
25. Tesfaye, K., Sonder, K., Cairns, J.E., Magorokosho, C., Tarekegn, A., Kassie, G.T., Getaneh, F., Abdoulaye, T.A., Abate, T.B., Erenstein, O.:. Targeting drought-tolerant maize varieties in Southern Africa: a geospatial crop modeling approach using big data (2016)
26. Fuchs, A., Wolff, H.: Drought and retribution: evidence from large scale rainfall index insurance in Mexico. s.l., s.n., pp. 13–14 (2011)
27. Schnase, J.L., Duffy, D.Q., Tamkin, G.S., Nadeau, D., Thompson, J,H., Grieg, C.M., McInerney, M.A., Webster, W.P.: MERRA analytic services: meeting the big data challenges of climate science through cloud-enabled climate analytics-as-a-service. Comput. Environ. Urban Syst. **61**, 198–211, ISSN 0198-9715 (2017). https://doi.org/10.1016/j.compenvurbsys.2013.12.003
28. Ma, C., Zhang, H.H., Wang, X.: Machine learning for big data analytics in plants. Trends Plant Sci. **19**(12), 798–808, ISSN 1360-1385 (2014). https://doi.org/10.1016/j.tplants.2014.08.004

29. Mucherino, A., Papajorgji, P., Pardalos, P.: Data mining in agriculture. Springer Science and Business Media **34** (2009)
30. Kempenaar, C., Lokhorst, C., Bleumer, E.J.B., Veerkamp, R.F., Been, T., van Evert, F.K., Boogaardt, M.J., Ge, L., Wolfert, J., Verdouw, C.N., van Bekkum, M., Feldbrugge, L., Jack, P.C., Waaij, B.D., van Persie, M., Noorbergen, H.: Big data analysis for smart farming, s.l.: Wageningen University and Research **655** (2016)
31. Armstrong, L., Diepeveen, D., Maddern, R.: The application of data mining techniques to characterize agricultural soil profiles. Proc. AusDM'07 Proc. Sixth Austr. Conf. Data Min. Anal. **70**, 85–100 (2007)
32. Meyer, G.E., Neto, J.C., Jones, D.D., Hindman, T.W.: Intensified fuzzy clusters for classifying plant, soil, and residue regions of interest from color images. Comput. Electr. Agric. **42**(3), 161–180, ISSN 0168-1699 (2004). https://doi.org/10.1016/j.compag.2003.08.002
33. Ishimwe, R., Abutaleb, K., Ahmed, F.: Applications of thermal imaging in agriculture—a review. Adv. Rem. Sens. **3**, 128–140 (2014). https://doi.org/10.4236/ars.2014.33011
34. Lucas, M.T., Chhajed, D.: Applications of location analysis in agriculture: a survey. J. Oper. Res. Soc. **55**(6), 561–578 (2004)
35. De Leeuw, J., Vrieling, A., Shee, A., Atzberger, C., Hadgu, K.M., Biradar, C.M., Keah, H., Turvey, C.: The potential and uptake of remote sensing in insurance: a review. Rem. Sens. **6**(11), 10888–10912 (2014)
36. Vibhute, A., Bodhe, S.K.: Application of image processing in agriculture: a survey. Int. J. Comput. Appl. **52**, 34–40 (2012). https://doi.org/10.5120/8176-1495
37. Hartung, F., Schiemann, J.: Precise plant breeding using new genome editing techniques: opportunities, safety and regulation in the EU. Plant J. **78**(5), 742–752 (2014)
38. Blue River Technology.: [Online] Available at http://www.bluerivert.com/. Accessed 2019 (2011)
39. Kramer, K., Hengeveld, G.M., Schelhaas, M.J., van der Werf, D.C., de Winter, W.: Genetic adaptive response: missing issue in climate change assessment studies (2013)
40. Nabuurs, G.J., Schelhaas, M.J., Pussinen, A.: Validation of the European forest information scenario model (EFISCEN) and a projection of Finnish forests (2000)
41. Sykuta, M.: Big data in agriculture: property rights, privacy and competition in ag data services. Int. Food Agribus. Manage. Rev. **19** (2016)
42. Carbonell, I.: The ethics of big data in big agriculture. Int. Policy Rev. **5**(1), 1–13 (2016)
43. Nandyala, C.S., Kim, H.-K.: Big and meta data management for U-agriculture mobile services. Int. J. Softw. Eng. Its Appl. (IJSEIA) **10**(1), 257–270 (2016)
44. Shin, D.-H., Choi, M.J.: Ecological views of big data: perspectives and issues. Telematics Inform. **32**(2), 311–320 (2015)
45. Sawant, M., Urkude, R., Jawale, S.: Organized data and information for efficacious agriculture using PRIDE model. Int. Food Agribus. Manag. Rev. **19**(A) (2016)
46. Atzberger, C.: Advances in remote sensing of agriculture: Context description, existing operational monitoring systems and major information needs. Remote Sens **5**(2), 949–981. aWhere Inc., 2015. [Online] Available at http://www.awhere.com/. Accessed 2018 (2013)
47. Becker-Reshef, I., Justice, C., Sullivan, M., Vermote, E., Tucker, C., Anyamba, A., Small, J., Pak, E., Masuoka, E., Schmaltz, J., Hansen, M., Pittman, K., Birkett, C., Williams, D., Curt, R., Doorn, B.: Monitoring global croplands with coarse resolution earth observations: the global agriculture monitoring (GLAM) project. Remote Sens **2** (2010). https://doi.org/10.3390/rs2061589
48. Wari, E., Zhu, W.: A survey on metaheuristics for optimization in food manufacturing industry. Appl. Soft Comput. **46**, 328–343 (2016)
49. Erenturk, S., Erenturk, K.: Comparison of genetic algorithm and neural network approaches for the drying process of carrot. J. Food Eng. **78**(3), 905–912 (2007)
50. Barrett, B., Nitze, I., Green, S., Cawkwell, F.: Assessment of multi-temporal, multi-sensor radar and ancillary spatial data for grasslands monitoring in Ireland using machine learning approaches. Remote Sens. Environ. **152**(2), 109–124 (2014)

51. Waldhoff, G., Curdt, C., Hoffmeister, D., Bareth, G.: Analysis of multitemporal and multisensor remote sensing data for crop rotation mapping. Int. Arch. Photogramm. Remote Sens. Spat. Inf. Sci. **25**(1), 177–182 (2012)
52. Song, M.-L., Fisher, R., Wang, J.-L., Cui, L.-B.: Environmental performance evaluation with big data: theories and methods. Ann. Oper. Res. 1–14 (2016)
53. Furbank, R.T., Teste, M.: Phenomics—technologies to relieve the phenotyping bottleneck. Trends Plant Sci. **16**(12), 635–644 (2011)
54. PEAT UG.: Plantix [Online] Available at http://plantix.net/. Accessed 2017 (2016)
55. González-Recio, O., Toro, M., Bach, A.: Past, present and future of epigenetics applied to livestock breeding. Front. Genet. **6**, 305 (2015)

Generic IoT Platform for Analytics in Agriculture

Balasubramani Pradeep, R. Balasubramani, Jason Elroy Martis and M. S. Sannidhan

Abstract Recently, it has become trend to use Internet of things for our day-to-day life, wherein we use the setup of various devices for the large collection of data and its transmission in various circumstances. It is actually made possible and success due to the involvement of different sensors that are capable of collecting the type and amount of data required. But many a times when using the setup of Internet of Applications for any kind of developed applications or an application to be developed, it is observed that the setup is built for a specified purpose and provides a one-dimensional application which basically restricts the usage of the developed system for some other purposes or does not provide a flexibility of extending the application to add further modules. In regard to this, our research work aims in creating a universal platform used especially in the field of agriculture for data transmission and reception from various sensors and plotting the statistical graphs on a dashboard. Apart from this as purpose of providing easy usage and extensibility smart devices can also be installed which provides the plug and play capability of our system.

Keywords Internet of things · Smart sensors · Real-time data streaming · Agriculture

B. Pradeep
University of Texas at Dallas, Richardson, USA
e-mail: pradeep.b2505@gmail.com

R. Balasubramani (✉) · J. E. Martis · M. S. Sannidhan
NMAM Institute of Technology, Nitte-54110, Udupi, Karnataka, India
e-mail: balasubramani.r@nitte.edu.in

J. E. Martis
e-mail: jason1987martis@nitte.edu.in

M. S. Sannidhan
e-mail: sannidhan@nitte.edu.in

P. K. Pattnaik et al. (eds.), *Internet of Things and Analytics for Agriculture, Volume 2*, Studies in Big Data 67,
https://doi.org/10.1007/978-981-15-0663-5_11

1 Introduction

The Internet of things (IoT) is a processing idea that combines the possibility of standard physical items being virtually linked with the Internet and having the capability to organize and collaborate themselves to live gadgets. The IoT buzzword is gargantuan on the principal grounds, especially in agriculture that can speak to itself by carefully moving toward an option that is more prominent than being independent from anyone else [1]. Never again does the query of data relate just to all clients, however, it is linked with encompassing articles and knowledgeable information. At present when numerous devices or gadgets combinatory as one, are known to have a so called "surrounding knowledge". Internet of things (IoT) does the work of all the physical gadgets that we can envision in our everyday life for gathering and communication of information through different stimuli and responses. This makes life manageable and automated with minimal to no human interruption [2].

Currently, the field of agriculture is more hit upon as people do not get much access to better and affordable technologies. It is also envisioned that if the trend continues the field of agriculture will drop down to a point of no return. Countries whose GDP depends much upon agriculture will be the most affected since 80% of the country depends upon agriculture. People, nowadays, talk a lot about bringing the word "smart" to the agriculture and this "Smart Agriculture" has certainly revolutionized the domain. It could only be done with the help of sensing external stimuli and communicating effectively to make a knowledgeable decision even at a microlevel scale. Internet of things immediately comes to the rescue which aids in the creation of these sensors and various interfaces. The major problem of Internet of things lies in the crux of interface for communication. Nowadays, developers tweak and develop application from scratch, which makes the life of software cycle much greater and too cumbersome. Therefore, there exists an urgent need in creation of a generic platform for detail communication of measurements from various sensors [3–6].

These sensors effectively communicate to the main controller which plots the graphs for analytical reasoning. This article focuses on the process of communication of data from the sensors effectively to the client for analytics and decision making. It also includes processes to set up a "Smart Switch and plug in which a device can be plugged in and controlled using a dashboard application after authentication. The data transmission is also kept confidential and secure to prevent intrusion and preserve clarity of informational integrity that is being exchanged between an authenticated client and server [7].

1.1 Internet of Things

The Internet of things (IoT) is the expansion of internet connectivity into numerous gadgets and everyday use items. The objects are embedded with electrical chips, connectivity protocols, and other stimuli sensing hardware (such as sensors), and these lively gadgets can interact and interpret with other machines over the cloud network, and they can be casually supervised and managed.

The definition of the acronym IoT has substantially improvised and has been perfected due to the convergence of multiple technologies, instant data analytics, artificial intelligence, commodity stimuli, and embedded system sensors. The age-old technology of embedded systems, wireless sensor networks, control and automation systems, and others cohesively help in the contribution to the Internet of things. In the current market scenario, Internet of things technology is tangibly linked with merchandises pertaining to the concept of the “smart”, gadgets and products (such as lighting fixtures, thermal conductive stats, home systems security and movement detection cameras, and other home appliances) are extensively connected, and can be coordinated via wide range of extra venous devices associated with that ecosystem, such as smartphones and intelligent voice activators and assistants [8].

The IoT concept has constantly being criticized closely, especially in regard to secrecy and confidentiality and privacy concerns related to these devices and their intention of pervasive presence. It is seen that Internet of things extends to very magnanimous set of applications for embedded devices. This set of devices are divided into consumer, industrial, and infrastructure spaces.

1.1.1 Consumer Applications

A very large portion of these smart devices is invented for the sole reason for consumer sale. They include a very wide array of devices ranging from connected vehicles to house management devices. Some examples are “SMART HOME”, “GOOGLE HOME”, etc.

1.1.2 Industrial Applications

The IIoT refers to a wide array of connected sensors, sensitive instrumentation devices, and other industrial devices inter-networked together with industrial appliances that include manufacture processes and power generation. This internetworked scenario allows for data assimilation, exchange, and analysis, potentially influencing advancements in productivity and speed of service as well. The industrial Internet of things is an evolving distribution-based controlling system which is abbreviated as DCS. It gives a scope for a higher degree of influential control of cloud computing for refinement of industrial processes [9, 10]. The IIoT is backed up by mechanisms such as computer secrecy and security, cloud computing, edge computing devices, mobile communication technologies, machine-to-machine interactions, three-dimensional printing, advanced interaction robotics, cognitive computing, and human–computer interfacing.

1.1.3 Agriculture

There are innumerable applications in the field farming for Internet of things. They include stimuli collection such as data on temperature, rains, humidity transitions, wind conditions, pest control and management, and soil nature content. This information is utilized for the purpose of automation on farming methodologies, analyze and calculate well-informed decisions for the dynamic improvement in quality and quantity and to minimize risk and waste, and reduce effort required to manage crops. For example, farmers can now effectively supervise soil temperature and moisture content of air from long distances and even apply calculated acquired data to perform micro-precision fertilization programs. In August month of the year 2018, Toyota introduced a dynamic partnership with Microsoft to create aquatic farming tools using their web application suite for Internet of things technology. This helped in generation of methodologies related to water management [11].

1.2 Internet of Things Architecture

See Fig. 1.

Following is the architecture of the Internet of things. The bottom layer proceeds through a wide variety of environmental stimuli also known as "things" in the IoT world. These things give a lot of information about the environment as well as help in building a large dataset of linked informational context. Stage 1 form the primary stage of the IoT architecture. It includes the sensors which sense the information omitted by the "IoT things" and converts them into digital data for further processing.

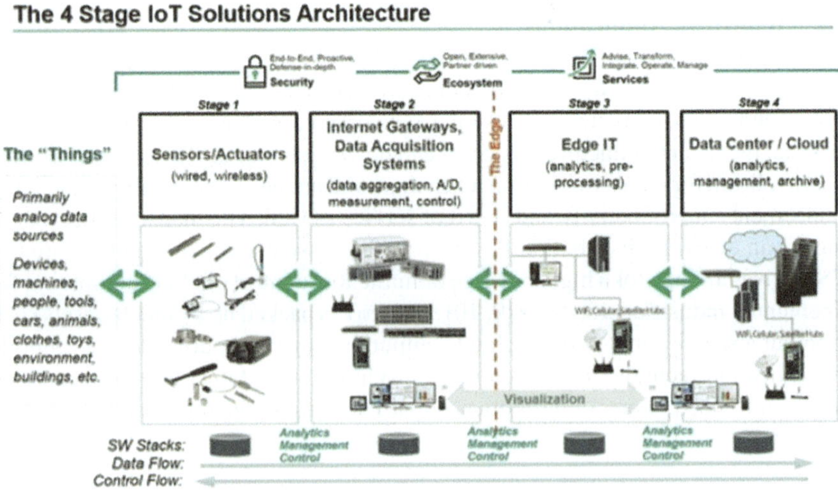

Fig. 1 Architecture of internet of things

Stage 2 includes data acquisitions and gateways which assimilate all the data gathered from various sensors and converts them into appropriate metrics for measurement and control. The data assimilation and control helps the gateways to visualize data in plots, sheets, and graphs for further processing. Stage 3 converges to a term called as "Edge". The data acquisition from Stage 2 from various gateways is transmitted to the edge infrastructure, which forms to be a backbone of the IoT framework. Data to the edge nodes are transferred mostly via (Wi-fi, LoRa, or ZigBee Transfer) wireless transfer for ease of access. Pre-analytical processing is done if required for further stages. Stage 4 is the topmost stage. The main processes in the last stage of IoT architecture converge in the cloud. It directly dives deep into the processing perspective, having a short follow-up revision for feedback. The data is then brought back to the world analyzed and well prepared for controlled and detailed assistance and management [12–14].

Our entire work is divided into two major categories, namely the hardware and the software section. The prior involves arranging the circuit setup for the data collection, and we involve a small subset of sensors for the proposed work. The software section creates a framework or an IoT platform for the user using multi-platform web interfacing. It provides much functionality to the user like user access and authentication, device management and control stats.

2 Related Works

In their research article [15], Heo and Kim proposed an IoT framework that specially focused on proving the security services. Article intended its work on security layer to provide security management system. As a part of implementation, researchers proposed a novel security architecture for an IoT system entitled M2M security architecture.

Article [16] articulated by Ammar.et al. proposed an IoT framework that focused on providing simplicity in implementing various IoT applications. In their implementation, they even focused on providing a strong security aspect for sensitive data used in their application sets. As an outcome of their work, they proposed a novel architecture for their framework which included all the essential elements for any of the application development over an IoT along with a cross-compatible hardware and essential security concepts.

In their article [17], authors have stressed the importance of security systems for any of the IoT framework. Apart from this, they have even emphasized the need of heterogeneous interoperability feature for any IoT framework. In this regard, they have succeeded in identifying the loopholes involved in proposing a standard security system. As a contribution to the research community, researchers in their article provided a set of guidelines that needs to be adopted by any IoT systems in terms of providing an efficient security aspect.

Anindya Dey, Kim Stuart, Matthew E. Tolentino in their research article [18] entitled "Characterizing the Impact of Topology on IoT Stream Processing" proposed a real-time data collecting sensor using a real-time platform system for an IoT named as "Namatad". As an outcome of this article, researchers were successful in identifying a topology for routing in an edge-based server with minimal latency and maximum QoS. They also succeeded in implementing a machine learning model for their work that can identify an alternative topology with respect to scaling and sensor deployment.

In their article [19], authors have proposed the implementation of SoC virtual platform for the IoT terminals. In this case, authors have focused on providing flexible software development for an IoT platform through SoC virtual platform without having to involve an actual hardware system that is required. This even provided the advantage of using any developed software immediately on the board as they are developed in a uniform environment. Result comparisons were successfully carried out and verified using a Raspberry Pi board which proved the same result as they were deployed on the board with a flexibility of development.

Article [20] authored by Dong-Hwan Park et al. proposed on the improvement of interoperability between any of the IoT devices making it easy for the usage. The focus was to provide ease of development in semantic IoT services. As a result, they targeted on developing an open IoT service platform for an easy access, easy connection and also more importantly an easy development. As an outcome, proposed system offered various services like proving a sleep satisfaction, sensor observation system in aggregation, and analysis of sleep pattern system. Ultimately proposed system provided the advantage of both interoperability and scalability of services on the IoT platform.

In their research article [21], paper discussed the proposal of implementing an Android-based IoT platform that could predict the security threats occurring in future. As a part of implementation, they focused on relying upon the permissions that arrive before and after update of any software system. Through this, a system was developed that could identify the security threats occurring based on the differences in permissions.

In an article [22], authors have discussed about challenges and limitations that are imposed when implementing a security aspect for an IoT system. According to their proposed research, securing of an IoT by covering the aspects of storage system, energy level and a protocol system of lightweight concept is more suitable. The aspects covered in modifying an existing system as it provided limited level of resources. Hence this research proposed a security protocol technique Physical Unclonable Functions (PUF's) considering the limitation in resources.

In their article [23] entitled "Light Weight Security Framework for IoT" proposed by Thenmozhi et al. have concentrated on implementing a security system based on lightweight cryptography. The technique aimed at proposing the lightweight cryptography system due to the availability of limited physical resource configurations like memory systems, processors, and also power performances. As an outcome of their research article, they succeeded in implementing a lightweight security system with a limited amount of data that ensure the data confidentiality. Experimental results have proved that the proposed system provided a better and more efficient result when compared to the existing system in terms of security aspects. Apart from this as a part of ensuring power dissipations, LIFI system was used in place of Wi-fi.

In a research article [24], researchers have MSSDL life cycle which basically focused on increasing the reliability of software systems in terms of the security aspects with respect to the bugs that usually encounter. The proposed system intended in providing a great security system for the internal architecture of the Microsoft systems. But the major drawback identified in the proposed system was its nature of being heavyweight to the security aspects.

In their research article [25], authors have closely observed the defects encountered in the research work have proposed a technique called CLASP which functions very similar to MSSDL in functionality but a better aspect of the proposed system is that the system is much more lightweight than MSSDL.

Research article [26] proposed another better technique which is an optimized version of CLASP. Authors in this research article focused on modifying CLASP to provide a framework that helps to develop strategic security for application based on security implementations for business risks.

In a research article [27] entitled "IoTSM: An End-to-end Security Model for IoT Ecosystems "authored by Joseph Bugeja et al. analyzed various security implication challenges with respect to IoT systems and to cover different aspects in tandem, authors have proposed a novel IoT framework that can provide a strategic security systems based on the organizations requirements. Proposed technique also focuses on providing an able matured model for the IoT systems that have the capability of expanding the security practices based on empirical model system.

In their research article [28], the authors have proposed a technique for small and medium-scale IoT solutions based on the domain-level requirement. In the course, they were successful in developing an IoT framework that works on the principle of rapid deployment technique for an industry.

Abbass et al., in [29], have bridged the application of deep learning system to access the security constraints for IoT intelligently. Basically, they have implemented a CNN that can improve the ability of security risk evaluation on the basis of continuous learning system. As a positive outcome of the research work, the proposal was successful in accessing the security threats via deep learning model and proved ably a better performance when compared to traditional techniques. As a part of negative aspect, CNN was found to be a black box model and consumed horrendous amount of time in data collection.

3 Methodology

We propose a generic IoT platform designed especially for agriculture helps us to get the data from various setups and conversely control the devices from distant places using the Internet as a medium. The readings of different sensors can be used to control an air humidifier. In turn it helps in the conservation of resources, which in turn drives the development of new technologies. For example, a smart bulb can help you save and monitor your energy consumption by showing you energy statistics. Moreover, it will help us to regulate the intensity and heat at which your device operates which will indeed help you save power as well as give nutrition to plants. Traditional system designs have been stand alone in their approach providing a specific type of application to the end user based on the requirements posed. Hence, the design approach followed the assembly of hardware components and the corresponding software interface to be highly specific. Our proposed system aimed at being flexible in terms of design is also beneficial for budding IoT enthusiasts as it would help them to gain an in-depth understanding of IoT and encourages to develop a multi-application system which one can use in their day-to-day life and also worth for the society in the field of agriculture (Fig. 2).

The above diagram shows an in-depth view of the proposed generic IoT platform. The area marked in dot-dashed lines marks our area of interest. The framework works in tandem with the IoT framework but separates it in part on with the authentication and confidentiality. The data from the sensors are collectively accumulated and transferred to our platform. This causes the preliminary part in our platform. The device is authenticated using authentication algorithms in order to verify the authenticity of the data and is tagged with a subscription Id and stored in the data store. The filter component is used to verify and strip off the unwanted residual subscription tag as well as only notify the data which is necessary to the client. The encryption and web interfacing wing encrypts the message using numerous encryption algorithms and transfers it to web viewing html format for ease of access viewing to the client in subscription. The data transfer component is a dynamic query processing for in term registration and deletion of the device. It also helps in aiding the filtration process to filter the quality of data given to the client.

The components will be explained in detail under further subsections.

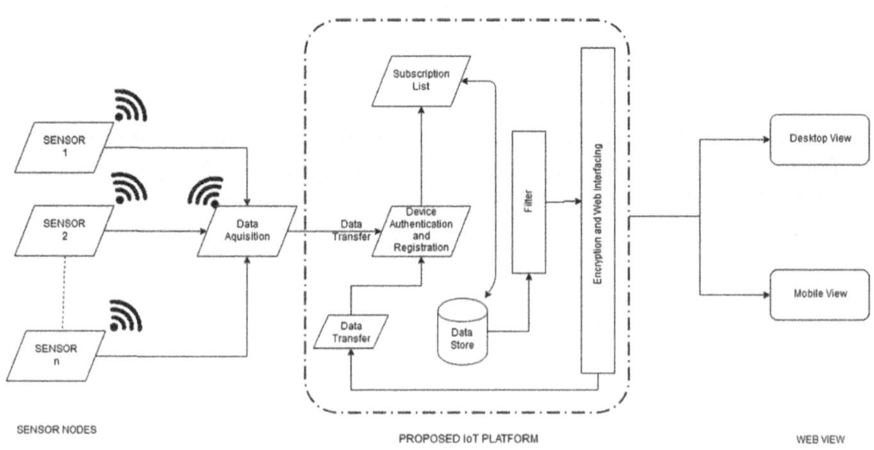

Fig. 2 Architecture for the proposed generic IoT platform for agricultural system

3.1 Device Authentication and Registration

This component is a very active component in our platform. Every sensor node that communicates to our platform has to communicate to the client through it. It ensures complete confidentiality and sees that no unwanted device can tap through data to the client. Every sensor is attached an unique Id to it. Initially, we could have used MAC-ID which is 48 bit in size, but nowadays every device has more than one mac-id, which poses a problem. In order to overcome this, the platform gives a unique registration code once the device communicates with the platform. The device authentication is done by using MAC and SHA-1 algorithms. SHA-1 expanded as Secure Hash Algorithm-1 as according to its name is an algorithm relied over a hashing technique. The technique is used for cryptographic applications to ensure secure transmission. Basically takes some plain text as its input entity and produces an output which is a hash value of considered input. It takes an input message of length n bits and produces an output message of length 120 bits. Generated output is basically known as a message digest [30].

3.2 Subscription List

This component includes a large list of subscribed clients to whom the sensor data readings are sent. The issue arises when multiple users subscribe to the same sensor readings. These are achieved by a well-known protocol called Message Queuing Telemetry Transport (MQTT). MQTT is basically a M2M protocol designed to be a lightweight method of transport. It is also ideal for mobile applications because of its small size, low power usage, minimized data packets, and efficient distribution of information to one or many receivers [31–35]

The main advantages of this protocol are:

1. **Better Error Reporting**—Very precisely, a logical code is being imbibed to required responsiveness for publications (PUBACK/PUBREC). A protocol called MQTT actually came into existence with lots of applicability's alike sensors works on a pipeline model system—in case of a failure occurring in the status of publication to be transmitted and delivered, in that awful case, sensor in action is not going to consider any kind of action from its side. Nonetheless, the applicability for the protocol under discussion is actually much expansive and an application on any cell-phone may also have an eagerness to warn the end user of the system if data required to be transmitted successfully fails to do so. It is very unusual nowadays to have a so-called return codes on every acknowledgement that the protocol system generates.
2. **Shared Subscriptions**—In case the rate of message transmission is considerably immense, subscriptions which are considered for sharing can actually be adopted for the sake of balancing the load, the messages across a number of receiving clients.

3. **Message Properties**—Being represented as a Metadata part of an header system of a message that is meant for the transmission. Ones that are considered are actually made use to develop other set of features in the list that is under consideration along with that it has to even permit the end users to make use of those properties that are being defined by the end user, e.g., to help in the encryption of plaintext by actually intimating the receiving entity about the type of the key to be applied for the purpose of decryption.
4. **Message Expiry**—One of the very effective methods that needs to be employed in which case a preference to abandon the transmitted messages in case if they fail to get delivered into the destination within a specific slice of time as defined by the end user.
5. **Session Expiry**—In case a scenario occurs where a client involved in the communication fails to form a connection within a period of time as defined by the end user, in such cases evidently state (e.g., messages that are buffered and subscriptions) can be disposed off without even cleaning up.
6. **Topic Alias**—One of the systems that actually permits strings in connection to the topic in communicating messages to be replaced with an individual number, cropping the amount of data bytes that actually needs to be transmitted in case if a person in publishing iteratively makes use of the identical topics.
7. **Will Delay**—This system permits an actual communicating to be published if a client that is involved in the communication system loses out of a connection for a time that is higher than the one that is being defined by the end user. Permits indications regarding the disruption of important applications in connection to the client end without being beset by positives encountered out of false.
8. **Allowed Function Discovery**—In the beginning of forming a connection, restrictions alike maximal size of a packet and the amount of (QoS > 0) messages in transmission can be transmitted to notify the client end about its responsibility in terms of communication.

Aforementioned list of subscriptions takes the responsibility of all sensor nodes and the request/responses at the client end by preserving the required contents from the data house and reading out the data during the point when it is actually required.

3.3 Filter

This component deals with the chipping off unwanted contents and verification of the client subscription from the sensor reading and also verifies whether the proper sensor readings are transferred to the user who has genuinely subscribed to it. It also forms an intermediate layer between the data store and the web interface layer.

3.4 Encryption and Web Interfacing

This layer completely deals with the security aspect of data transmission. It receives incoming packets of data from the filter component and encrypts the packet for confidentiality, to prevent masquerade and man in the middle attacks on data transmission. Encryption is done which increases the performance and throughput of encryption and decryption. Web viewing is used to transmit data to the device from which the client wants to view. It can be any device and can have multiple displays. The job of the web view ensures that any resolution display will not hinder the viewing details of the user. It is done by using NodeJS, JQuery, and Bootstrap. The web interface is very dynamic and fully fledged to introduce components to fulfill ease of access.

Logically the platform to be truly generic must be strictly classified into three categories.

1. **Sensor nodes**: This logical subdivision only acquires data from external stimuli and transmits them to the IoT Platform. The data transfer is done mainly via wireless modes of communication. It ensures flexibility and ease of deployment. The wireless modes of transfer are Wi-fi, LoRa, ZigBee.
2. **IoT Platform**: This is the main logical subdivision of the three-tiered M2M architecture of our IoT platform. It includes the data analytics and control mechanisms for data acquisition and control. The data store used is implemented in MongoDB, mainly concentrated on noSQL architecture.
3. **Web View**: The final sub division of the architecture is the web view. It contains codes in JQuery, JavaScript, Bootstrap to utilize full functionality of the client and to reduce the load on the server, bringing in the concept of Fat client. Only data is pumped to the client and displayed beautifully on the browser. It also manages multiple views to bring in flexibility.

4 Results and Discussions

We have tested our generic platform in the field of agriculture by sensing the most common features that farmers depend upon like light ambience, temperature, and humidity. An accelerometer is also attached to show if the sensor setup has fallen or toppled over by extreme wind. These sensor's data are progressively collected by using a microcontroller like Arduino UNO and then passed on to the NodeMCU 12-E to transmit the data wirelessly to the IoT platform. The schematic of the circuit is shown below (Fig. 3).

The results are obtained in the form of charts at our mobile end from the web view as shown below. Note that the user has signaled that the platform should show the web view in the form of plots (Fig. 4).

We have also added another set of IoT device called an actuator which will perform a certain action for a certain event. This is used to switch the light on and off for

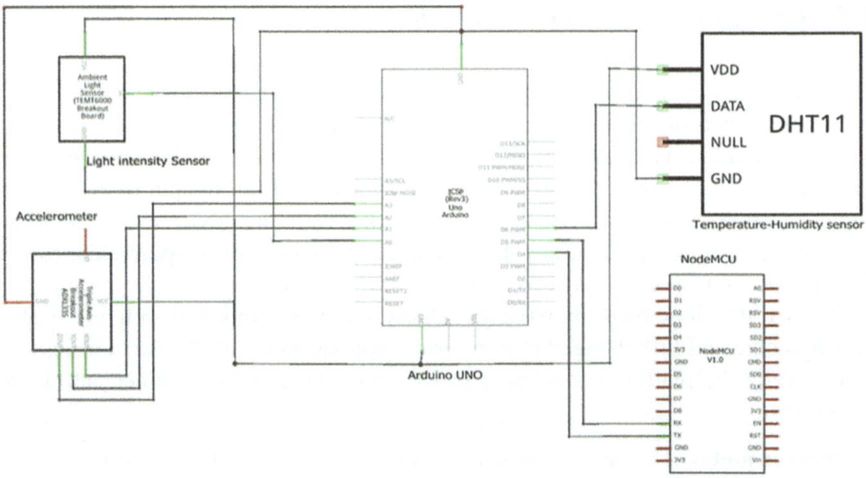

Fig. 3 Schematic diagram for sensors showing the connection to the microcontroller

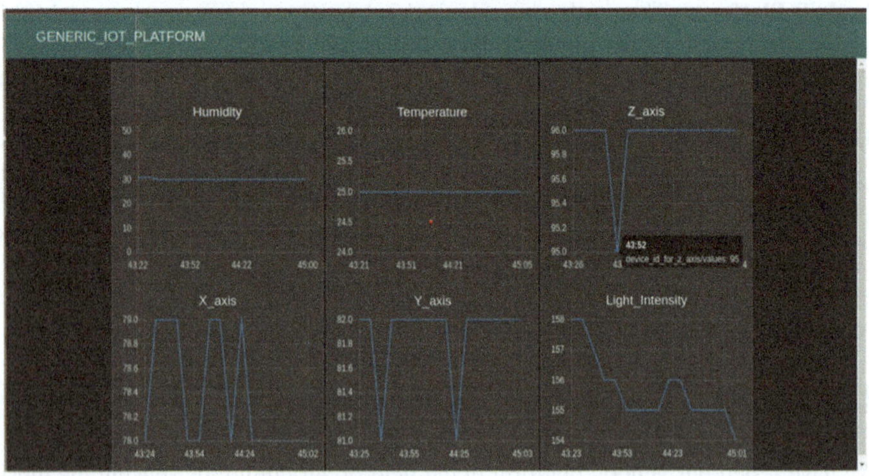

Fig. 4 Dashboard output on our IoT platform

regulation of light intensity of the plants. We call this device "Smart Switch". The schematics of the connection are shown below (Fig. 5).

We know that the NodeMCU cannot directly control AC devices; because they are not configured to do that and the voltages used by AC devices are too high. So, we uncouple the devices by using a device called a relay. The relay used here is a solid state relay called S116s02. The NodeMCU acting as a microcontroller signals the relay by passing a voltage of 5v through an npn transistor which in turn switches

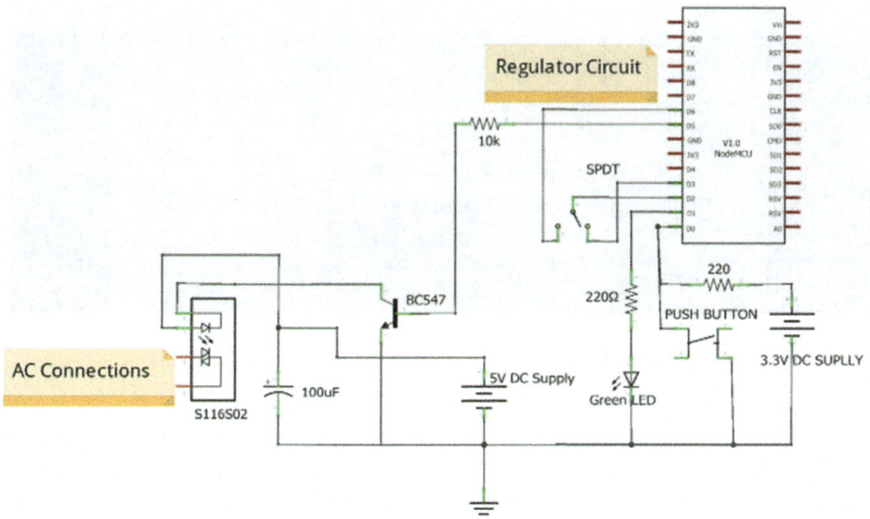

Fig. 5 Schematic diagram for "smart switch"

the light. We can also control the intensity of the light as shown in the schematic below (Fig. 6).

The control of the Smart Switch is given in the dashboard as shown (Fig. 7).

Note that we have configured out platform to set the view of the Smart Switch. It displays control for On/Off as well as adding features for timers and also measures the energy consumed by the bulb. The user can directly manage the bulb from a remote place. It can also be configured if it crosses any threshold kWh.

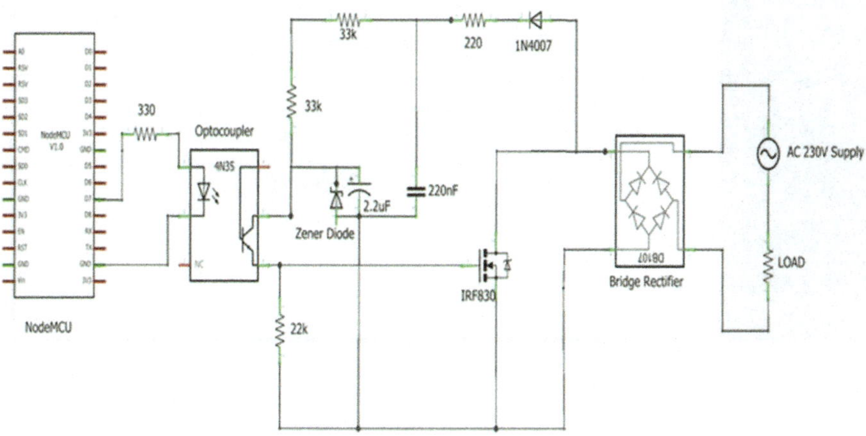

Fig. 6 Schematic diagram for "smart switch" light intensity regulation

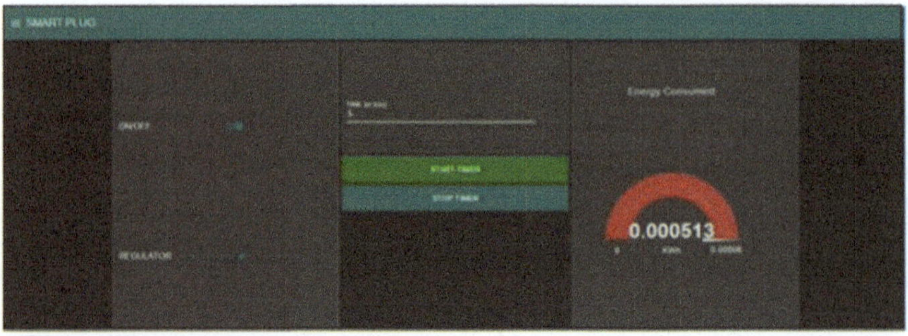

Fig. 7 Dashboard view for "smart switch"

As you know, the sensor and the user have to register before using our platform as well as the sensors have to be authenticated. We have given a sign up page which sends an email with authentication key of the sensor. The front view of the user dashboard is shown below.

Analysis of the data can also be done from our platform as shown in Figs. 8 and 9, respectively, depicted below.

Analysis of the data can also be done from our platform as shown in Fig. 10.

The figure depicted above shows how the temperature analysis is done from a start data to end date. The data from the sensors from our date and time is filtered accordingly from our platform database and then given to the web view as shown.

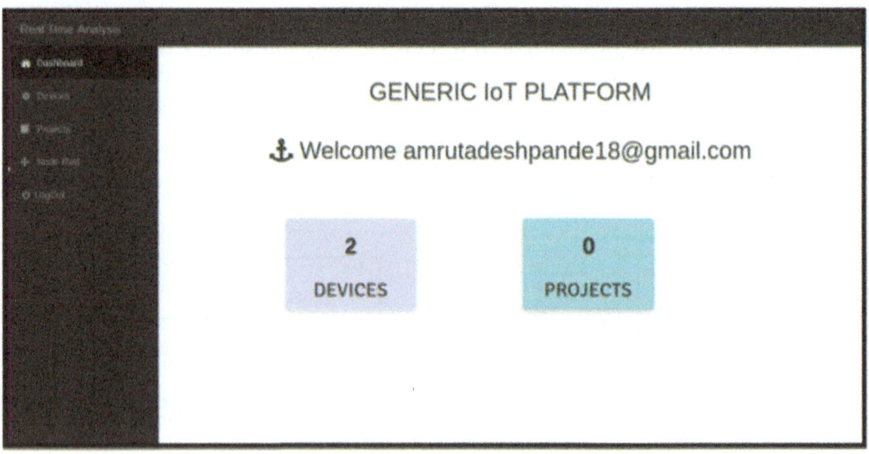

Fig. 8 Dashboard view for user

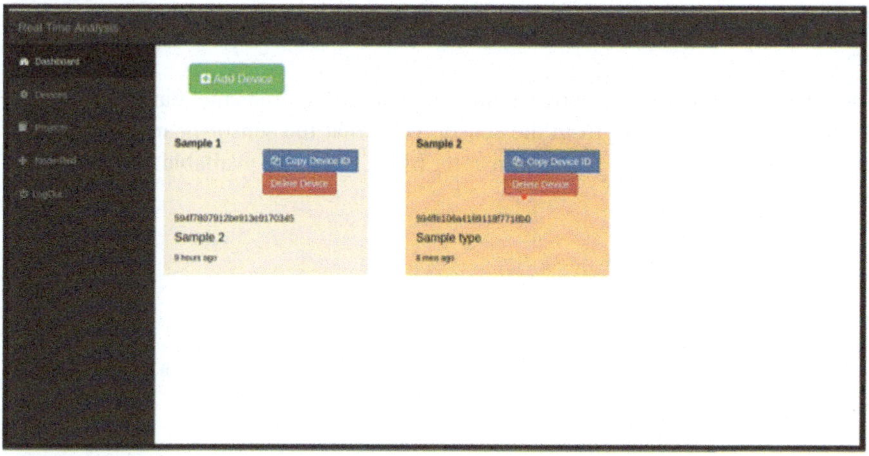

Fig. 9 View for user devices registered and their user id's

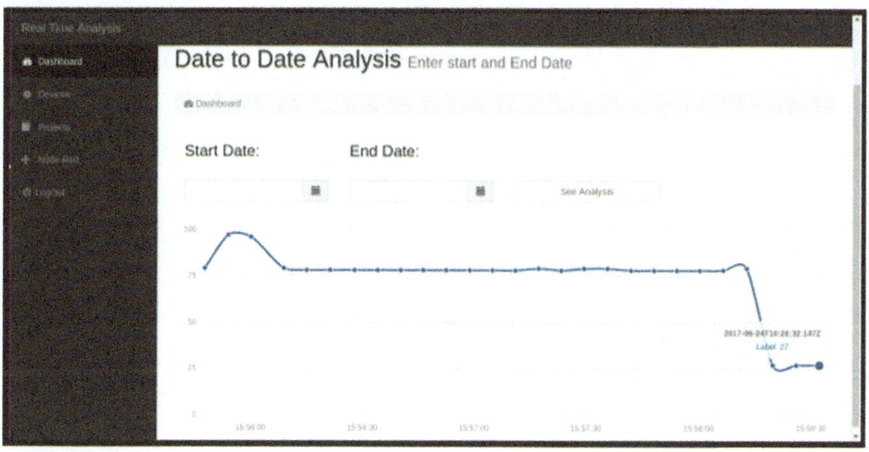

Fig. 10 Analysis view for temperature

5 Program Code and Flow

This section shows below the codes for all different components used in our setup. This section is divided into two parts, i.e., the hardware part code for the sensor and the software part of the IoT platform.

5.1 Light Intensity Sensor and Data Transmission

The code shown below symbolizes how the nodeMCU transmits data and how the Arduino collects the data from the sensor. Note that the sensor is analog in nature and requires an analog-to-digital converter (ADC), already available in the Arduino Board.

```
void intensity( ){

/*

Read the data from analog pin and map them from 0 to 255


*/

Inten=map(analogRead(A0),0,1024,0,255);

/*

map function maps A0 from 1024 values to 255 values only

*/

Serial.println("Light Intensty is: ");

/*

Serially prints the light intensity with a message

*/

Serial.println(inten);

/*

Prints the light intensity value that is being mapped


*/
```

Code that is used to write the data into the nodeMCU captured by Arduino

```
Void writedata( )
{

/*
Module to write the data
*/
Serial.write(temp);

/*
 serially transmit the temperature from Arduino to
nodeMCU
*/
Serial.write(hum);
/*
 serially transmit the humidity from Arduino to nodeMCU
*/

Serial.write(tinten);

/*
 serially transmit the light intensity from Arduino to
nodeMCU
*/

}
```

Code for nodeMCU transmitting the data to the platform

```
while(Serial.avalable( ))
      {
            /*
            check for serial availability from the
            Arduino
            */
            x=Serial.read();
            /*
            Convert the data to a specified format for
            our proposed platform
            */

            snprintf(msg.75,"%d",x);
```

```
        if(i==0)
        {

        /* Encrypt the message using encrypt message
        function along with authentication key
        */
    encryptMessage(msg,"594a734/4569872156/89values");
        }


    }
```

5.2 *Smart Switch*

The Smart Plug performs the following functionalities

i. ON/OFF devices (~100 W) from mobile app
ii. Regulation of light intensity from mobile app
iii. Timer functionality for turning the switch ON/OFF
iv. Energy consumption graph in kWh
v. Indicator or Smart Switch for the following conditions

- Connected
- Connecting
- Disconnected
- Reconfiguration of the NodeMCU in the Smart Switch.

The flowchart of the Smart Switch goes as follows (Fig. 11)

5.3 *Data Store*

The data store is implemented in Mongo DB and interacted using NodeJS script. Schemas are created for device authentication and user identification. The schemas are given below.

5.3.1 Schema for Device Authentication

Below is the code depicted for the design of the schema pertaining to the authentication required for an IoT device.

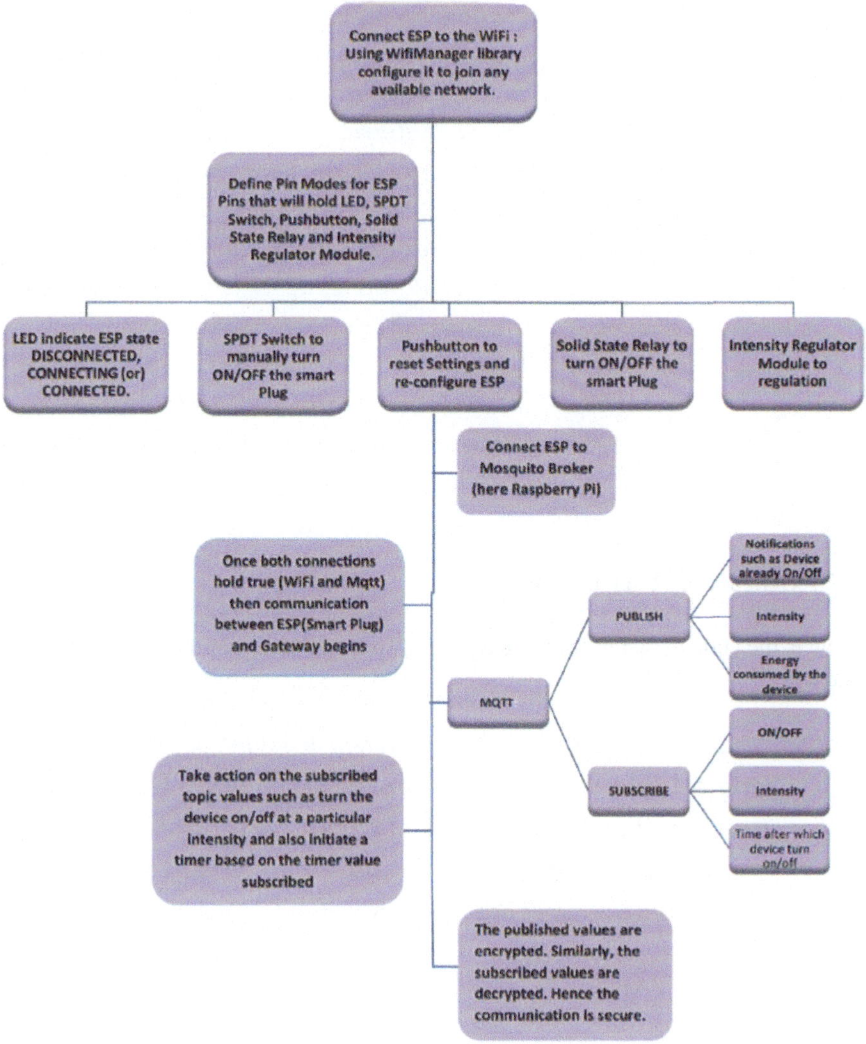

Fig. 11 Flowchart for the "Smart Switch" working

```
var authSchema = mongoose.Schema( {local : {
            api_key : String, (a unique string to
            identify each user(generated server-side)
            email : String, (email id of the user:
            verification email will be sent on this mail)
            password : String, (password of the user: for
            logging in after verification, hashed)
            url : String, (url generated server-side to
            be sent to the email for verification)
            state : Boolean, (If verified: state = true -
            > schema is deleted from collection, else
            false)},{collection:'authSchema',safe:true});
```

5.3.2 Schema for User Details

Below is the code depicted for the purpose of storing the required user details in the database.

```
local : {api_key : String,a unique string to
      identify each user(generated server-side)
      email : String, email id of the user:
      verification email will be sent on this mail
      password : String, password of the user: for
      logging in after verification, hashed using
      b-crypt node});
```

5.3.3 Schema for Notification Details

Below is the code depicted for the purpose of sending the notification to the device that is registered with the proposed system.

```
var notification = mongoose.Schema({time : {type :
      Date, default: Date.now}, time of receiving
      the notification from the device
      message: String message conveyed by the
      notification});
```

5.3.4 Schema for Sensor Details

Below we have depicted a code that is able to store the data collected from the sensor nodes into our proposed platform data store.

```
var sensorSchema = mongoose.Schema({device_name :
      String,// user inputted string to identify
      the device
      device_id : { type: String, unique: true },
      server generated string to uniquely identify
      a device
      description : String, user inputted
      description of the device
      last_access : {type: Date, default:
      Date.now}, time of creation of the device
      (name is misleading)
      device_type : String, user inputted : type of
      device
      timer : {type:String,default:"false"},
      boolean, if timer is set or unset
      timerDate : String, if timer is set, the date
      to which it is set to
      project_id : {type:String,default:""}, if the
      device belongs to a project, this field
      stores the id of the project it belongs to,
      else stores nothing
```

```
intensity : {type:String,default:"0"}, last
set intensity of the device
state : {type: String, default: "false"},
state is true if device is on, else false
activity : [notification],// array of all the
notifications     sent by the given device to
the application.
values : [sensorDataSchema],// stores all the
time-value pairs used to plot the data sent
by the device
energy : [ ]// Array of numbers, stores the
energy values for the last 30 days.
// The indices of the array correspond to the
days of a month.
},{collection:'sensor',safe:true});
```

5.3.5 Schema for Sensor Data

The code that is put up is used for the web view of our proposed platform.

```
var sensorData = mongoose.Schema({
    time : {type : Date, default:
    Date.now},timestamp of the instant when the
    data value to be plotted is received by the
    application
    value : Number, data value to be plotted,
    sent by the device});
```

6 Conclusion and Future Work

Internet of things has become a mandatory requirement in our day-to-day lives, and it is integrating into almost everything that we use every day. The only problem that arises with this is the communication platform. For this, we propose a novel solution that is completely generic on both ends, i.e., the objects as well as the users. This solution provides a straightforward approach which tackles the communication platform scenario as well as introduced security and feasibility. We have also tested this platform on our own built IoT devices such as Smart Switch and agricultural IoT sensory devices. The major advantages of this are ease of connectivity for analytics which even motivates an end user with very minimal knowledge on IoT frameworks.

In future, this work can be extended on all platforms that intensively depend on IoT devices. We can further incorporate machine learning and deep learning approaches for automatic analyses which make the system having minimal dependence on end users.

References

1. Sannidhan, M.S., Martis, J.E. and Bhandary, A.: A cost effective approach for detecting electricity theft using raspberry pi board. In: 2017 International Conference on Current Trends in Computer, Electrical, Electronics and Communication (CTCEEC) (2017)
2. Chanthakit, S., Rattanapoka, C.: MQTT based air quality monitoring system using node MCU and node-RED. In: 2018 Seventh ICT International Student Project Conference (ICT-ISPC), Nakhonpathom, pp 1–5 (2018)
3. Shenoy, J., Pingle, Y.: IOT in agriculture. In: 2016 3rd International Conference on Computing for Sustainable Global Development (INDIACom), New Delhi, pp. 1456–1458 (2016)
4. Prathibha, S.R., Hongal, A., Jyothi, M.P.: IOT based monitoring system in smart agriculture. In: 2017 International Conference on Recent Advances in Electronics and Communication Technology (ICRAECT), Bangalore, pp. 81–84 (2017). https://doi.org/10.1109/icraect.2017.52
5. Jaiganesh, S., Gunaseelan, K., Ellappan, V.: IOT agriculture to improve food and farming technology. 2017 Conference on Emerging Devices and Smart Systems (ICEDSS), Tiruchengode, p. 260–266 (2017). https://doi.org/10.1109/icedss.2017.8073690
6. Heble, S., Kumar, A., Prasad, K.V.V.D., Samirana, S., Rajalakshmi, P., Desai, U.B., Sabeena, S.: A low power IoT network for smart agriculture IEEE 4th World Forum on Internet of Things (WF-IoT). Singapore **2018**, 609–614 (2018). https://doi.org/10.1109/WF-IoT.2018.8355152
7. Chaczko, Z., Braun, R.: Learning data engineering: creating IoT apps using the node-RED and the RPI technologies. In: 2017 16th International Conference on Information Technology Based Higher Education and Training (ITHET), Ohrid, pp. 1–8 (2017)
8. Rajalakshmi, A., Shahnasser, H.: Internet of things using node-red and alexa. In: 2017 17th International Symposium on Communications and Information Technologies (ISCIT), Cairns, QLD, pp. 1–4 (2017)
9. Abinaya, P., Geetha, C.: Dynamic detection of node replication attacks using X-RED in wireless sensor networks. In: International Conference on Information Communication and Embedded Systems (ICICES2014), Chennai, pp. 1–4 (2014)
10. Lekić, M., Gardašević, G.: IoT sensor integration to node-red platform. In: 2018 17th International Symposium INFOTEH-JAHORINA (INFOTEH), East Sarajevo, pp. 1–5 (2018)
11. Kodali, R.K., Anjum, A.: IoT based home automation using node-red. In: 2018 Second International Conference on Green Computing and Internet of Things (ICGCIoT), pp. 386–390. Bangalore, India (2018)
12. Zhong, C., Zhu, Z., Huang, R.: Study on the IOT architecture and gateway technology. 2015 14th International Symposium on Distributed Computing and Applications for Business Engineering and Science (DCABES), Guiyang, pp. 196–199 (2015). https://doi.org/10.1109/dcabes.2015.56
13. Saadeh, H., Almobaideen, W., Sabri, K.E.: Internet of things: a review to support IoT architecture's design. In: 2017 2nd International Conference on the Applications of Information Technology in Developing Renewable Energy Processes & Systems (IT-DREPS), Amman, pp. 1–7 (2017). https://doi.org/10.1109/it-dreps.2017.8277803

14. Navani, D., Jain, S., Nehra, M.S.: The Internet of things (IoT): a study of architectural elements. In: 2017 13th International Conference on Signal-Image Technology and Internet-Based Systems (SITIS), Jaipur, pp. 473–478 (2017). https://doi.org/10.1109/sitis.2017.83
15. Heo, S., Kim, H.W.: An analysis of IoT security requirements and communications of the Korean, pp. 16–22 (2017)
16. Ammar, M., Rusello, G., Crispo, B.: Internet of things: a survey on the security of IoT frameworks. J. Inf. Secur. Appl. **38**, 8–27 (2018)
17. Hwang, I., Kim, Y.-G.: Analysis of security standardization for the internet of things. In: 2017 International Conference on Platform Technology and Service (PlatCon) (2017)
18. Dey, A., Stuart, K., Tolentino, M. E.: Characterizing the impact of topology on IoT stream processing. In: 2018 IEEE 4th World Forum on Internet of Things (WF-IoT) (2018)
19. Lee, H.-R., Lin, C.-H., Park, K.-H., Kim, W.-J., Cho, H.-J.: Development of SoC virtual platform for IoT terminals based on OneM2M. 2017 International SoC Design Conference (ISOCC) (2017)
20. Park, D.-H., Bang, H.-C., Pyo, C. S., Kang, S.-J.: Semantic open IoT service platform technology. 2014 IEEE World Forum on Internet of Things (WF-IoT) (2014)
21. Park, I.-K., Kwak, J.: Implementation of permission management method for before and after applications the update in android-based IoT platform environment (2017)
22. Idriss, T., Idriss, H., Bayoumi, M.: A PUF-based paradigm for IoT security. 2016 IEEE 3rd World Forum on Internet of Things (WF-IoT) (2016)
23. Thenmozhi, S., Ranjitha, J.B., Anu, S., Soniya, A.J.: Light weight security framework for IoT. In: 2019 5th International Conference on Advanced Computing and Communication Systems (ICACCS) (2019)
24. Howard, M., Lipner, S.: The security development lifecycle: SDL: a process for developing demonstrably more secure software, vol. 8. Microsoft Press (2006)
25. Graham, D.: Introduction to the CLASP process (2006) [Online]. Available https://goo.gl/wducjb [Online]
26. OWASP: Software assurance maturity model. OWASP. [Online] (2018). Available https://goo.gl/9cCA4h [Online]
27. Bugeja, J., Vogel, B., Jacobsson, A., Varshney, R.: IoTSM: an end-to-end security model for IoT ecosystems. 2019 IEEE International Conference on Pervasive Computing and Communications Workshops (PerCom Workshops) (2019)
28. Vakaloudis, A., Oleary, C.: A framework for rapid integration of IoT Systems with industrial environments. In: 2019 IEEE 5th World Forum on Internet of Things (WF-IoT) (2019)
29. Abbass, W., Bakraouy, Z., Baina, A., Bellafkih, M.: Classifying IoT security risks using deep learning algorithms. In: 2018 6th International Conference on Wireless Networks and Mobile Communications (WINCOM) (2018)
30. Chen, H.W., Lin, F.J.: Converging MQTT resources in ETSI standards based M2M platform. In: 2014 IEEE International Conference on Internet of Things (iThings), and IEEE Green Computing and Communications (GreenCom) and IEEE Cyber, Physical and Social Comp
31. Tantitharanukul, N., Osathanunkul, K., Hantrakul, K., Pramokchon, P., Khoenkaw, P.: MQTT-topics management system for sharing of open data. 2017 International Conference on Digital Arts, Media and Technology (ICDAMT), Chiang Mai, pp. 62–65 (2017)
32. Mishra, B.: TMCAS: An MQTT based collision avoidance system for railway networks. In: 2018 18th International Conference on Computational Science and Applications (ICCSA), Melbourne, VIC, pp. 1–6 (2018)
33. Bryce, R., Shaw, T., Srivastava, G.: MQTT-G: a publish/subscribe protocol with geolocation. In: 2018 41st International Conference on Telecommunications and Signal Processing (TSP), Athens, pp. 1–4 (2018)
34. Sasaki, Y., Yokotani, T., Mukai, H.: Comparison with assured transfer of information mechanisms in MQTT. 2018 International Japan-Africa Conference on Electronics, Communications and Computations (JAC-ECC), Alexandria, Egypt, pp. 95–98 (2018)
35. Harsha, M.S., Bhavani, B.M., Kundhavai, K.R.: Analysis of vulnerabilities in MQTT security using Shodan API and implementation of its countermeasures via authentication and ACLs. In:2018 International Conference on Advances in Computing, Communications

Mission SDGs Through Food Waste Management: Nature and Approaches

Arindam Chakrabarty and Uday Sankar Das

Abstract The global fraternity has been embarrassed to understand how to feed the elephantine population of this planet. The greed of comfort, convenience, and technology ecosystem has mesmerized our life in such a fashion that the human society has been up-rooted from the nature. The wave of infrastructure development results in heavy encroachment of fertile land as well as yielding to regressive fertility of land. On the contrary, the society has not learned to optimize the utilization of resources whatever forms it may be. These compel the United Nations to formulate specific target-oriented Sustainable Development Goals that need to be achieved by 2030. Food waste is still a menace of mankind. It may be of many forms and dimensions. Food waste exists in every phase of supply chain. Most surprisingly, the stigma of this menace reaches to our household also. The irony of life is that, on one hand, we are habituated to accept that wastage in foodstuff as part of our livelihood, and on the other hand, the United Nations reveals its concern for poverty, hunger, and many other unaccomplished goals. The world is now on the verge of Fourth Industrial Revolution. The IoT-based ecosystem has been emerged as an inseparable entity of the modern societies. This paper has attempted to assess and account the loss of economy for wastage of food items from global canvas to national perspective. This study has also focused on how to use IoT platform so that the food wastage can be reduced up to a considerable amount both in the supply chain and even in household practices. This research work is based on secondary information like research papers, reports, and results of other relevant studies. The paper has attempted to develop and devise a conceptual and strategic model where the IoT ecosystem can be incorporated to ensure real-time solutions and to curb on massive food wastage practices. If the model is implemented and practiced with appropriate case specific modifications

A. Chakrabarty (✉) · U. S. Das
Department of Management, Rajiv Gandhi University (Central University), Rono Hills: Doimukh, Itanagar, Arunachal Pradesh 791112, India
e-mail: arindam.management@gmail.com; arindam.chakrabarty@rgu.ac.in

U. S. Das
e-mail: udaysankar2011@gmail.com

P. K. Pattnaik et al. (eds.), *Internet of Things and Analytics for Agriculture, Volume 2*, Studies in Big Data 67,
https://doi.org/10.1007/978-981-15-0663-5_12

and customizations, it would make the economy more efficient and address different perspectives and dimensions of UN Sustainable Development Goals (UNSDGs) to a larger extent primarily in Indian Context.

Keywords Food waste management · Supply chain · Fourth Industrial Revolution IoT ecosystem · UNSDGs · India

1 Introduction

The journey of human civilization started with the ignition of fire, invention of wheels, and creations of indigenous tools and techniques through the passage of various ages. Today, we are in the age of information automation and moving toward Fourth Industrial Revolution. The success of our human civilization is based on key primary needs, i.e., food, clothing, and shelter. However, this is the irony of life that in spite of achieving various developmental indicators, the human society is still unable to fulfill the basic needs like food which is reflected in various literatures, research outcomes and from the experiences of reality. To cater the critical issues, the United Nations move ahead from achieving Millennium Development Goals (MDGs) to Sustainable Development Goals (SDGs) where poverty and hunger were given primary thrust.

1.1 Food Waste Across the Globe

Around 1.3 billion tonnes of food produced gets wasted globally which account to around one third of the food produced. A whopping US$680 billion and US$310 billion are lost in the industrialized countries and developing countries, respectively. A 670 and 630 million tonnes of food are wasted for industrialized countries and developing countries. Perishable items like fruits and vegetables account for the highest losses post-harvest. Food loss and food waste per year are roughly account to 20% of oilseeds, 35% for combined for meat, dairy and fish, 30% for cereals, and 40–50% for root crops. Per capita food lost or wasted every year is between 95 and 115 kg/year in Europe and North America. While 6–11 kg a year is wasted in countries of South-Eastern Asia and sub-Saharan Africa. A 40% loss of food items occur at post-harvest or at processing level in developing nations, and same amount is lost at retail and consumer levels in industrialized nations (Table 1; Fig. 1).

Table 1 Per capita food losses and waste (kg/per year), at consumption and pre-consumptions stages, in different regions [1]

Region	Production to retailing	Consumer
Europe	187	94
North America and Oceania	181	115
Industrialized Asia	165	70
Sub-Saharan Africa	159	7
North Africa, West and Central Asia	181	31
South and Southeast Asia	115	13
Latin America	198	24

Adapted from key finding of Food and Agriculture Organization of the United Nations

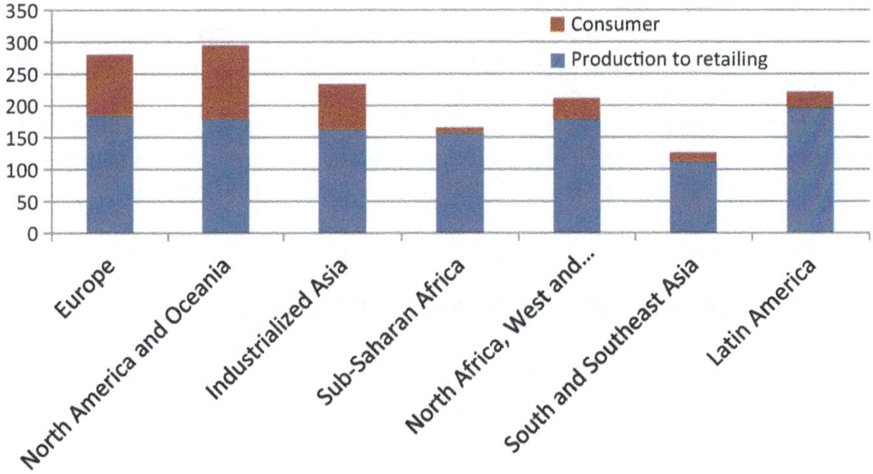

Fig. 1 Per capita food losses and waste (kg/per year), at consumption and pre-consumptions stages, in different regions (adapted from key finding of Food and Agriculture Organization of the United Nations) [2]

1.2 Food Waste in India

India has adorned the developing nation tag since the dawn of independence and is at the frontiers of food loss and food waste. India is ranked at 103 in the latest Global Hunger Index [3] report published in 2018 while a large section of the population still live below poverty line (21.92% of the total population as per RBI data published on Sep 16, 2015) [4] bureaucratic loopholes and intentional ignorance towards the issue has kept the problem alive till date. FCI, a premier food distribution corporation and a state owned utility is yet to fix the grain drain problem. It is reflected in numerous news reports that the food loss by FCI was in the tune of 1.94 lakh MT food grain that had been wasted between 2005–2013 [5]. Apart from this, India still lacks organized cold storage facilities or processing plants within proximity of farmers producing perishable food items like fruits and vegetables.

1.3 Socio-economic Impact of Food Waste in the Context of Sustainable Development Goal

The Global Hunger Index [3] of 2018 point a staggering 124 million people across the globe suffer from acute hunger due to malnutrition, internal displacement, refugee status and poverty. Four of the total 17 goals prescribed by the United Nations SDGs focus on Poverty, Hunger, Good health and wellbeing, and responsible consumption and production (Goal 1: No poverty; Goal 2: Zero hunger; Goal 3: Good health and well-being; Goal 12: Responsible consumption and production). Food loss [6] is the primary reasons why most of the developing world still suffers from malnutrition and hunger which is caused by poor storage and processing in the post production phase and developed nations are highest contributor towards food waste [6] due to fast pace life, irresponsible and ignorant behavior towards the issue. While the former can be solved by application of technology and infrastructure management the later can be solved by improving upon the social physiological behaviors of the masses.

1.4 Introduction to Fourth Industrial Revolution

The term Fourth Industrial Revolution or 4IR [7] was coined by Professor Klaus Schwab, founder and executive chairman of the World Economic Forum in his book "The Fourth Industrial Revolution" based on the technologies of the artificial intelligence, machine learning, quantum computing, 3D printing, and the IoT. Around the year 1760, First Industrial Revolution started in Great Britain, which was powered by the invention of steam engines. Almost a century later, Second Industrial revolution started which was characterized by mass production (from craft-oriented production to mass production) in various industries like steel, oil, and electric. Some of the greatest inventions like internal combustion engine and light bulb came through in the same period. The Third Industrial Revolution or "The Digital Revolution" started somewhere around 1960s powered by the greatest inventions of the century, semiconductor chips, which gave rise to personal computing and eventuality made the Internet a real possibility. Now, 4IR is knocking on the doorsteps of humanity which will eventually change the way we eat, live, think, or nearly everything we do to sustain our life.

2 Literature Review

The unprecedented developments in the fields of digital, physical, biological technology are the three major drivers of the Fourth Industrial Revolution. There is an attempt to understand the impact of these technologies on various global, industrial, economic, and social developments [8].

One quarter of the food supplied for human consumption is wasted across the food supply chain. High-income countries generate food waste at all levels including household which is the highest. The study gathers data from 1062 Danish respondent measuring the intention and attitude not to waste food. Food waste can be controlled by making perceived behavioral control [9].

Sustainable Development Goals (SDGs) are successor of Millennium Development Goals (MDGs). It proposes 17 goals with 169 targets with numerous indicators [10].

Large-scale food waste in the global food supply chain has attracted attention due to its environmental, social, and economic impacts. There has been an attempt to understand the difference between food surplus, avoidable and unavoidable food waste by various specialist, to manage the waste, and to identify the most appropriate mechanism to create a sustainable supply chain management. There is also an attempt to understand and distribute the food surplus to poor people or to convert it as an animal feed [11].

The UN organization, Food and Agriculture Organization estimates a 32 percent loss of food produce in 2009 based on weight while a 24 percent in terms of calorie. Food wastage has negative impact on economy, and it represents a wasted investment, consumer expenses, and farmers' income loss. While food loss refers to losses incurred due to spills and spoils, food waste refers to losses due to infrastructure limitations or post-consumer waste which is generally fit for consumption. Food waste is generally a conscious decision to throw away the food [12].

Wireless sensor network (WSN) surrounds all living beings in the modern era and influences day to day living. A communication between all these wireless-enabled network creates the Internet of Things (IoT) to form a seamless environment to create a common operating picture (COP). The IoT has evolved from the static web2 (social networking web) to web3 (ubiquitous computing web) increasing the data demand [13].

Continuous population growth will keep a continuous demand for food supply for another 40 years approximately, while there will be a decrease in the capacity to produce food due to overexploitation of land and other natural resources including a threat from climate change. However, a more efficient ways of food production can be explored [14].

In order to feed the nine billion human populations by 2050, a review of food waste in the global supply chain is discussed. Data of post-harvest losses of grains are outdated, and current global losses are unknown. The impacts of food waste in the development of BRIC economies are also unknown while developing nations face food wastage post-harvest due to its perishability developed countries contribute to high post-consumer food wastage which suggest a scope for behavioral change to reduce wastage in affluent population [15].

There is an attempt to understand the link between inflation in food prices and riots or food riots. Demonstrators of the riot pointed political repression injustice and inequality which mobilized and bought together various political coalitions to promote human dignity [16].

Electronic nose has been a trending technology for the last two decades largely due to numerous applications built around the sensors. Recent changes in the computing power have given the electronic nose a new possibility of various applications. It has provided a numerous benefits in the fields of biomedical, agricultural, environmental, food, cosmetics, manufacturing, military, pharmaceutical, and various scientific researches. Now, electronic noses can monitor all phases of industrial manufacturing [17].

An introduction to radio frequency identification systems and their strength, weaknesses along with deployment challenges is discussed along with various extensions that offer read/write memory and environmental sensing along with social issues [18].

Electronic noses (e-noses) are sensors that can detect various volatile organic compounds. A wide range of applications can be designed based on pattern recognition with the help of artificial intelligence or neural network. The future trend of the sensors is also explored [19].

3 Objectives of the Study

1. To explore varied forms of agriculture food loss at different stages up to the phase of consumption.
2. To provide an IoT-based solution for reducing food waste to a considerable amount primarily in closed system or household condition.

4 Research Methodology

This paper has attempted to understand, study food waste and its impact in the global context, provide solution through existing technological framework (IoT and electronic nose), and achieve related SDGs. This paper has been developed using secondary information collected from various relevant sources and documents. The paper has focused on how a technological solution would help achieve sustainable development goals of the United Nations by 2030.

5 Analysis and Interpretation

Analysis I

The various forms of agricultural food loss can be expressed by using the flow chart diagram as mentioned below.

Loss of Agricultural Produce at Farming Stage

- Agriculture food loss starts at the very beginning when farmers produce crop without adequate planning and sharing information among the producer's community and other stakeholders, and as a result of that, food loss occurs with overproduction of crops [20].
- Farmers harvest their crop prematurely for his personal consumption or to earn money. This accounts both the economic and nutritional loss of food grains [20].
- Massive use of pesticides and fertilizers diminishes the nutritional value of the food grains, and at times, it could be deterrent to human life.

Loss of Agricultural Produce at Transportation/Supply Chain

- During the movement of crops or food grains from field to warehouse or market, there are several supply chain/transportations loopholes like lack of refrigeration system, using FIFO technique or VED analysis (Valuable, Essential, and Desirable) (Fig. 2).

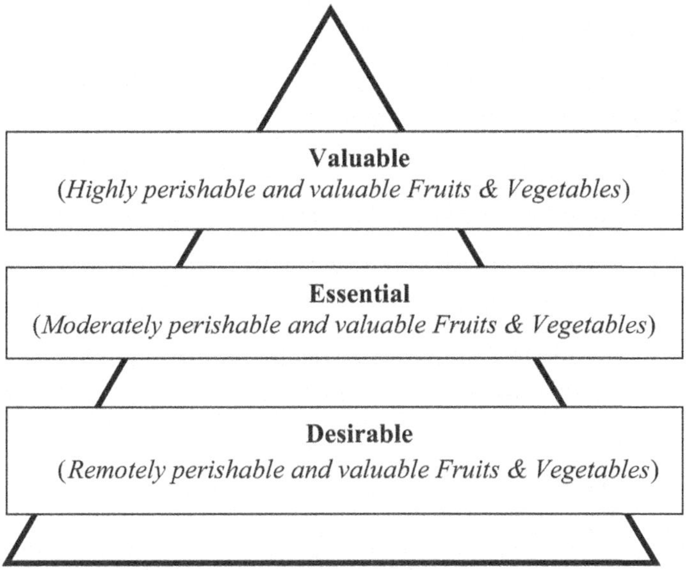

Fig. 2 Schematic diagram of application of VED analysis

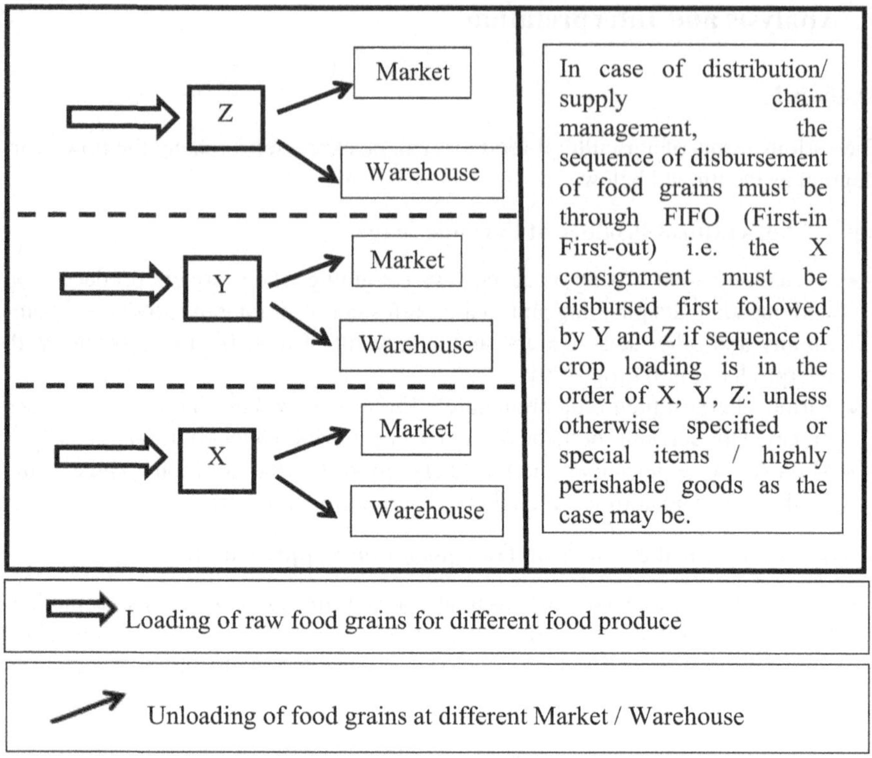

Fig. 3 Schematic diagram: application of FIFO method

Loss of Agricultural Produce at Warehouse/Cold Storage

See Fig. 3.

Loss of Agricultural Produce at Market Place

- In course of selling of food grains in the market, it is difficult to predict actual demand for each variety of food grains or crops. As a result of that, there is substantial loss of food grains in the market places which could not be sold at all or may be sold at a tendered expiry mode.
- In fact, accurate demand forecasting for every set of crops is next to impossible due to uncertainty of market dynamics and impulse purchasing behavior of the consumer.
- Lack of adequate refrigeration and cold storage system.
- Lack of Warehouse facilities.
- Due to excessive overproduction of crops, it results in higher storing cost and stagnation of crops which yield loss of food value, nutrition, and economy (Fig. 4).

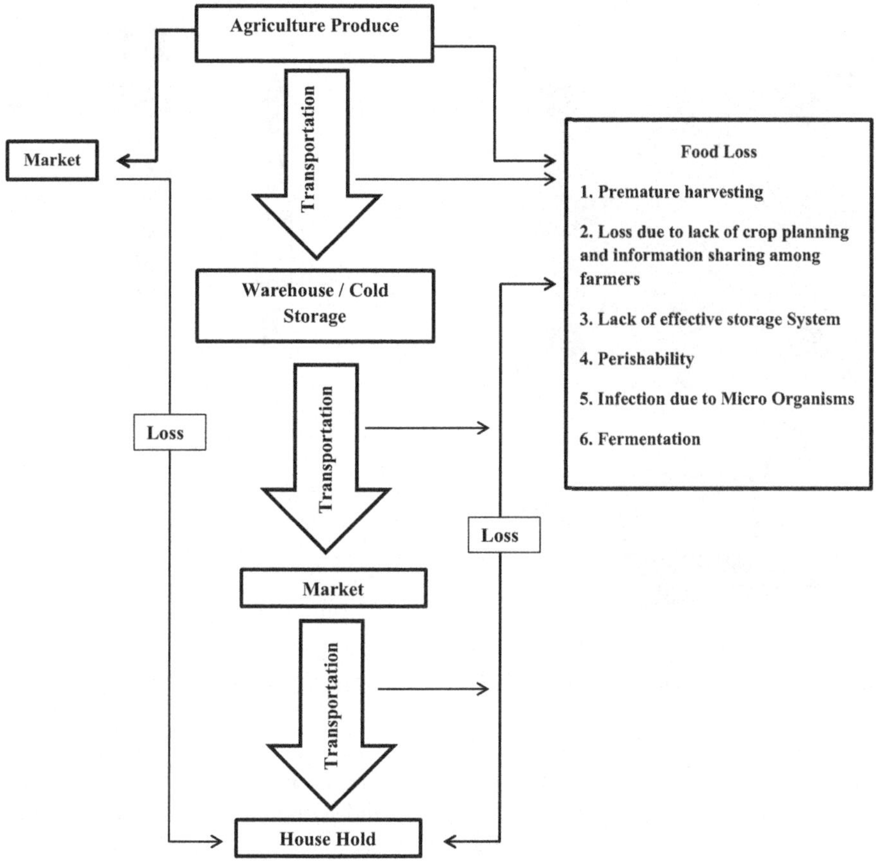

Fig. 4 Schematic diagram of food wastes at various stages

Loss of Agricultural Produce at Household

- Various ways of food waste take place at the household level both in uncooked and cooked format.
- The loss of uncooked produce is primarily because of oversupply of vegetables or food grains due to lack of ready information about the food reserves in the household. From lower middle class and above category, the family depends on household refrigerators for the purpose of preservation of food grains, vegetables, etc.
- The loss of cooked food occurs due to overestimation of perceived demand of food for day to day domestic consumption. This loss can be managed by experiential learning of the household and commitment of the members toward sustainable use of resources.

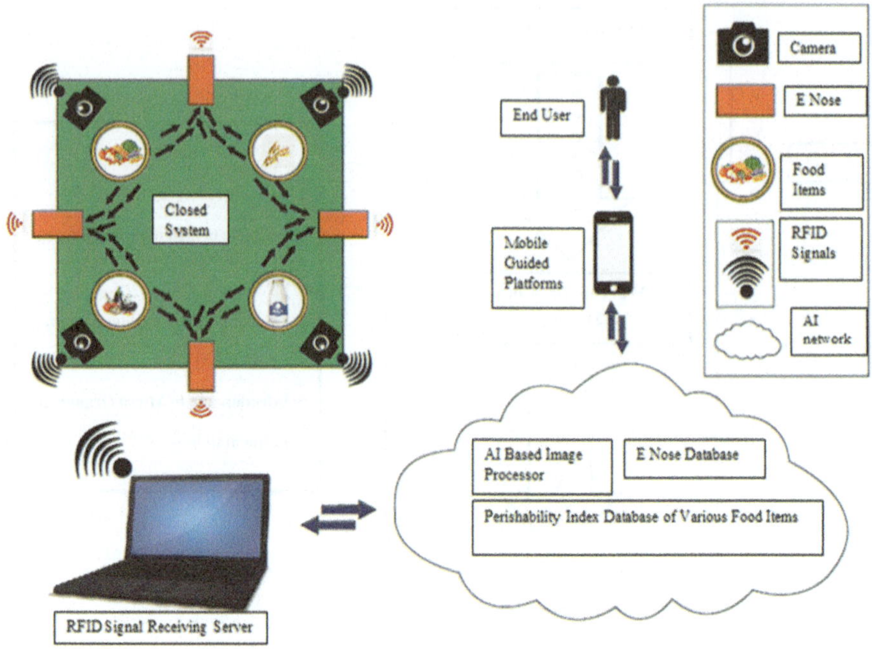

Fig. 5 IOT-based solution for reducing food waste primarily in closed system or household condition

- In fact, the loss of uncooked produce at the household level may be solved with appropriate smart in-house storage system and economic usage mechanism led by IoT intervention.

Analysis II

See Fig. 5.

Indicative algorithm for the model

1. The confined storage system (e.g., household refrigerator) needs to be augmented by multiple IoT devices that would act as E-Nose and E-iris as a means of sensory devices.
2. The IoT, E-nose and E-Iris would be provide adequate database for image processing for with pigment support (e-iris) and aromatic database (e-nose) on each specific food grains and vegetables generally stored in the system.
3. The sensory device will receive appropriate stimuli both aromatic and pigment and that would process using IoT, intelligence system (AI) by matching the preset database and finally the system would generate its predictive analytics about the nature of food produce stored in the system.
4. The processed results would be transferred at all the connected portals with the IoT ecosystem on real-time basis so that the user can be aware of the quantity and quality of each of the variants food products.

5. Based on this input, the user can take most appropriate purchase decision for further procurement of food grains, i.e., the set of variants to procure or not to procure and at what quantity. This real-time information would enable the household to prioritize which vegetables to cook immediately on priority basis in order to minimize both the nutritional and economic losses.

6 Conclusion

Food waste management has multiple implications on society as well as its economy. The world is highly apprehensive on the issues of food security, fertility of land, and nutritional benefits for upbringing off the human civilization. The sustainable development goals firmly advocate that poverty, hunger, and responsible production and consumption are the few integral dimensions which need to be improved if the world would like to achieve the essence of sustainability. Whatever the policy is formulated at macro level, finally the success depends on what extent it has been implemented and practiced at the very micro household level. This paper has emphasized on developing a sustainable solution to address the food loss issues at the household level with the intervention of IoT led smart technology.

Limitation of the Study

This paper has designed and developed a strategic solution in order to combat household syndrome. However, if the model is efficiently implemented, the degree of minimization of food loss could be explored so that the impact of the model could be studied, and necessary improvement on the structural morphology could be incorporated.

Working definition:

e-nose: An electronic nose is used to identify odors by detecting the "fingerprint" of a chemical compound using pattern recognition software [21].

References

1. Food and Agriculture Organization of United Nation: Data on food loss and waste (2019) https://stats.unctad.org/Dgff2016/planet/goal12/data/Fig12_1.xls
2. Food and Agriculture Organization of United Nation: Key facts on food loss and waste (2019). http://www.fao.org/save-food/resources/keyfindings/en/
3. Global Hunger Index (GHI): Global hunger index by severity and country wise ranking (2018). https://www.globalhungerindex.org/results/
4. Reserve Bank of India: Number and percentage of population below poverty line (Sep, 2015). https://www.rbi.org.in/scripts/PublicationsView.aspx?id=16603
5. The Hindu: FCI admits 1.94 lakh MT food grain wasted between 2005 and 13 (May, 2016). https://www.thehindu.com/news/national/fci-admits-194-lakh-mt-foodgrain-wasted-between-200513/article5680994.ece

6. Food and Agriculture Organization of United Nation: Technical platform on the measurement and reduction of food loss and waste (2019). http://www.fao.org/platform-food-loss-waste/en/
7. World Economic Forum: The Fourth Industrial Revolution or 4IR (Jan, 2016). https://www.weforum.org/agenda/2016/01/the-fourth-industrial-revolution-what-it-means-and-how-to-respond/
8. Li, G., Hou, Y., Wu, A.: Fourth industrial revolution: technological drivers, impacts and coping methods. Chin. Geogra. Sci. **27**(4), 626–637 (2017)
9. Stancu, V., Haugaard, P., Lähteenmäki, L.: Determinants of consumer food waste behaviour: two routes to food waste. Appetite **96**, 7–17 (2016)
10. SDG, U.: Sustainable development goals (2015)
11. Papargyropoulou, E., Lozano, R., Steinberger, J.K., Wright, N., Bin Ujang, Z.: The food waste hierarchy as a framework for the management of food surplus and food waste. J. Clean. Prod. **76**, 106–115 (2014)
12. Lipinski, B., Hanson, C., Lomax, J., Kitinoja, L., Waite, R., Searchinger, T.: Reducing food loss and waste. World Res. Inst. Working Paper, pp. 1–40 (2013)
13. Gubbi, J., Buyya, R., Marusic, S., Palaniswami, M.: Internet of things (IoT): a vision, architectural elements, and future directions. Future Gen. Comput. Syst. **29**(7), 1645–1660 (2013)
14. Godfray, H.C.J., Beddington, J.R., Crute, I.R., Haddad, L., Lawrence, D., Muir, J.F., Toulmin, C.: Food security: the challenge of feeding 9 billion people. Science **327**(5967), 812–818 (2010)
15. Parfitt, J., Barthel, M., Macnaughton, S.: Food waste within food supply chains: quantification and potential for change to 2050. Philos. Trans. Royal Soc. B: Biol. Sci. **365**(1554), 3065–3081 (2010)
16. Bush, R.: Food riots: poverty, power and protest 1. J. Agrar. Change **10**(1), 119–129 (2010)
17. Wilson, A., Baietto, M.: Applications and advances in electronic-nose technologies. Sensors **9**(7), 5099–5148 (2009)
18. Want, R.: An introduction to RFID technology. IEEE Perv. Comput. **1**, 25–33 (2006)
19. Scott, S.M., James, D., Ali, Z.: Data analysis for electronic nose systems. Microchim. Acta **156**(3–4), 183–207 (2006)
20. Stuart, T.: Waste: uncovering the global food scandal. WW Norton & Company (2009)
21. IEEE Spectrum: Electronic nose or e-Nose (Oct, 2018). https://spectrum.ieee.org/the-human-os/biomedical/devices/meet-the-enose-that-actually-sniffs
22. Tjoa, A.M., Tjoa, S.: The role of ICT to achieve the UN sustainable development goals (SDG). In: IFIP World Inf. Technol. Forum (pp. 3–13). Springer, Cham (2016, September)
23. Schaller, E., Bosset, J.O., Escher, F.: 'Electronic noses' and their application to food. LWT-Food Sci. Technol. **31**(4), 305–316 (1998)
24. Economic Times: India wastes Rs. 244 crore worth of food a day: Report (Jan, 2018). https://economictimes.indiatimes.com/wealth/personal-finance-news/india-wastes-rs-244-crore-worth-of-food-a-day-report/articleshow/62391489.cms?from=mdr

Frost Prediction in Highland Crops Management Using IoT-Enabled System and Multiple Regression

J. M. Mendez and D. D. Dasig

Abstract The magnitude of the effects of climate change in the agriculture-driven growth, countries, and regions in the world has expurgated the crop production and a decrease in agricultural productivity and eventual price surge. In 2014, it was accounted that agriculture is one-third of the global gross domestic product which was crucial to the country and global economic growth. With the recent problems in the agricultural sector, agriculturists and engineers have explored the advent of the cyber-physical system and automation age by using Precision Agriculture to address these issues and problems particularly in frost prediction in highland crops management. This chapter will discuss the concepts of Precision Agriculture and applications of the Internet of things (IoT) in agriculture, the design and challenges of an IoT-enabled system for highland crops management, and utilized multiple regression as a frost prediction technique. The remote sensing device deployed in the farm collects the frost climatic events including air velocity, temperature, and humidity, and transmitted to the web server. The system provides early warnings and frost forecast for the farmers using SMS. The system helped farmers as a useful resource to conduct frost protection activities on the farm, thereby reducing the frost harm to the crops.

Keywords Frost prediction · Internet of things · Multiple regression · Microclimate

---

J. M. Mendez
College of Computer Studies and Engineering, LORMA Colleges, Carlatan Campus, San Fernando, La Union 2500, Philippines
e-mail: janelli.mendez@lorma.edu

D. D. Dasig (✉)
College of Science and Computer Studies, De La Salle University—Dasmarinas, DBB-B, 4115 West Ave, Dasmariñas, Cavite, Philippines
e-mail: dhuniel2006@yahoo.com; dddasig@dlsud.edu.ph

P. K. Pattnaik et al. (eds.), *Internet of Things and Analytics for Agriculture, Volume 2*, Studies in Big Data 67,
https://doi.org/10.1007/978-981-15-0663-5_13

1 Introduction

The agricultural sector has confronted issues and difficulties on how it could bolster food production and feed the changing and increasing demand of the society. Agricultural engineers, scientists, and technocrats have advocated on the development of intelligent agricultural production to countervail food production concerns called "smart" system. Agribusiness has been modified and changing its landscapes in a few years. More and more farmers must adapt to new technologies to obtain products with greater value and cost-effective, because of the current disruption and high competition of the market and the high demands on the quality of products. Moreover, linked to these determining factors, agriculture is suffering the damages caused by climate change such as floods or constant temperature changes. In the same way as the industry, agriculture is forced to modernize its work methodologies, to automate its protocols, to reduce production costs, to increase production by improving the quality of its products, and to fight against meteorological inclemency [1].

Atmosphere fluctuation and change can unfavorably influence human and natural ecosystem as well as the ecological systems. Understanding the adaptation and mitigation capacities of agricultural producers is essential for improving climate-related support to the sector. Farmers need be able to engage in a farm management and assortments of approaches and maneuvers to accomplish and countervail climate-related risks, as well as uncertainties from other sources [2]. Understanding how climate affects your growing conditions is essential to successful gardening. Climate is the extent of middling disparity in general climate configurations over time in a very eclectic topographical region or zone.

In a farm field, climate strongly influences growing conditions: air and soil temperatures, precipitation, and heat accumulated over the growing season. When air temperatures fall below 0 °C, sensitive crops can be injured, with significant effects on production. For example, in Benguet, there are more economic losses to frost damage than to any other weather-related phenomenon. Therefore, impacts on affected farmers and the local economy are often devastating. Although it is important, information on how to protect crops from freezing is relatively limited. Consequently, there is a need for a widely available, simplified source of information to help farmers address this serious problem.

Technically, "frost" alludes to the arrangement of frost precious stones on surfaces, either by solidifying of dew or a stage change from vapor to frost; notwithstanding, the word is broadly utilized by people, in general, to depict a meteorological occasion when yields and different plants experience freezing damage. Producers frequently utilize the expressions "frost" and "freeze" reciprocally, with the vague definition being "an air temperature not exactly or equivalent to 0 °C". Instances of frost definitions in the published works and literatures include: the event of a temperature not exactly or equivalent to 0 °C estimated in a "Stevenson-screen" cover at a stature somewhere in the range of 1.25 and 2.0 m; the event of an air temperature under 0 °C, without characterizing the safe house type and tallness; when the surface temperature dips under 0 °C; and the presence of a low air temperature that makes harm or passing the plants, without reference to frost arrangement and formations.

In the investigation of [3], frost was defined and falling into two categories "advective" and "radiative." Advective frosts are related to enormous scale attacks of cold air with a well-blended, blustery atmosphere and a temperature that is regularly below zero, notwithstanding during the daytime. Radiative frosts are related to cooling because of energy loss through the radiant trade during clear, quiet evenings, and with temperature reversals (e.g., temperature increments with heights). Now and again, a mix of both advective and radiative conditions will happen. This might be trailed by a few days of clear, quiet conditions that are helpful for radiation frosts. Moreover, the researchers have posited conditions that are considered as "micro-scale shift in weather conditions frost" or micro-scale advection frosts. These happen when the area is presented to radiation-type frost conditions; however, nearby chilly air waste prompts fast drops in temperature from a more minor perspective inside the radiation frost territory.

Similarly, freezing and frosts bibliotheque definitions in lexicons and the writing are variable and befuddling; be that as it may, on an overall premise, the term frost protection is more ordinarily utilized than freeze protection. A "frost" is attributed to as an event of an air temperature of 0 °C or lower, estimated at a stature of somewhere in the range of 1.25–2.0 m above soil level, inside a suitable climate shelter. The water inside plants could conceivably solidify during a frost episode, contingent upon a few shirking factors (e.g., supercooling and convergence of frost-nucleating microscopic organisms).

A "freezing" happens when extracellular water inside the plant freezes (e.g., changes from a fluid to frosts). This could conceivably prompt harm of the plant tissue, contingent upon resistance factors (e.g., solute substance of the cells). A frost occasion turns into a freeze occasion when extracellular frost frames within the plants. Freeze damage happens when the plant tissue temperature falls beneath a basic worth where there is an irreversible physiological condition that is helpful for death or breakdown of the plant cells. This harming plant tissue temperature is connected with air temperatures called "basic temperatures" estimated in standard instrument covers. Below-zero air temperatures are brought about by decreases in reasonable warmth substance of the air close to the surface, basically coming about because of (1) a net vitality deficit through radiation from the surface to the sky (e.g., radiation frosts); (2) wind blowing in below-zero air to supplant hotter air (e.g., shift in weather conditions frosts); or (3) a mix of the two procedures.

A "shift in weather conditions frosts" or "freeze" happens when the air temperature falls beneath 32 °F (0 °C) during the night, yet the breeze speed is sufficiently high (normally more prominent than 5 mph) with the goal that it blends the air and there is practically no reversal arrangement. In a "solidify," the temperature often remains beneath 32 °F (0 °C) notwithstanding during the day, and it can proceed for a few days straight. Then again, "radiation frosts" or essentially a "frosts" is regular in the Province of Benguet, and these happen when there are quiet breezes and clear skies, a reversal structures, and the temperature fall underneath 32 °F (0 °C). As a rule, the temperature will transcend 32 °F (0 °C) during the day following a frost occasion; be that as it may, frost occasions can happen a few evenings in succession.

Both freeze and frost occasions are related to low stickiness. Now and again, the atmosphere conditions begin as a "solidify" on the principal night and change to a "frosts" on ensuing evenings. Most insurance strategies are compelling against "frosts" occasions yet not against "solidify" occasions. Since the harm is brought about by frosts development inside the plant tissue bringing about cell drying out and cell divider harm, the terms frosts and freezes harm are exchangeable. Frost occasions can be partitioned into two classes: radiation frosts and advective frosts. The previous will, in general, happen at the meteorological attributes of clear skies, no wind, and low dew-point temperature. The last typically happens under the meteorological states of overcast skies, moderate-to-solid breezes, no temperature reversal, and low moist. Frost occasions are progressively basic in the Province of Benguet, so the analyst utilized the term frosts harm when depicting the damaging impact of frosts arrangement on vegetables.

Frosts hazard is a significant unavoidable truth for agribusiness. Frigid temperatures limit the length of the developing season and are responsible for decreases in yield and nature of harvests. Delicate harvests can endure genuine harm at temperatures above solidifying also. In either case, the base temperature is basic. Of all the climate elements influencing farming, the base temperature is among the most factors here and there. Contiguous fields and nursery destinations can have altogether different frosts-free periods and are along these lines fit various employments. Hazard can be diminished via cautious administration of the land and successful response to figures. Pre-summer and late-summer frosts more often than not include least temperatures only marginally beneath solidifying or freezing; it is in these circumstances that preventive measures can be best. By effectively turning away a solitary solidifying circumstance, the developing season might be reached out by weeks.

The area of Benguet in the self-ruling locale of Cordillera speaks to the most part rain-fed upland agroecological zone dedicated to high-esteem crop generation and exchange. Benguet's rocky landscape and immense reliance on occasional precipitation make the rural partners in the territory especially touchy to regular climatic oddities. The region would suitably fill in as study site when investigating the estimation of atmosphere-related data especially for delicate on-farmers and off-farmers decisions [4].

Situated in the northern part of the Philippines, alongside the Cordillera's mountain ranges, and a huge and largely agribusiness-based Benguet area, poverty is prevalent in the region. Benguet is no more unusual to penury. In 2000, the National Statistics Coordination Board [5] announced that somewhat over 30% of the commonplace populace lives in poverty. Land corruption and degradation, peripheral and marginal profitability and inadequate or lack of foundation and infrastructure are a portion of the issues being looked by the province. These serious interrelated issues are basic in numerous highland and upland areas and territories as a result of unfavorable natural conditions and an absence of resources that these regions are named "Less Favored." It is comparable given these reasons that Benguet can be named a less favored zone. A few man-made and natural resources impede regions and locales economic development.

From January 30 to February 2, 2019, Atok, Benguet, hit a 9 °C temperature that made a cover of frost secured a portion of this present town's homesteads, yet neighborhood authorities eased apprehensions, and that this would affect the supply and costs of vegetables in the Province of Benguet [6]. At the point when Atok encountered a temperature of 7°, farmers mixed to the patches of farmland still grasped by frost. Sitios (sub-towns) Lower and Upper Engladad in Barangay (village), Paoay in Atok endured pockets of frost, which additionally influenced the Kibungan towns of Nilicob, Proper Madaymen, Taliboy-oc, Cagam-is, and Masala, and the town of Cada in Mankayan town, as indicated by agriculture information discharged by the Philippine Information Agency's office in the Cordillera.

A portion of the farmers depended on indigenous learning with regard to frost forecasting. It was accepted that an unmistakable sky with no wind development toward the evening is an indication that a frost scene will happen around the evening or early morning the next day. At the point when frost occurs, farmers assume that there is nothing that should be possible about it.

This study was conducted to develop and deploy a frost prediction in highland crops management using IoT-enabled system and multiple regression. The system will also monitor the farm field temperature, humidity, and dew as the most important climatic parameters. The system was deployed and tested on the select municipalities in the Province of Benguet, Philippines as illustrated in Fig. 1.

2 Background of the Study

Agriculture has been the conventional and still the real wellspring of employment for the individuals in the province. At least 54% of the labor force is engaged in vegetable/cut flower farming. Vegetable products from the province supply 80% of vegetable needs of the metropolis. Vegetables grown are cabbage, Chinese cabbage, broccoli, cauliflower, potatoes, carrots, and other temperate vegetables. Cut flower production is also gaining attention because of its promising profitability. These are roses, chrysanthemums, lilies, carnation, Shasta daisies, and others. The "Rose Garden of the Philippines" is situated in the area especially Barangay Bahong, La Trinidad. Tough and slanting territory and valleys describe the province.

The province is a leading producer of vegetables; it also has potential for growing oranges, pears, and other temperate fruits. Also, just recently, Benguet has been dubbed as the term "Strawberry Country." In Benguet, vegetable farmers brought their produce either directly or through the assembler wholesalers, commission agents, or wholesalers to the La Trinidad Trading Post and Hangar Market. The traders at the trading post initially got a large proportion of the farmers' produce. Subsequent trading activities, however, occurred where vegetables sold at the trading post went to the Hangar market. Some of these vegetables were sold in the domestic market through local wholesaler-retailers and retailers while the major proportion

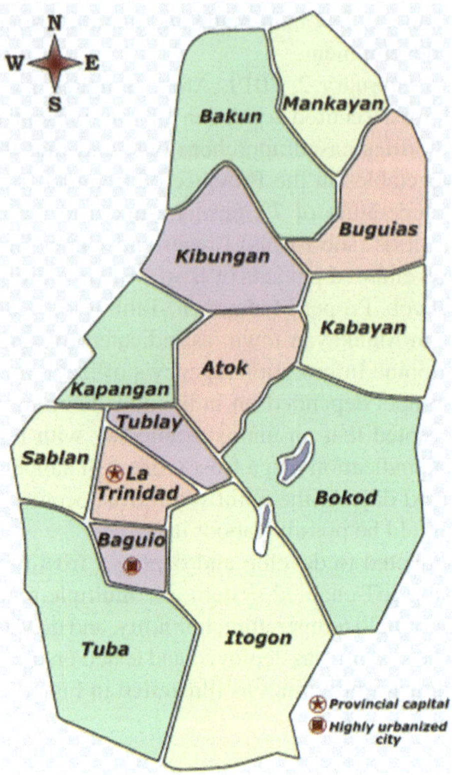

Fig. 1 Geographical map of the Province of Benguet, Philippines

of about 73 and 17% were brought out of the region through the large assembler wholesalers and wholesalers [7].

Data from the office of the Provincial Agriculture in Benguet revealed that in terms of average production in metric tons from 2010 to 2016, potatoes (18,666 mt) are the top commodity produced followed by cabbage (14,790 mt) and chayote at 9472 mt. Carrot is the top four with 7219 mt, and completing the top five is Chinese cabbage with 4222 mt' production from 2002 to 2015. Even looking at the five-year average production from 2011 to 2015, we can still infer that the same vegetables are in the top five in Benguet. There is about 19.5 thousand metric tons for potato, chayote (17.5 mt), cabbage (16.9 mt), carrots (8.8 mt), and completing the top five is Chinese cabbage at 4447 mt. In terms of average production from 2011 to 2015, Broccoli is included in the top 10 produced vegetables in Benguet (Fig. 2).

Frost was considered as a regular and accepted phenomenon in certain municipalities of the province, until recently when its occurrence and perceived impact gained more interest from social media, traditional news outlets, and the general public. As of now, there is no available frost forecast or advisory in the province. In the study [1], it characterized the occurrence of frost in some areas in Benguet and explained that

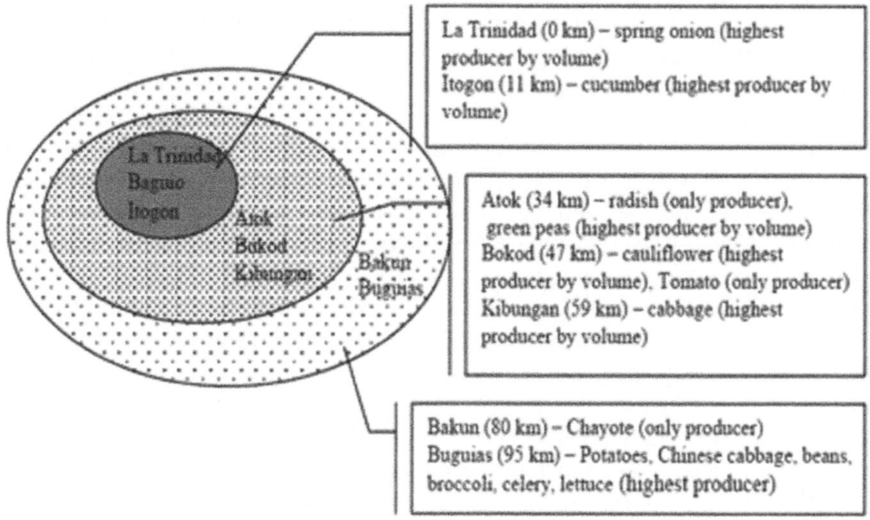

Fig. 2 Crops produced by the Province of Benguet, Philippines

the areas with high risk have a high frequency of the cloudless nights with temperatures falling below within a minimum temperature threshold. This data applicable as cloud cover is one of the central points for radiative frost occurrences. During the passage of a cold surge, highly elevated mountainous areas with clear skies like Atok and Kibungan are more prone to frost. Apart from the meteorological conditions, the geology assumes an imperative job in the frost arrangement and frost formations. Elevation and morning potential insolation (MPI) have the most influence on frost risk. The seasonal variability of the solar radiation in the morning contributes to the soil energy balance, local diurnal wind development, and risk of radiative frost [1].

Plants are not damaged by freezing temperature but by frost formation inside plant tissues. The frost crystals, which are huge concerning plant cells, structure in the space between cells. The frost crystals develop by coaxing the water out of the cells prompting parchedness and dehydration of the cell [8, 9]. Along these lines, when the frost liquefies, the cell divider is harmed. In this manner, anything that decreases the opportunity of frost development inside the plant tissue maintains a strategic distance from freeze injury [10, 11].

In addition, when a basic change in one of the estimations happens, at that point the farmers and the agriculture specialists were recognized employing SMS. Extreme climate conditions have been recognized as one of the main reasons for the decline of agrarian efficiency and agricultural productivity in numerous nations around the world. In the agriculture industry, advanced decision support through Internet of things (IoT) technologies are gaining attentions as it enables precision farming. Smart agriculture based on microclimate data analysis and advanced decision support systems helps improve productivity, avoid unnecessary costs related to harvesting, and use proper pesticide and fertilizers to enable precision farming [12, 13].

Temperature is one of the essential elements affecting plant development and geological appropriation toward compositional quality [14]. Similarly, frosts will have an adverse effects on crop production and even environmentally, and economically. The high humidity, frost as well as hail have adverse effects to crops yield and quality [15]. The minimum temperature which damage is caused varies with plant species and with growth stage for a given plant. Precision agriculture (PA) utilizes the information and communication technologies and engineering application in order to manage crop production and the lifecycle of crop management. Part of the Internet of things (IoT) technology the ambiental parameters needed (e.g., temperature, moistness, dew) can be estimated utilizing sensors. Wireless sensor networks (WSN) consist of random or planned spatially distributed nodes, also called nodes, each of which is equipped with sensors, memory for storage, a microprocessor for computing their decisions, batteries for energy supply and can communicate wirelessly with other nodes in a short-range. Today, it is possible to build multifunctional sensor device, which are smaller, cheaper, that can facilitate better communication, and operationally, a low-power device, in contrast with the traditional data loggers.

Notwithstanding the noteworthy advances in farming innovation in the course of the most recent couple of decades, agricultural sector and production remain dependent on weather and climate. It is a clear reality that climatic variability will play an even greater role than in the past, as sufficient food supplies will not be available to feed the world population adequately at its present rate of increase, decelerating advances in technologies, natural resources and security with up-to-date information on crop conditions [16–18]. These variations can be addressed and treated using data-driven decisions and programs.

Apart from the different techniques developed to fight against frost, there still exists an unresolved issue: to foresee when the frost will happen and on the off chance that it is relied upon to happen, to determine the exact period that is more effective to run the anti-frost technique. This is a problematic concern if the administration will deliberate about that in this similar portion of terrestrial land, the temperatures may be different. Climatic changes observed and felt by the community, generally, agree with observed changes by PAGASA noted as still "normal" although the temperature and rainfall pattern is pronounced as some of the manifestations of climate change.

These communities are also replete with mechanisms both at the farm and household levels to negotiate changes in the climate. Such mechanisms, however, are not necessarily sustainable.

According to [19, 20], the majority of the farmer in Benguet cultivates medium-scale (48.2%) and small-scale farms (42.1%). Small-scale farms predominate in Kibungan (58.8%) and Kapangan (74.3%), while medium-scale farms are in Atok (71.9%) and Bugias (56.3%). These results validate NSCB 2007 data where a combined area of 6743 ha from the total 29,983 ha of Benguet farms is less than a hectare [5]. Of these farms, nearly three-fourths (70.1%) are owned by the farmers, while 28.5% are either renting (17.8%) or caretaking (10.7%). Most (54.3%) of the farmers inherited their landholdings.

Before the modern way of weather forecasting, the indigenous people (IPs) of Benguet have their system of forecasting weather mainly based on the natural phenomenon (biotic and abiotic) on their environment. Under this system are periodical changes on the behavior of local animals most especially birds, drop and rise of temperature, rainfall, and other weather elements. Birds are scientifically proven to have demonstrated to be able to foresee or predict that there will be a shift in weather/climate through detecting changes in the barometric (air) pressure [21, 22]. For example, nearby people in the region or province, they are guided by the behavior of local birds. In that capacity, they named every month with the neighborhood vernaculars as agricultural calendar.

In the past, the IPs of Benguet plans their agriculture activities on planting and harvesting which help strengthened their community and household food security. Usually, land preparation and clearing of the garden for the first cropping are conducted from January to February. The yield would be planted in March and April and would be harvested in June or July. In the past, March and April are rainy months; thus, farmers of Loo and Paoay could plant in these months. But with climate change, gardens with no irrigation/water source are not farmed except for farmers with water pump. The usual first cropping crops would be carrots and cabbage. In Bayabas and Taloy Sur, the month of March would be the start of "pinagkakaingin" (swidden farming).

The stormy season as a rule starts in May. Be that as it may, it has been observed that the standard cycle for the dry and rainy season has been disturbed, thereby changing of planting and harvesting schedules. Farmers now would take the risk of planting knowing that the timing of rainy and dry seasons has been distorted. The next cropping will start from October to December, and the month of August is known to have the strongest typhoons; therefore, gardens are fallowed in this month; Lan preparations start in September and planting in frost October or November. In Atok, cropping starts before the expected time of frost formation. Most frost occasions happen during clear and quiet evenings.

3 Methodology

In this section, the methods, tools, instrument, and data gathering techniques are presented. The researchers employed the Design Science Research Methodology (DSR). This methodology guided the researchers to propose, develop, and deploy an IoT-based technology for highland crop management as a new development in precision farming to improve the farmers' livelihood techniques and their social aspect through modifications on the processes and technologies [23, 24].

Table 1 Distribution of respondents

Respondents	Total population		Sample size
Farmers	853		156
Land owners	226		41
Local government officials (PAGASA, DA)	**20**		4
Total	1099		201

3.1 Research Locale and Sampling Technique

The study was conducted for the Province of Benguet of Cordillera Administrative Region (CAR) in the Philippines. Benguet is one of the highland provinces in the Philippines, and also known as the Salad Bowl of the Philippines being the huge producers of upland vegetables [25–27]. During the pilot testing, tests were conducted in the crops and vegetable farms in Atok, Benguet, 16° 35′ N 120° 42′ E with 7400 ft elevation and most frequently affected by frost and climactic variations. Atok has a mountainous land area, with many steep cliffs. Accordingly, the Philippine Highway System's highest point is Atok's Barangay Paoay at 2255 m (7400 ft) above sea level [25, 26].

A systematic sampling was utilized as a more efficient sampling technique to simplify the process of selecting a sample or to ensure ideal dispersion of sample units throughout the population [28–30]. Lynch formula was used to determine the research respondents or participants;

$$\frac{\mathrm{SS} = Z^2 * (p) * (1 - p)}{C^2} \tag{1.1}$$

where $Z = Z$ value, $p =$ percentage picking a choice, expressed as decimal and $C =$ confidence interval.

The population of the study is one thousand ninety-nine (1099) and was shown as illustrated in Table 1 according to their group and common characteristics as farmers—853, landowners—226, and local government officials (PAGASA, DA)—20.

3.2 Project Development Methodology

The researchers utilized a Modified Iterative and Incremental Project development methodology as illustrated in Fig. 3. This methodology aimed at mitigating pitfalls in the projects due to unrealistic goals and planning [31–33]. The requirements include the functional and non-functional requirements. The whole project was broken into

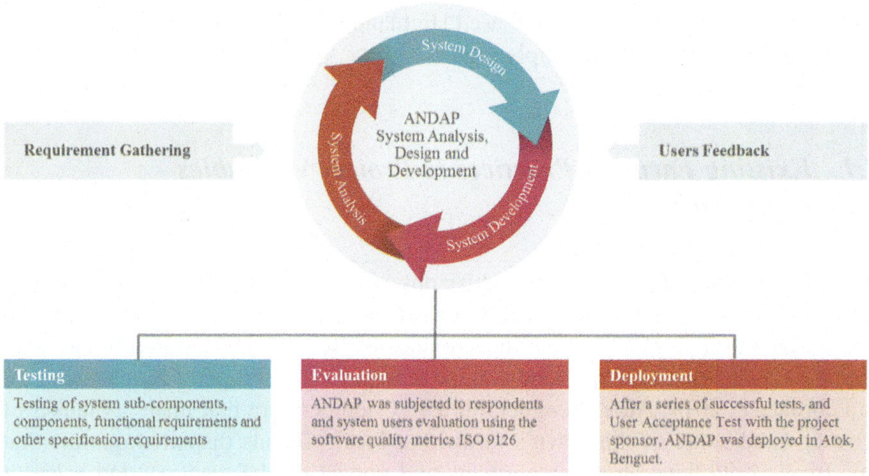

Fig. 3 Modified iterative and incremental methodology

smaller chunks with respective milestone and project stakeholders, project costing, and timeline and quality metrics. The requirements were analyzed and categorized into user requirements, process requirements, software requirements, hardware requirements, and technical requirements [32, 34].

During the analysis phase, two areas were treated, the data analysis and requirements analysis. During the system design phase, the block diagrams, schematic diagrams, database design, web design, mobile app design, and among others were designed and synchronized with other requirements. During the system development phase, the researchers embarked into codifying the designs, bread boarding, and implementing the schematic diagrams. Each design is presented to the users to encourage user engagement in the SD phases, and to accommodate changes to the requirements [31, 32, 34].

Testing of system subcomponents, components, functional requirements, and other specification requirements was also conducted to identify bugs, glitches, and other system issues to be remediated early on. Anti-ANDAP was subjected to respondents and system users' evaluation using the software quality metrics ISO 9126. After a series of successful tests, and user acceptance test with the project sponsor, anti-ANDAP was deployed in Atok, Benguet.

4 Results and Discussion

This section will detail the existing farming practices to protect vegetables from frost, parameters considered in determining frost occurrence. It will also discuss the architecture and design of the IoT-based Highland Crop Management, and the frost prediction technique using multiple regression.

4.1 Existing Farming Practices to Protect Vegetables from Frost

Based on the data gathered from the active observation, informal and semi-structured interview, and secondary data analysis, the researchers concurred with the following results. Due to its high elevation, the Province of Benguet has a mild climate with about 10 °C compared to the temperature at Baguio City Proper. From January 30 to February 2, 2019, Atok, Benguet, hit a 7 °C temperature that made a blanket of frost covered some of this town's farms, but local officials dispelled qualms that this frost would have an adverse effects and impact on the resource and price of vegetables in the Province of Benguet. When Atok experienced a temperature of 7°, farmers scrambled to the patches of farmland still embraced by frost. Sitios Lower and Upper Engladad in Barangay Paoay in Atok suffered pockets of frost, according to agriculture data released by the Philippine Information Agency's off frost in the Cordillera.

According to the farmers, frost was considered as a regular and accepted phenomenon in Atok, until recently when its occurrence and perceived impact gained more interest from social media, traditional news outlets, and the general public. As of now, there is no available frost forecasting or advisory in the province as stated by Edward Haight owner of the Sakura Park in Barangay Atok; also there are a few sorts of research that were directed in their homestead; however, a portion of the exploration was not focused on the open field.

Frost is most likely to happen during calm, clear nights, especially if the humidity is low. Under these conditions, there are no clouds to reflect heat to the earth, and no wind to mix the descending cold air with ascending warm air. Temperature changes all the more immediately when mugginess is low, which are the reason temperatures drop so quickly after sunset in desert areas. Most of the time if the farmers feel that frost will occur, they usually get "dayami" or straw grass from the lowland areas and they fog their farm using the straw grass, but the risk of it was the transportation of straw grass to Atok and frost usually occur at dawn; most of the time was 2:00 a.m. to 6:00 a.m. The farmers cannot go to their homestead during those seasons of the day. This data is important as overcast spread is one of the central points for radiative frost occurrences. During the passage of a cold surge, highly elevated mountainous areas with clear skies like Atok and Kibungan are more prone to frost according to Mr. Haight.

Also, to understand the cold weather, the farmers considered and observed if the sun warms the world's surface during the day and the warmth is emanated again into the climate during the night. The chilliest temperatures happen about sunrise. Mists around evening time assimilate and reflect warmth to the earth. Quiet, starry evenings represent the most serious peril of frost since there is no wind to blend the rising warm air with the diving cold air and no mists to transmit warmth back to the dirt.

Humidity eases back temperature changes noticeable all around. Cold air settles descending, streaming like water, to the absolute bottom. Tourist rises. Cold breeze mixes temperature misfortune.

The dew point is the temperature at which the air discharges moisture. It is an element of temperature and relative stickiness. At the point when the dew point is above frosty, it leaves the proof as haze noticeable all around or dew on the ground. During the evening, vegetable field (from radiation misfortune) to the dew point, product buildup, or dew and after that may keep on cooling to the point of solidification or lower. The dew at that point solidifies into frost.

Farmers of high-value crops that are vulnerable to frost consistently check climate figures for the daily dew point. The dew point is considered the "storm cellar" or low temperature for the night because the buildup of humidity or arrangement and formation of frosts eases back or freezes the pace of temperature drop. On the off chance that the dew point is above frigid, most plants are probably not going to be harmed by frost. In any case, subsolidifying dew-point flags the need to think about defensive measures. As illustrated in Fig. 4, the low humidity and humid air and the moist soil can create frost to their vegetable fields.

Accordingly, low temperature (e.g., chilling and solidifying) damage can happen in all plants; however, the components and kinds of harm differ impressively. Numerous organic products, vegetable, and decorative yields of tropical starting point experience physiological harm when exposed to temperatures underneath about + 12.5 °C, henceforth well above frigid gum-based paint. Nonetheless, harm over 0 °C is chilling damage as opposed to solidify damage. Freeze damage happens in all plants because of frost development. Harvest plants that create in tropical atmospheres, frequently experience serious frost harm when presented to a temperature marginally underneath zero like Atok, Benguet. A few vegetables, which began in a temperate atmosphere, yet can be harmed at temperatures close to 5 °C and some subtropical organic products trees that can withstand temperatures to 2–3 °C. Species

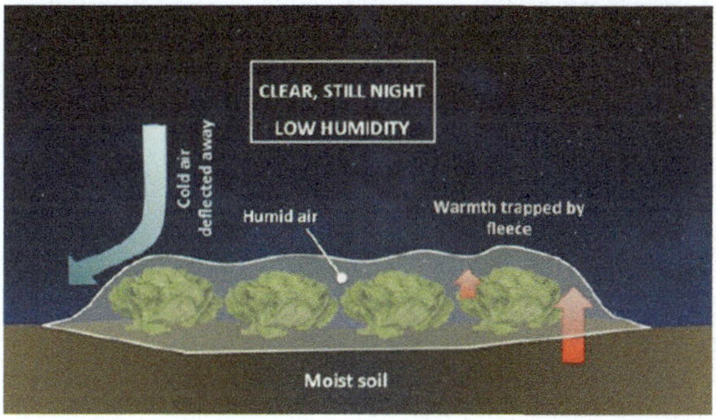

Fig. 4 Occurrence of frost

or assortments show diverse frost harm at the equivalent temperature and phenological arrange, contingent upon precursor climate conditions, and their adjustment to cold temperatures before a frost night is designated "solidifying."

In the meeting with CHARM farmers (Cordillera Highland Agricultural Resource Management), an association of farmer and landowners of Benguet, it was exemplified that during cold periods, plants will, in general, freeze against freeze damage, and they lose the freezing state after a warm spell. Freezing is most presumably identified with an in-wrinkle in solute substance of the plant tissue or diminishes in ice-nucleation dynamic (INA) microorganism fixations during cold periods or a mix.

Frost harm to harvests has been an issue for people since the primary yields were developed. Regardless of whether all parts of yield creation are very much overseen, one night of frosty temperatures can prompt complete crop misfortune.

In Atok, Benguet, harming occasions are ordinarily advective, frost occasions with quiet conditions and more grounded reversals are progressively run of the mill. The harming occasions ordinarily begin with shift in weather conditions of virus air pursued by a couple of evenings of radiation frost. In calm atmospheres, frost periods are shorter in term and happen more every now and again than in different towns of Benguet.

To an enormous degree, the potential for frost harm relies upon neighborhood conditions. Subsequently, it is hard to show a topographical assessment of potential harm.

The normal length of the frost-free period, which keeps going from the event of the last below zero temperature in the spring to the first in the pre-winter, is here and used to topographically portray the potential for harm.

Shielding plants from frost harm more often than not is practices by catching the warmth that is emanating starting from the earliest stage clear plastic, by moving the plant to a protected area, or by straightforwardly giving warmth by means of lights. Each measure gives around 2–3 degrees of protection. Keep plants well-watered to upset the drying up of frosts (Fig. 5).

Similarly, during frost at night, if the temperature is low enough, extracellular water freezes, drawing water out of the plant cells. If cell desiccation is limited, the

Fig. 5 Traditional way of protecting the crops from frost

water will move back into the cells as temperatures rise the next morning without causing damage. During this process, the leaves will take on a black water-soaked appearance, but they regain a normal appearance as warming takes place. If desiccation is severe, cell wall damage causes cell death. The lower the temperature, the longer the exposure, and the faster the temperature drops, the greater the damage to the plant.

If an unexpected light frost catches unprepared, the farmers simply drape a plastic tarp, blanket, or newspapers over plants before nightfall to trap the heat. It is better to drape the cover over a frame or wire cages so that it does not touch the plant leaves. Leaves touching the covers are likely to be frost damaged. People tend to leave coverings on for weeks, but the coverings should be removed during the day to allow heat in and then recovered at night. Other farmers turn on sprinklers to decrease the temperature by drizzling the farm field.

Based on the interview responses, and active observations, the farmers relied on indigenous knowledge in frost forecasting. It was believed that a clear sky with no wind movement in the afternoon is a sign that a frost episode will occur at night or early morning. When frost happens, the farmers simply drape a plastic tarp, blanket, or newspapers over plants before nightfall to trap the heat. Some of the farmers also use light bulbs to heat the plants, and others turn on the sprinkler as manual frost protection.

4.2 Parameters Considered in Determining Frost Occurrence

Internet of things (IoT) is a set of device and software elements, sensors that allow association, accumulation, representation and information sharing, which through their properties improve vitality productivity and reduce human effort while also bringing economic benefits [35–37]. Nowadays, different organizations and government departments are taking interest in implementing the technology for agriculture parameters measurement. Manual collection of parameters for desired factors can be erratic, not continuous, and produce variations from measurements results in difficulty in controlling agriculture environment. IoT and remote sensor hub can decrease the time and endeavors required for checking the horticulture condition [35, 36, 38].

The IoT likewise maintains a strategic distance from the loss of horticulture parameter database and spares it into a capacity gadget or cloud for long life [36, 37, 39]. Also, it will allow monitoring in critical place without the need to put personnel in hazardous situations. A monitoring system should have quicker response time, better quality control of the crop, and lower labor cost [38–40]. In this study, several parameters such as soil moisture, precipitation, dew, the temperature are considered as variables needed in detecting frost occurrence; this parameter can monitor on a real-time basis using IoT [35, 36, 41].

Sensors are set at various areas to gather the information to anticipate and predict the behavior of a particular area in a Barangay in Atok, Benguet. The nodes are monitored remotely using the Internet, and the data gathered from the sensors are put in the cloud and to extend the evaluated pattern on the web browser. The data collected concerning the normal or specified ranges of particular parameters. The embedded system is an integration of sensor device, wireless communication which enables the user to remotely access the various parameters and store the data in the cloud [35, 37, 39, 41].

4.3 The IoT-Based Highland Crop Management

4.3.1 IoT Architecture the System Block Diagram

The IoT-based Highland Crop Management utilized Internet-based technical architecture "IoT architecture" as illustrated in Fig. 6. In this architecture, the anti-ANDAP system was developed and deployed using the three layers including the IoT device layer, IoT gateway layer, and the IoT platform layer.

The researchers concur that the IoT architecture was able to facilitate the wireless sensor network (WSN) gateway so that ANDAP can perform its functions such as measuring, processing, and analyzing environmental indicators related to frost [42].

The IoT-based Highland Crop Management System is illustrated in the block diagram in Fig. 7. The sensor unit includes the temperature, humidity, dew, precipitation, and soil sensors. In the IoT architecture, the sensor unit is the IoT device layer. It is also called the sensor node in a wireless sensor network context. The sensor node is connected to the Arduino Mega 328P microcontroller and power by a solar panel. The GSM/GPS module with Wi-Fi shield is the IoT gateway layer. This layer serves as the bridge between the IoT device layer and IoT platform layer. It uses the Wi-Fi Access Protocols IEEE 802.11 standard, while the IoT platform layer includes the Online Crop Management System and the mobile application for anti-ANDAP, web, and other cloud computing services.

In this system, three are three (3) sensor nodes deployed in the locale (Atok, Benguet). The sensor nodes as connected together as a wireless sensor network. The WSN is connected to a gateway node (GSM/GPS and Wi-Fi shields), which allows the IoT device to send sensory data from the environment and other ambiental information need for frost forecasting to the platform layer, such as the Online Crop Management System illustrated in Fig. 8.

Fig. 6 IoT architecture

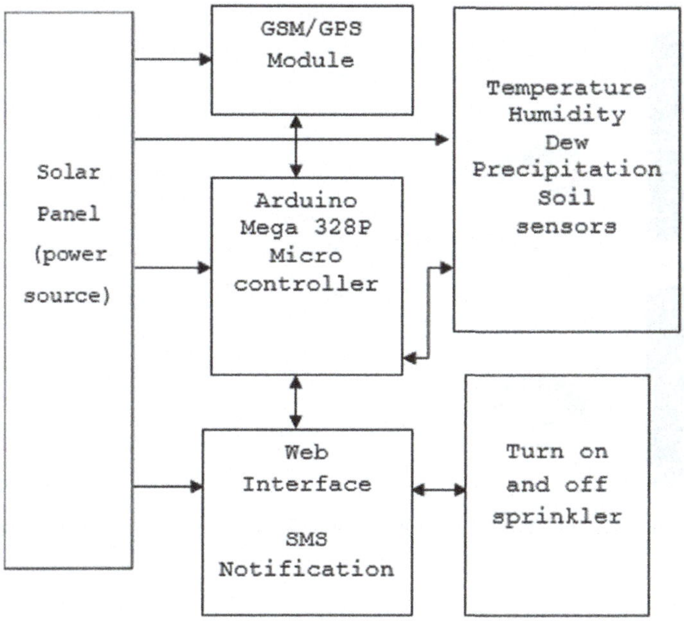

Fig. 7 IoT-based highland crop management block diagram

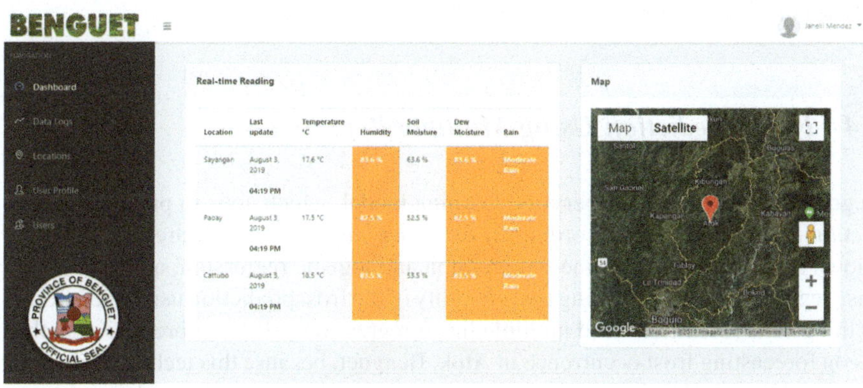

Fig. 8 Graphical user interface of online highland crop management system

4.3.2 The System—Online Crop Management System

Using the online application and mobile application, farmers can check to monitor the frost prediction parameters themselves even at home.

Due to scarce in electricity supply in the locality, farmers need to monitor their distant sprinklers during irrigation at night. The system interface has a feature-rich pump starter and sensors status monitoring. Farmers can also switch ON and OFF

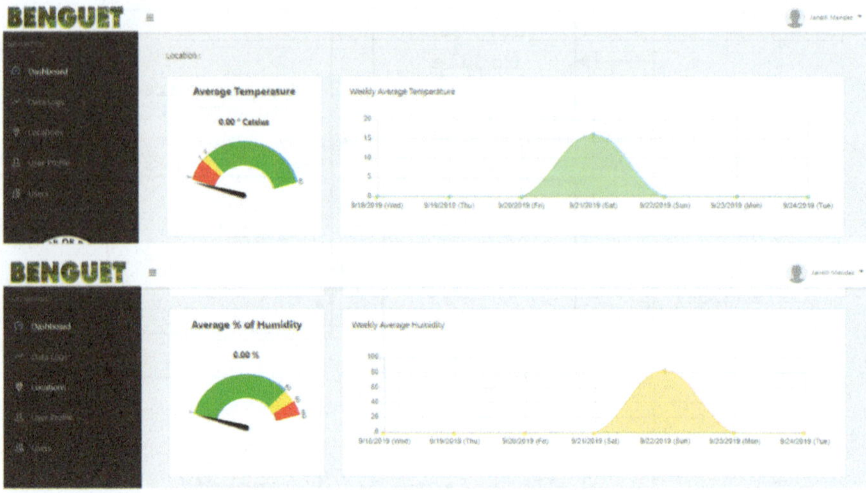

Fig. 9 Graphical user interface of environmental and ambiental condition monitoring for frost prediction

the sprinklers when deemed necessary. Figures 8 and 9 show the graphical user interface of Online Highland Crop Management System. The GUI also shows the data analytics that are generated based on the climatic variations and sensor nodes data sent to the web-based and mobile applications.

4.4 Frost Prediction Using Multiple Regression

Regression constitutes a supervised learning model, which aims to provide the prediction of output varies according to the input variables, which are known. Most known algorithms include linear regression and logistic regression, as well as stepwise regression. In determining the reliability of the frost prediction using the Internet of things, the researcher used multiple linear regression (MLR) as predictive analytics on forecasting frost occurrence in Atok, Benguet, because this technique is much simpler than other techniques and can be more useful for operational activities to prevent frost damages to agricultural crops.

Regression attempts to determine the strength of the relationship between one dependent variable usually denoted by Y and a series of other changing variables known as independent variables. In multiple linear regression (MLR), there are more than two variables among which one is the dependent variable and all others are the independent variables, and the equation looks like this:

$$Y_i = \beta_0 + \beta_1 x_{i1} + \beta_2 x_{i2} + \beta_3 x_{i3} \cdots \beta_p x_{ip} \quad (2)$$

where Y_i is the i-th value of the dependent variable, β_0 is the intercept, β_1 is the i-th angular coefficient, and x_i is the i-th vector of observations (predictors).

For this study, there are six (6) predictors that believed have significant impact on frost detection and are used in applying multiple regression such as average temperature (°C), maximum temperature (°C), minimum temperature (°C), average humidity (%) (measured by 8:30 am and afternoon humidity measured by 5:30 p.m.), dew (mm), precipitation (%). For cross-validation, local temperature data taken from Atok, Benguet, was taken as dataset. The dataset from PAGASA for 3 years was used for the experiment during the month and year from January 2017 to May 2019 which has a total number (k) of 29 observations taken. To measure the quality of the MLR equation, the predicted temperature amount is compared with the actual temperature gathered by the nodes on the field.

Before the MLR was undertaken, the correlation of the predictors using correlation entity was conducted with the results illustrated in Table 2. The results imply that minimum temperature was highly correlated with average temperature and maximum temperature with minimum temperature. Thus, precipitation is not correlated with maximum temperature, and humidity is not correlated with precipitation because of the negative values shown on their correlation.

Table 3 describes the summary output of the regression statistics taken from

Table 2 Correlation of predictors

Predictor	Ave	Min	Max	Dew	Precip. (mm)	Hum. (%)
Ave C	1					
Min C	0.993917954	1				
Max C	0.876113185	0.864295666	1			
Dew (mm)	0.533599548	0.535485727	0.417316696	1		
Precip. (mm)	−0.65237895	−0.65832911	−0.47075748	−0.7103738	1	
Hum. (%)	0.4437614	0.468292939	0.357942558	0.06579885	−0.7723735	1

ave C average temperature in °C; *min C* minimum temperature in °C; *max C* maximum temperature in °C; dew dew in mm; *precip. (mm)* precipitation in mm; *hum. (%)* humidity (%)

Table 3 Regression statistics

Summary output	
Regression statistics	
Multiple R	0.751513272
R square	0.564772198
Adjusted R square	0.446073707
Standard error	6.337164685
Observations	29

Table 4 Multiple regression table

Predictors	Coefficient correlation	Standard error	p-value
Intercept	−67.3909377	65.55815905	0.015
Average temperature (°C)	2.97551657	5.29959839	0.080
Minimum temperature (°C)	−4.725115525	4.948244714	0.050
Maximum temperature (°C)	1.67216714	1.243787101	0.092
Dew (mm)	4.735493886	0.989893894	0.061
Precipitation (%)	7.163066034	3.479092271	0.051
Humidity (%)	−0.203586847	0.35390595	0.070

the dataset which has R square of 0.56; hence, it shows that the regression square has a 56% low variability of statistic and 0.75 multiple R signifies a 75% good multiple regression. The adjusted regression square was computed with 0.446 output to compare between predictors, and standard error was 6.337 quite high.

Further, all predictors considered in the multiple regression have significant impacts in the fitting of the intercept as illustrated in Table 4. Maximum temperature, average temperatures, and dew and precipitation have positive effects on temperature, while the effects of minimum temperature and humidity are negative. On the p-value interpretation, all of the predictors are significant since they are all below the 0.15 standard error for p-value.

Applying multiple regression approaches on that dataset finds out predictable equation between temperature and other predictors. So MLR was computed given below.

$$\begin{aligned} Y_i = & -67.3909377 + 2.97551657 * X_1 - 4.725115525 * X_2 \\ & + 1.67216714 * X_3 + 4.735493886 * X_4 \\ & + 7.163066034 * X_5 - 0.203586847 * X_6 \end{aligned} \tag{3}$$

where

X_1 average temperature
X_2 minimum temperature
X_3 maximum temperature
X_4 dew
X_5 precipitation and
X_6 humidity.

Based on the dataset, example numbers are $X_1 = 17$, $X_2 = 13$, $X_3 = 19$, $X_4 = 0.3052$, $X_5 = 9.8$, and $X_6 = 72$; the value of the multiple regression would be $Y_i = 10.522$, which signifies that there are 10.52% frost that will occur with that specific numbers of predictors.

Based on the residual output of the dataset, the predicted value was shown and plotted as shown in Table 5.

Table 5 Residual output

Observation	Predicted t
1	10.522585
2	10.35941
3	8.8045597
4	10.220742
5	12.662168
6	13.622304
7	15.304227
8	11.030896
9	5.9691042
10	11.31716
11	9.1312452
12	9.6276945
13	10.957603
14	11.361045
15	11.178337
16	11.543338
17	15.226635
18	17.720764
19	14.660084
20	11.913541
21	15.0971
22	27.351176
23	13.473796
24	28.093354
25	17.358863
26	28.106629
27	22.230981
28	21.220673

It signifies that during the 24th month which falls last December 2018, there is a high tendency that frost occurs and during 26th month which falls on February also a high tendency that frost will occur. Results show that frost did occur during December 2018 and February 2018.

The researchers plot graphs as illustrated in Figs. 10 and 11 that depict the relationship between the actual value of temperature data and predicted value of temperature and the actual value of the average temperature and the relative humidity using multiple linear regression equation, and from the graph, it is observed that MLR method

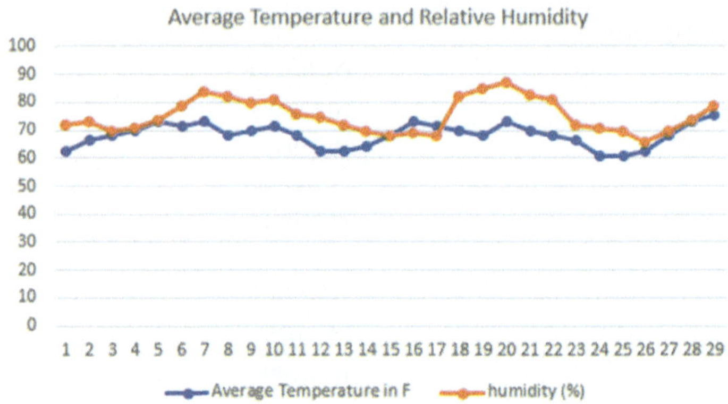

Fig. 10 Comparison of the average temperature and relative humidity

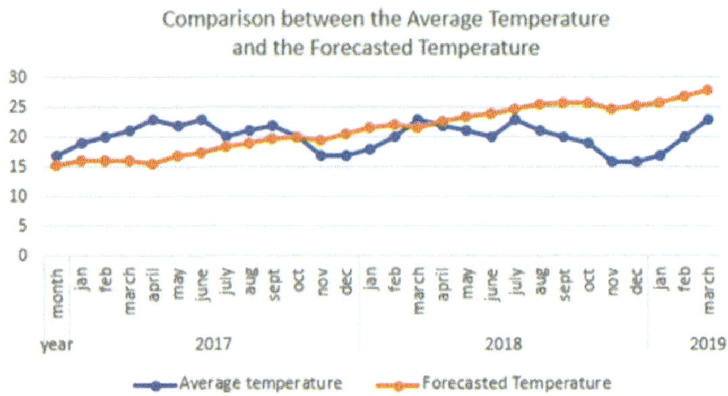

Fig. 11 Comparison of the actual temperature and the forecasted temperature

for prediction of temperature achieves closer values between actual and predicted temperature values.

Also, the residual plots of the predictors during the testing were taken as shown in Fig. 12.

The line fit plot of the residual is compared with the predicted value based on the residual output of the dataset. It shows that all of the predictors are in-line with the real-time values as with the predicted values as shown in Fig. 13.

Another figure describes the comparison of the real-time dew and the forecasted dew based on the multiple linear regressions taken from the dataset as illustrated in Fig. 14.

The comparison between the real-time precipitation and the forecasted is also shown in Fig. 15.

Figure 16 shows the comparison of the humidity based on the dataset taken as during the testing.

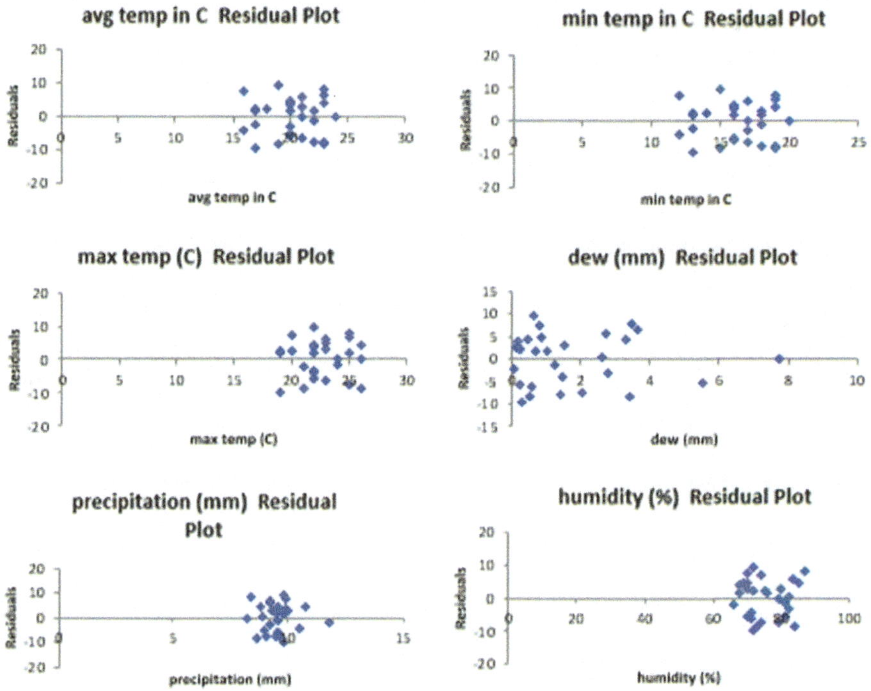

Fig. 12 Residual plots of the predictors

The comparison between the real-time minimum temperature and the forecasted minimum temperature signifies that the line graph was correlated with each other also illustrated in Fig. 17.

On the basis of the results, the multiple correlation and regression could be successfully used to analyze the relationships between the temperature, humidity, dew, and precipitation. Frost could be easily predicted using the multiple linear regression function. The graphs above prove the efficiency of the regression. With larger dataset, better and more precision can be achieved.

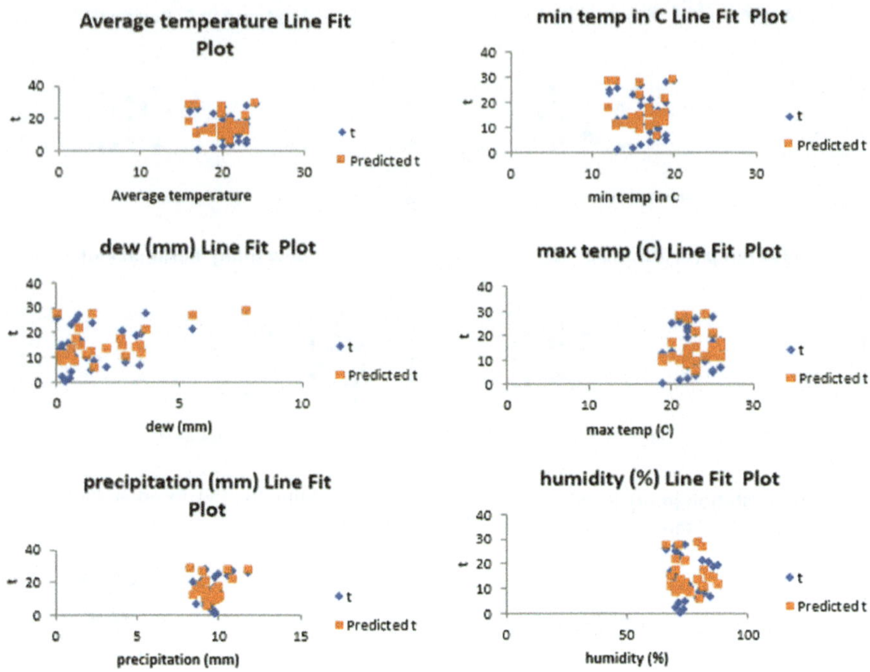

Fig. 13 Residual line fit

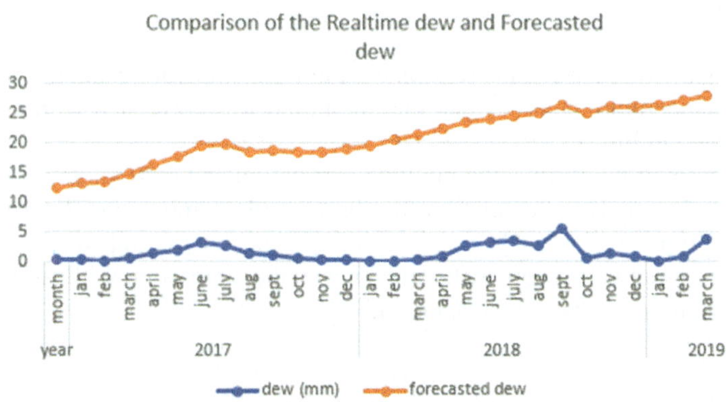

Fig. 14 Comparison of the real-time dew and forecasted dew

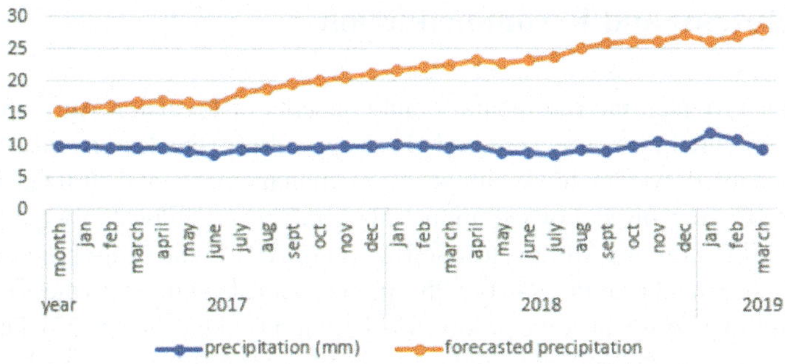

Fig. 15 Comparison of the real-time precipitation and the forecasted precipitation

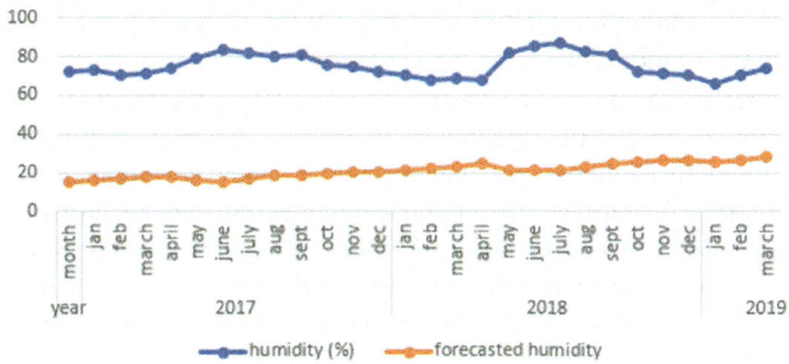

Fig. 16 Comparison of real-time humidity and forecasted humidity

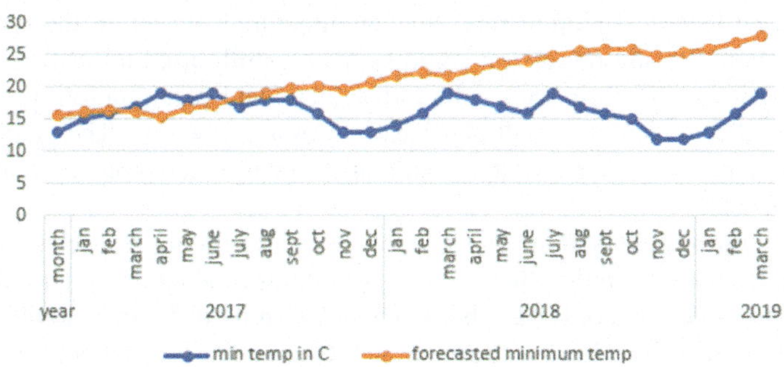

Fig. 17 Comparison of the real-time minimum temperature and forecasted temperature

5 Conclusions and Recommendation

This chapter provided the rudimentary and concepts of Precision Agriculture, and applications of the Internet of Things (IoT) in agriculture, the design and challenges of an IoT-enabled system for highland crops management, and utilized Multiple Regression as a frost prediction technique. The study was conducted for the Province of Benguet of Cordillera Administrative Region (CAR) in the Philippines. Benguet is one of the highland provinces in the Philippines, and also known as the Salad Bowl of the Philippines being the huge producers of upland vegetables [25–27]. During the pilot testing, tests were conducted in the crops and vegetable farms in Atok, Benguet, 16° 35′ N 120° 42′ E with 7400 ft elevation and most frequently affected by frost and climactic variations. Atok has a mountainous land area, with many steep cliffs. Accordingly, the Philippine Highway System's highest point is Atok's Barangay Paoay at 2255 m (7400 ft) above sea level [25, 26].

The researchers employed the Design Science Research Methodology (DSR). This methodology guided the researchers to propose, develop, and deploy an IoT-based technology for highland crop management as a new development in precision farming to improve the farmers' livelihood techniques and their social aspect through modifications on the processes and technologies; as well as employed Modified Iterative and Incremental Project development methodology. Based on the interview responses, and active observations, the farmers relied on indigenous knowledge in frost forecasting.

To help the vegetable farmers in frost forecasting and mitigate its adverse effects to their farms, the IoT-based Highland Crop Management was developed using the Internet-based technical architecture "IoT architecture." Based on the results, the multiple correlation and regression could be successfully used to analyze the relationships between the temperature, humidity, dew, and precipitation. Frost could be easily predicted using the multiple linear regression function. The graphs above prove the efficiency of the regression. With larger dataset, better and more precision can be achieved.

Further studies could be done on IoT application in agriculture, such as the use of blockchain technology in the lifecycle of agricultural production system, financing and product marketing, applying other machine learning techniques to help predict climatic variations, and other bioinformatics and biomedical techniques for plant and crop management.

References

1. Marasigan, A.R.: Characterization of frost events in Benguet, Philippines. Published Master's Thesis for the degree in meteorology. Institute of Environmental Science and Meteorology, University of the Philippines, Diliman (2017)
2. Hansen, J., Marx, S., Weber, E.: The role of climate perceptions, expectations, and forecasts in farmer decision making. The Argentine Pampas and South Florida. International Research Institute for Climate Prediction (IRI) the Earth Institute, Columbia University (2014)

3. Kalma, J.D., Laughlin, G.P., Caprio, J.M., Hamer, P.J.C.: The Bioclimatology of Frost: Its Occurrence, Impact and Prediction. Springer, Berlin (2012)
4. National Statistics Offfrost: Cut flower industry: an update. J. Philippine Statist. **59**(2) Second Quarter 2008. Manila, Philippines. Retrieved from https://psa.gov.ph/sites/default/files/2ndQ2008.pdf, Accessed 28 Aug 2018
5. National Statistical Coordination Board, October 2011. Philippine Statistical Yearbook (2011). Retrieved from https://psa.gov.ph/sites/default/files/2011_PSY_0.pdf, 28 Aug 2018
6. Fagyan, A.: 8 villages in Benguet susceptible to climate change. SunStar Baguio. 07 June 2017. Retrieved from http://www.sunstar.com.ph/baguio/opinion/2017/06/07/8-villages-benguetsusceptible-climate-change-546212, 4 July 2018
7. Piadozo, M.E.S.: Efficiency of Benguet vegetable prfrost linkages. J. Int. Soc. Asian Agric. Sci **19**(2), 104–119 (2013)
8. Hoffmann, N.E., Bischof, J.C.: The cryobiology of cryosurgical injury. Urology **60**(2), 40–49 (2002)
9. Petzold, G., Aguilera, J.M.: Frost morphology: fundamentals and technological applications in foods. Food Biophys. **4**(4), 378–396 (2009)
10. Snyder, R.L., Melo-Abreu, J.D.: Frost protection: fundamentals, practfrost and economics. Frost Prot. Fundam. Practfrost Econ. **1**, 1–240 (2005)
11. Battany, M.C.: Vineyard frost protection with upward-blowing wind machines. Agric. For. Meteorol. **157**, 39–48 (2012)
12. Moon, A., Kim, J., Zhang, J., Liu, H., Son, S.W.: Understanding the impact of lossy compressions on IoT smart farm analytics. In: 2017 IEEE International Conference on Big Data (Big Data), pp. 4602–4611. IEEE (2017, December)
13. Vuran, M.C., Salam, A., Wong, R., Irmak, S.: Internet of underground things in precision agriculture: architecture and technology aspects. Ad Hoc Netw. **81**, 160–173 (2018)
14. Lee, S.K., Kader, A.A.: Preharvest and postharvest factors influencing vitamin C content of horticultural crops. Postharvest Biol. Technol. **20**(3), 207–220 (2000)
15. Rosenzweig, C., Iglesias, A., Yang, X.B., Epstein, P.R., Chivian, E.: Climate change and extreme weather events; implications for food production, plant diseases, and pests. Global Change Human Health **2**(2), 90–104 (2001)
16. Mermut, A.R., Eswaran, H.: Some major developments in soil science since the mid-1960s. Geoderma **100**(3–4), 403–426 (2001)
17. Joshi, P.K., Gulati, A., Birthal, P.S., Tewari, L.: Agriculture diversification in South Asia: patterns, determinants and policy implications. Econ. Polit. Week. 2457–2467 (2004)
18. Piao, S., Ciais, P., Huang, Y., Shen, Z., Peng, S., Li, J., Friedlingstein, P.: The impacts of climate change on water resources and agriculture in China. Nature **467**(7311), 43 (2010)
19. Napaldet, J.T., Buot Jr., I.E.: Floral diversity assessment of Balili river as potential phytoremediators. Wetlands Biodiv. **7**, 17–28 (2017)
20. Napaldet Jones, T.: Diversity of aquatic macrophytes in Balili River, La Trinidad, Benguet, Philippines as potential phytoremediators. Biodiver. J. Biol. Div. **20**(4), 1048–1054 (2019)
21. Jenouvrier, S.: Impacts of climate change on avian populations. Glob. Change Biol. **19**(7), 2036–2057 (2013)
22. McMichael, A.J., Campbell-Lendrum, D.H., Corvalán, C.F., Ebi, K.L., Githeko, A., Scheraga, J.D., Woodward, A.: Climate change and human health: risks and responses. World Health Organization (2003)
23. Peffers, K., Tuunanen, T., Rothenberger, M.A., Chatterjee, S.: A design science research methodology for information systems research. J. Manag. Inf. Syst. **24**(3), 45–77 (2007)
24. Baskerville, R., Pries-Heje, J., Venable, J.: Soft design science methodology. In: Proceedings of the 4th International Conference on Design Science Research in Information Systems and Technology (p. 9). ACM (2009, May)
25. San Luis, G.D., Balangcod, T.D., Abucay Jr., J.B., Wong, F.M., Balangcod, K.D., Afifi, N.I.G., Apostol, O.G.: Phytochemical and Antimicrobial Screening of Indigenous Species That Have Potential for Revegetation of Landslides in Atok. Benguet, Philippines (2014)

26. Barcelo, J.M., Barcelo, R.C., Alvarez, A.: Ochratoxin A, fungal contamination and antioxidant property of defective Arabica coffee in Benguet, Philippines. Emirates J. Food Agri. 10–17 (2017)
27. Simongo, D.K., Gonzales, I.C., Sagalla, E.J.D.: Evaluation of potato entries for yield and fry quality grown in different elevations of Benguet, Philippines. J. ISSAAS Int. Soc. Southeast Asian Agric. Sci. **17**(2), 117–127 (2011)
28. Stehman, S.: Comparison of systematic and random sampling for estimating the accuracy of maps generated from remotely sensed data. PE RS-Photogramm. Eng. Rem. Sens. **58**(9), 1343–1350 (1992)
29. Deeks, J.J., Macaskill, P., Irwig, L.: The performance of tests of publication bias and other sample size effects in systematic reviews of diagnostic test accuracy was assessed. J. Clin. Epidemiol. **58**(9), 882–893 (2005)
30. Onwuegbuzie, A.J., Leech, N.L.: Sampling designs in qualitative research: making the sampling process more public. Qual. Report **12**(2), 238–254 (2007)
31. Larman, C., Basili, V.R.: Iterative and incremental developments: a brief history. Computer **36**(6), 47–56 (2003)
32. Chomicki, J.: Iterative modification and incremental evaluation of preference queries. In: International Symposium on Foundations of Information and Knowledge Systems, pp. 63–82. Springer, Berlin (2006, February)
33. Daniel, D.D.: An evaluative study on project management tools for managing engineering projects. Ascendens Asia J. Multidisc. Res. Abstr. **1**(1) (2017)
34. Nyssen, C.: An efficient and accurate iterative method, allowing large incremental steps, to solve elasto-plastic problems. In: Computational Methods in Nonlinear Structural and Solid Mechanics, pp. 63–71. Pergamon (1981)
35. Ashton, K.: That 'internet of things' thing. RFID J **22**(7), 97–114 (2009)
36. Atzori, L., Iera, A., Morabito, G.: The internet of things: a survey. Comput. Netw. **54**(15), 2787–2805 (2010)
37. Dasig, D.D.: A study on the sectors of economy servfrostd by pre-industry system developers among companies in Metro Manila: A tool for business reengineering. arXiv preprint arXiv: 1409.7277 (2014)
38. Gubbi, J., Buyya, R., Marusic, S., Palaniswami, M.: Internet of things (IoT): a vision, architectural elements, and future directions. Fut. Gen. Comput. Syst. **29**(7), 1645–1660 (2013)
39. Dasig Jr., D.: A frontier in organizational and business process innovation in servfrost management through lean six sigma Kaizen project implementation. J. Adm. Bus. Stud. **3**(6), 263–283 (2017)
40. Dasig, D.D., Benosa, M.C., Pahayahay, A.B., Asejo, N.R., Tadeo, C., Ellaine, M., Sabado, D.C.: Globalization of technology capability in these technological constellations. Int. J. Bus. Econ. Affairs, 106–115
41. Dasig, D.: User experience of embedded system students on Arduino and field programmable gate array (FPGA). In: Proceedings of the Second International Conference on Advances in Applied Science and Environmental Engineering-ASEE, pp. 124–128 (2014, February)
42. Gubbi, J., Buyya, R., Marusic, S., Palaniswami, M.: Internet of Things (IoT): a vision, architectural elements, and future directions. Future Gen. Comput. Syst. **29**(7), 1645–1660 (2013)

The manufacturer's authorised representative in the EU is Springer Nature Customer Service Centre GmbH, Europaplatz 3, 69115 Heidelberg, Germany. If you have any concerns regarding our products, please contact ProductSafety@springernature.com

Printed and bound by CPI Group (UK) Ltd, Croydon, CR0 4YY
30/07/2026
02181667-0008